Alkali-Activated Cements and Concretes

T0175706

Also available from Taylor & Francis

Sulfate Attack on Concrete
J. Skalny, J. Marchand & I. Odler

Spon Press   Hb: 0–419–245502

Determination of Pore Structure Parameters
K. Aligizaki

Taylor & Francis   Hb: 0–419–228004

Diffusion of Chloride in Concrete: Theory and Application
Leif Mejlbro, Ervin Poulsen

Taylor & Francis   Hb: 0–419–25300–9

Aggregates in Concrete
Mark Alexander, Sidney Mindess

Taylor & Francis   Hb: 0–415–25839–1

Information and ordering details

For price availability and ordering visit our website **www.tandf.co.uk/built environment**
Alternatively our books are available from all good bookshops.

Alkali-Activated Cements and Concretes

Caijun Shi

Professor of the School of Civil Engineering and Architecture, Central South University, China, CJS Technology Inc., Burlington, Ontario, Canada

Pavel V. Krivenko

Professor of the Kiev National University of Civil Engineering and Architecture, Kiev, Ukraine

Della Roy

Professor Emerita of Materials Science, Materials Research Laboratory The Pennsylvania State University, University Park, USA

CRC Press
Taylor & Francis Group
Boca Raton London New York

CRC Press is an imprint of the
Taylor & Francis Group, an **informa** business

CRC Press
Taylor & Francis Group
6000 Broken Sound Parkway NW, Suite 300
Boca Raton, FL 33487-2742

First issued in paperback 2019

© 2006 Caijun Shi, Pavel V. Krivenko and Della Roy
CRC Press is an imprint of Taylor & Francis Group, an Informa business

Typeset in Saban by
Integra Software Services, Pvt. Ltd, Pondicherry. India

No claim to original U.S. Government works

ISBN-13: 978-0-415-70004-7 (hbk)
ISBN-13: 978-0-367-86363-0 (pbk)

British Library Cataloguing in Publication Data
A catalogue record for this book is available from the British Library

Library of Congress Cataloging in Publication Data
A catalog record for this book has been applied for

Visit the Taylor & Francis Web site at
http://www.taylorandfrancis.com

and the CRC Press Web site at
http://www.crcpress.com

Contents

Acknowledgements

We would like to express our profound thanks and deep gratitude to those who make this book possible: Ms Chen Xiaoyang and Yuan Jiang of Chongqing University Press, China, for their enthusiastic help in reproducing the drawings and compilation of book index, Caijun Shi's daughter Joy Shi for reproducing and annotating many drawings for this book, Pavel Krivenko's colleague Dr Elena Kavalerova for searching and selecting relevant publications in Russian and translating them into English, Prof. Jushi Qian of Chongqing University, Chongqing, China, for collecting relevant Chinese publications, Dr Chuck Cornman of W. R. Grace, USA, for his review and comments on Chapter 2, Beatrix Kerkhoff of Portland Cement Association, USA, for her review and comments on Chapter 3, Dr Adrian Brough of Leeds University, UK, for his review and comments on Chapters 4 and 5, Dr Frank Collin of Maunsell Australia, Australia, for his review and comments on Chapters 6 and 7, Steve Kosmatka of Portland Cement Association, USA, for his review and comments on Chapters 8 and 13, Dr Ana Fernandez-Jimenez of Eduardo Torroja Institute (CSIC), Spain, for her review and comments on Chapters 8, 10 and 11, Prof. Jan Deja of University of Mining and Metallurgy, Poland, for his review and comments on Chapters 12 and 13.

We also want to thank the following individuals and organizations for their permission for the use of certain figures in this book: Drs Jan Deja, Ana Fernandez-Jimenez and Weiming Jiang, Technical Research Centre of Finland (VTT), Thomas Telford Ltd, John Wiley and Sons, American Concrete Institute, Elsevier Science, Chapman and Hall Ltd, Cement and Concrete Institute (South Africa) and National Council for Cement and Building Materials (New Delhi, India).

Caijun Shi
Pavel Krivenko
Della M. Roy

Chapter 1

Introduction

1.1 Historical developments of alkali-activated cements and concretes

The first use of alkali as a component of cementing material dates back to 1930, when Kuhl investigated the setting behaviour of mixtures of ground slag powder and caustic potash solution. Chassevent measured reactivity of slags using caustic potash and soda solution in 1937. Purdon (1940) did the first extensive laboratory study on clinkerless cements consisting of slag and caustic soda or slag and caustic alkalis produced by a base and an alkaline salt in 1940. The important historic developments of alkali-activated cements are summarized in Table 1.1.

In later 1957, Glukhovsky first discovered the possibility of producing binders using low basic calcium or calcium-free aluminosilicate (clays) and solutions of alkali metal (Glukhovsky 1959). He called the binders "soil cements" and the corresponding concretes "soil silicates". Depending on the composition of starting materials, the binders can be divided into two groups: alkaline binding system $Me_2O–Me_2O_3–SiO_2–H_2O$ and alkaline-earth alkali binding system $Me_2O–MeO–Me_2O_3–SiO_2–H_2O$. Extensive researches and developments on alkali-activated cements and concrete have been carried out since then. The Trief cements and F-cements from the Scandinavian countries (Forss 1983a, b) and alkali-activated blended cements are more recent examples (Davidovits 1988, Roy and Silsbee 1992).

In 1981, Davidovits of France produced binders by mixing alkalis with burnt mixture of kaolinite, limestone and dolomite (Davidovits 1981). He called the binders "geopolymer" since they have polymeric structure. He also used several trademarks such as Pyrament, Geopolycem and Geopolymite for the binder (Davidovits 1994). This type of materials virtually belongs to the alkaline binding system $Me_2O–Me_2O_3–SiO_2–H_2O$, as discovered by Glukhovsky.

Malek *et al.* (1986) identified alkali-activated cement type materials as the matrix formed in the solidification of certain radioactive wastes, while Roy and Langton (1989) showed some analogies of such materials with ancient concretes.

Table 1.1 Important historic developments of alkali-activated cements

Year	Name	Country	Work
1930	Kuhl	Germany	Investigated setting behaviour of slags in the presence of caustic potash.
1937	Chassevent	unknown	Measured reactivity of slags using caustic potash and soda solution.
1940	Purdon	Belgium	Investigated clinker-free cements consisting of slag and caustic soda or slag and caustic alkalis produced by a base and an alkaline salt.
1957	Glukhovsky	USSR	Synthesized binders using hydrous and anhydrous aluminosilicates (glassy rocks, clays, metallurgical slags, etc.) and alkalis, proposed $Me_2O–MeO–Me_2O_3–SiO_2–H_2O$ cementing system, and called the binder "soil cement".
1982	Davidovits	France	Mixed alkalis with a burnt mixture of kaolinite, lime stone and dolomite, and used several trademarks such as Geopolymer, Pyrament, Geopolycem, Geopolymite.

Krivenko (1994a) further showed that alkalis and alkali metal salts, as well as silicates, aluminates and aluminosilicates, exhibit reaction in alkaline aqueous medium when the alkali concentration is sufficient. Such interaction takes place with clay minerals, with aluminosilicate glasses of natural and artificial origin, in which calcium is absent, as well as with calcium binding systems under natural conditions, and form a water-resistant hardened alkali or alkali-alkaline-earth hydroaluminosilicates, which are analogous to natural zeolites and micas.

Kiev National University of Civil Engineering and Architecture organized two international conferences on alkali-activated cements and concretes in 1994 and 1999 (Krivenko 1994b, 1999). Davidovits in France organized two international conferences on geopolymer in 1988 and 1999 and Van Deventer from the University of Melbourne organized one conference on geopolymer in 2002. The fourth international conference and a workshop on geopolymer were held in France and Australia, respectively, in 2005.

1.2 Applications and specifications

Since the discovery of alkali-activated cements and concretes, they have been commercially produced and used for different purposes in a variety of construction projects in the former Soviet Union, China and some other countries. These commercial products and applications include:

- Structural concrete
- Masonry blocks
- Concrete pavements
- Concrete pipes
- Utility poles
- Concrete sinks and trenches
- Autoclaved aerated concrete
- Refractory concrete
- Oil-well cement
- Stabilization and solidification of hazardous and radioactive wastes

Over 60 specifications and standards related to alkali-activated slag cements and concretes were developed in the former Soviet Union. These specifications and standards cover raw materials, cements, concretes, structures and productions for alkali-activated slag cements and concretes. A lot of experience has been gained from design, production and applications during the past 40 years, which are invaluable for the further development and application of alkali-activated slag cement and concrete.

1.3 Future prospect of alkali-activated cements and concretes

Glukhovsky, as indicated above, in particular, emphasized the difference between the composition of traditional portland cements and the basic rock-forming minerals of the earth's crust. The major hydration products of the former are, of course, calcium silicate hydrate (C–S–H) and portlandite (Ca(OH)$_2$), whereas the latter cements are more represented poorly crystallized aluminosilicate gel, containing alkalis. The formation of the latter should give rise to the probability of enhanced durability. This is aside from the role of alkali-activated cements in helping protect the environment through (a) their utilization of by-product materials in their manufacture and (b) end applications in waste management.

During the past two decades, alkali-activated cements and concretes have attracted strong interests all over the world due to their advantages of low energy cost, high strength and good durability compared to portland cements. A major incentive for further development of such cements is generated by the annual output of fly ashes from power plants and other by-product materials, which is so enormous that there is a constant need to find new uses for them. In the US, approximately 49% of the utility wastes are simply landfilled, 41% are contained in surface impoundments, and about 10% are disposed of by discharging into old quarry operations. Increasingly, storage is conducted "on site" due to reduced costs to the utilities. This is to say nothing of the need to utilize SO$_3$-rich by-products.

Although much of the development of alkali-activated or alkaline cements has been based on activated slags, as will be seen, there is great potential for utilization of these other by-products.

In an early review paper, Roy (1999) summarized some challenges for alkali-activated cements and concretes. Based on laboratory and field experience with alkali-activated cements and concrete, many concerns and problems still need to be addressed, including the following:

- Alkali-activated cement and concrete often exhibit efflorescence problem due to the leaching out of alkalis, which react with CO_2 in the air to form alkali carbonates.
- Alkali-activated cement paste and concrete often exhibit larger drying shrinkage and may lead to cracking under dry conditions. The shrinkage usually increases as the lime content in the system decreases.
- Alkali-activated cement and concrete can be carbonated very slowly under wet conditions, but will crack and be carbonated very quickly under drying conditions.
- More research on alkali–aggregate reaction of alkali-activated cement and concrete is needed to understand the reaction and expansion mechanisms. The characteristics of by-products vary from source to source and affect the potential expansion of alkali–aggregate reaction. Cautions for potential alkali–aggregate reaction should be taken, especially when the aggregate may be known to be alkali-reactive.
- Most chemical admixtures on the market are mainly for portland cement-based concrete, they do not work well with alkali-activated cement concrete.
- The chemical and physical characteristics of raw materials may vary from time to time, from source to source. Quality control and quality assurance of alkali-activated cement and concrete is a big challenge.
- Considerable effort will be needed to gain more widespread development of the use of the materials through the evolution of more performance-based standards and not necessarily those directly derived from current portland cement standards. Although portland cement performance equivalence is one target to aim at, it is recognized that even portland cements are not "perfect" and that, in many cases, the properties of alkali-activated cements, especially durability, are superior.
- Development of database: Greater confidence will be gained in the manufacture and use of alkali-activated cements as a more extensive database is available to enhance the predictability of performance. This is particularly true because of some of the variability in natural or by-product/waste source materials. The current text should contribute significantly to this database.

1.4 Organization of this book

This is the first English book that intends to summarize the progresses in the past in alkali-activated cements and concretes. Chapter 2 discusses the production and properties of some commonly used chemical activators. Chapter 3 discusses the production and properties of cementing components used in alkali-activated cements and concrete. Chapter 4 discusses the hydration and microstructure of alkali-activated slag cements, Chapter 5 discusses the properties of alkali-activated slag cement pastes and mortars, Chapter 6 discusses the properties of alkali-activated slag cement concrete, Chapter 7 discusses the durability of alkali-activated cements and concrete, Chapter 8 presents the mix design of alkali-activated slag cement concrete, Chapter 9 discusses alkali-activated portland cement-based blended cements, Chapter 10 discusses alkali-activated lime-pozzolan cements, Chapter 11 discusses other alkali-activated cementitious systems, Chapter 12 presents applications and case studies, and Chapter 13 summarizes standards and specifications for alkali-activated cements, concretes and products.

This book can be used as a textbook for advanced courses in civil engineering materials, or as a reference book for students, laboratory workers, engineers and scientists.

Chapter 2

Alkaline activators

2.1 Introduction

Usually, caustic alkalis or alkaline salts are used as alkaline activators of alkali-activated cements and concretes. Glukhovsky *et al.* (1980) classified them into six groups according to their chemical compositions:

1. caustic alkalis: MOH;
2. non-silicate weak acid salts: $M_2CO_3, M_2SO_3, M_3PO_4$, MF, etc.;
3. silicates: $M_2O \cdot nSiO_2$;
4. aluminates: $M_2O \cdot nAl_2O_3$;
5. aluminosilicates: $M_2O \cdot Al_2O_3 \cdot (2-6)SiO_2$;
6. non-silicate strong acid salts: M_2SO_4.

Of all these activators, NaOH, $Na_2CO_3, Na_2O \cdot nSiO_2$ and Na_2SO_4 are the most widely available and economical chemicals. Some potassium compounds have been used in laboratory studies. However, their potential applications will be very limited due to their availability and costs. On the other hand, the properties of sodium and potassium compounds are very similar. This chapter discusses the production, properties and applications of NaOH, $Na_2CO_3, Na_2O \cdot nSiO_2$ and Na_2SO_4.

2.2 Caustic soda

2.2.1 Introduction

Caustic soda is one of the workhorses of chemical industry, sharing this position only with soda ash and sulphuric acid (The Occidental Chemical Corporation 1992a). There is scarcely a single chemical which does not require one or more of these three basic chemicals for its production, and furthermore, there is hardly an industrial product which does not depend upon caustic soda. Commercial grades of caustic soda are produced through electrolysis of brine. In addition to liquid, there are four types of commercially available solid caustic soda: solid, flake, compounder's and beads.

Solid caustic soda is obtained by cooling molten caustic soda, from which all the water has been evaporated, in metal drums. Flake caustic soda is made by passing molten caustic soda over cooled flaking rolls to form flakes of uniform thickness. The flakes are milled and screened into several forms of controlled particle sizes. The manufacture of caustic soda beads involves feeding molten liquor into a prilling tower under carefully controlled operating conditions, producing a spherical bead of uniform size. Solid caustic soda, flake caustic soda and caustic soda beads have the same chemical composition and differ only in particle size.

2.2.2 Properties of caustic soda

2.2.2.1 NaOH – H₂O System

Anhydrous caustic soda has a specific gravity of 2.13 at 20 °C and a melting point of 318 °C. Figure 2.1 shows the phase diagram of NaOH – H_2O system. It can be seen that NaOH has a solubility of approximately 53% at 20 °C. There are six hydrates – $NaOH \cdot H_2O$, $NaOH \cdot 2H_2O$, $NaOH \cdot 3.5H_2O$, $NaOH \cdot 4H_2O$, $NaOH \cdot 5H_2O$ and $NaOH \cdot 7H_2O$, which exist in

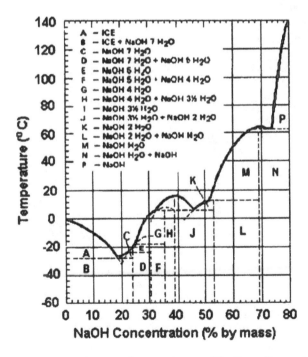

Figure 2.1 NaOH – H_2O phase diagram (The Dow Chemical Company 1994).

over-saturated solution. The combined water decreases with the increase of NaOH concentration.

The dissolution of NaOH releases a large amount of heat at low concentrations. The heat of dissolution slightly increases with concentration up to about 15%, then decreases with concentration.

The viscosity of NaOH solution increases with concentration. At 20°C, water has a viscosity of 1.00 centipoise, while it is 1.72 for 10% and 4.5 for 20% NaOH solutions. The viscosity of NaOH solution decreases slightly with temperature.

2.2.2.2   Dilution of concentrated NaOH solution

For anhydrous caustic soda, adequate amount can be dissolved to obtain a required concentration of NaOH solution. Usually, 50% NaOH solution is the most convenient and economical form of caustic soda. Thus, the concentrated solution has to be diluted to the required concentration. The dilution of a concentrated NaOH solution can be calculated as follows:

$$D = A\frac{B-C}{C} \qquad (2.1)$$

where
  A = specific gravity of strong solution;
  B = % NaOH in concentrated solution;
  C = % NaOH in desired solution;
  D = volumes of water added to each volume of strong solution.

Since the dilution of NaOH releases a large amount of heat, special cautions should be taken for diluting caustic soda solution: (1) caustic soda solutions are always added to water with agitation and water is never added to the caustic soda solution. If caustic soda is concentrated in one area, a rapid temperature increase can result in dangerous mists or boiling or spattering, which may cause immediate violent eruption; (2) the water should be lukewarm (30–40°C)(80–100°F) and should never be hot or cold water when starting. Figure 2.2 shows the temperature of final solution using 50% or 73% caustic solution for dilution.

2.2.3   Use of caustic soda in cement and concrete production

Caustic soda can be used as an accelerator of cement hydration. However, it results in a decrease of strength after 7 to 14 days of hydration. For most ultimate uses, caustic soda is used in solution form and the anhydrous solid caustic soda must be dissolved. Because of very high heat of solution, a

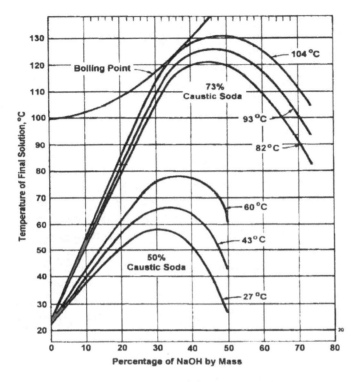

Figure 2.2 Dilution temperature of NaOH solution (The Dow Chemical Company 1993).

rapid temperature increase can result in dangerous boiling and/or spattering if caustic soda is added to a solution too fast, or if the solution is not sufficiently agitated or if added to a hot or cold liquid.

2.3 Sodium carbonate (soda ash)

2.3.1 Introduction

Sodium carbonate, or soda ash, is a white anhydrous powdered or granular material containing well above 99% sodium carbonate (Na_2CO_3). The accepted commercial standard for soda ash is expressed in terms of the equivalent sodium oxide (Na_2O) content. In other words, commercial 58% soda ash contains the Na_2CO_3 equivalent of 58% Na_2O.

According to physical characteristics such as bulk density and size and shape of particles, soda ash can be classified into light soda ash and dense soda ash. Light soda ash has a bulk density of 510 to 620 kg/m^3 and is produced by calcining the sodium sesquicarbonate precipitate recovered from the carbonation towers or vacuum crystallizers. Dense soda ash has a bulk density of 960 to 1060 kg/m³ and is produced by hydrating light soda ash followed by dehydration through calcination to produce denser crystals. Light soda ash and dense soda ash have the same other physical properties and chemical properties. Several publications have very detailed descriptions on sodium carbonate (Kostick 1993a, 1994, Garrett 1992, General Chemical Industrial Products Inc. 2003).

2.3.2 Sources of sodium carbonate

Sodium carbonate can be obtained from either natural sources or manufacture processes. Natural precipitates containing sodium carbonate-bearing minerals are located in shallow, non-marine alkaline lake sand marshes and usually co-exist with various chloride and sulphate salts in various countries through the world. These precipitates exist in five modes of occurrences: buried, surface or subsurface brines, crystalline shoreline or bottom crusts, shallow lake bottom crystals and surface efflorescence. Mannion (1983) and Kostick (1994) well compiled the world sodium carbonate occurrences. Table 2.1 lists some minerals that contain varying proportions of sodium carbonate.

The majority of synthetic soda ash is from the Solvay process and the remainder from the ammonium chloride, New Asahi and caustic carbonation processes. Because the Solvay process is energy, capital and labour intensive, and generates by-product calcium chloride and sodium chloride that cause environmental problems when discharged, the percentage of

Table 2.1 Sodium carbonate-bearing minerals (Kostick 1994)

Mineral	Chemical formula	%Na_2CO_3
Thermonatrite (monohydrate)	$Na_2CO_3 \cdot H_2O$	85.5
Wegscheiderite	$Na_2CO_3 \cdot 3NaHCO_3$	74.0
Trona (sesquicarbonate)	$Na_2CO_3 \cdot NaHCO_3 \cdot 2H_2O$	70.4
Nahcolite (sodium bicarbonate)	$NaHCO_3$	63.1
Bradleyite	$Na_3PO_4 \cdot MgCO_3$	47.1
Pirssonite	$Na_2CO_3 \cdot CaCO_3 \cdot 2H_2O$	43.8
Tychite	$MgCO_3 \cdot 2Na_2CO_3 \cdot Na_2SO_4$	42.6
Northupite	$Na_2CO_3 \cdot NaCl \cdot CaCO_3$	40.6
Natron (sal soda or washing soda)	$Na_2CO_3 \cdot 10H_2O$	37.1
Dawsonite	$NaAl(CO_3)(OH)_2$	35.8
Gaylussite	$Na_2CO_3 \cdot CaCO_3 \cdot 5H_2O$	35.8
Shortite	$Na_2CO_3 \cdot 2CaCO_3$	34.6
Burketite	$Na_2CO_3 \cdot 2Na_2SO_4$	27.2
Hanksite	$Na_2CO_3 \cdot 9Na_2SO_4 \cdot KCl$	13.6

synthetic soda ash decreased from 85% of total world production in 1971 to 70% in 1991.

2.3.3   Properties of sodium carbonate

2.3.3.1 $Na_2CO_3 - H_2O$ system

The phase diagram of $Na_2CO_3 - H_2O$ system is shown in Figure 2.3. There are three hydrated sodium carbonates: sodium carbonate monohydrate ($Na_2CO_3 \cdot H_2O$), sodium carbonate heptahydrate ($Na_2CO_3 \cdot 7H_2O$) and sodium carbonate decahydrate ($Na_2CO_3 \cdot 10H_2O$). The formation of which type of hydrated product depends on the concentration and temperature of the sodium carbonate solution.

The solubility of sodium carbonate in pure water can be derived from the phase diagram. It increases with increasing temperature until 35.4 °C, after which the solubility gradually decreases. A saturated solution at a temperature of 35.4 °C and above varies from 30 to 33% Na_2CO_3.

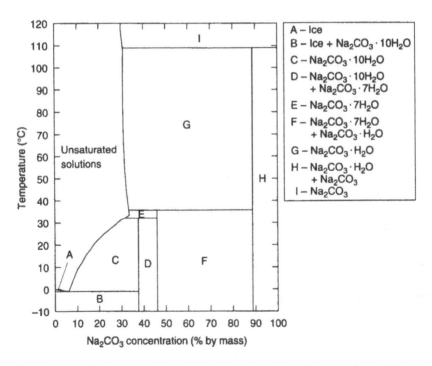

Figure 2.3 $Na_2CO_3 - H_2O$ phase diagram (based on General Chemical Industrial Products Inc. 2003).

Any soda ash added beyond saturation converts to the decahydrate form from −2.1 to 32 °C, to the heptahydrate form between 32 and 35.4 °C, and to the monohydrate form at temperatures above 35.4 °C. These processes are exothermic and generate heat that warms the solution. The presence of impurities may change the solubility of sodium carbonate. In that case, the phase diagram must be modified.

Sodium carbonate monohydrate ($Na_2CO_3 \cdot H_2O$), which contains 85.48% Na_2CO_3 and 14.52% crystallized water, precipitates as small crystals from saturated solutions above 35.4 °C, or may be formed simply by wetting soda ash with the calculated quantity of water at or above this temperature. Its solubility decreases slightly with increasing temperature. It loses water on heating and converts to Na_2CO_3 in saturated solution at 109 °C.

Sodium carbonate heptahydrate ($Na_2CO_3 \cdot 7H_2O$) contains 45.70% Na_2CO_3 and 54.3% crystallized water. It is of no commercial interest because of its narrow temperature stability, which extends from 32.0 to 35.4 °C.

Sodium carbonate decahydrate ($Na_2CO_3 \cdot 10H_2O$), commonly called sal soda or washing soda, contains 37.06% Na_2CO_3 and 62.94% crystallized water. It may be re-crystallized from saturated solution below 32.0 °C and above −2.1 °C or by wetting soda ash with calculated quantity of water within this temperature range. The crystals readily effloresce in dry air, forming a residue of low hydrates, principally the monohydrate.

When sodium carbonate anhydrous and hydrate are dissolved in water, they show different thermal properties. Sodium carbonate anhydrous and sodium carbonate monohydrate release heat but sodium carbonate heptahydrate and sodium carbonate decahydrate absorb heat. Table 2.2 is the heat of dissolution of one mole of sodium carbonate in 200 moles of water at 25 °C.

The heat of dissolution also depends on the concentration of solutions. Figure 2.4 is the relationship between the concentration and heat of dissolution of sodium carbonate anhydrous at 25 °C. It can be seen that the more concentrated the solution is, the greater the heat of dissolution. This means that dilution of a concentrated sodium carbonate solution will absorb heat and decrease the temperature of the solution. According to Figure 2.4, if a 20% solution is diluted to, say 10%, the temperature will decrease through the absorption of the following amount of heat: 70.4–63.6 = 6.8 kcal/kg of Na_2CO_3.

Table 2.2 Heat of dissolution of one mole of sodium carbonate in 200 moles of water at 25 °C

Formula	Na_2CO_3	$Na_2CO_3 \cdot H_2O$	$Na_2CO_3 \cdot 7H_2O$	$Na_2CO_3 \cdot 10H_2O$
Heat of Dissolution (kcal/mole)	5.87	2.69	−10.71	−16.26

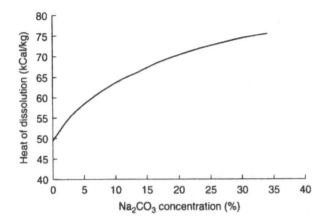

Figure 2.4 Effect of concentration on heat of dissolution of sodium carbonate at 25 °C (General Chemical Industrial Products Inc. 2003).

2.3.3.2 Specific gravity, concentration and temperature of sodium carbonate solutions

The specific gravity of a solid or liquid substance is the ratio of the weight of a given volume of the substance to the weight of an equal volume of water. The temperatures of both must be specified or understood, since volumes change with temperature. The specific gravity of a solution of sodium carbonate at 60 °/60 °F refers to the weight of a given volume of sodium carbonate solution at 60 °F as compared with the weight of an equal volume of water at 60 °F (General Chemical Industrial Products Inc. 2003).

In the United States, the National Bureau of Standards specified another scale, degrees Baume at 60 °F, to express specific gravities. For liquids heavier than water, the conversion of specific gravity and degrees Baume can be described as follows (General Chemical Industrial Products Inc. 2003):

$$\text{Degrees Baume at } 60\,°F = 145 - \frac{145}{\text{specific gravity at } 60°/60°F} \tag{2.2}$$

$$\text{specific gravity at } 60\,°F = \frac{145}{145 - \text{degrees Baume at } 60°/60°F} \tag{2.3}$$

The relationships between specific gravity, concentration and temperature are shown in Figure 2.5. This relationship is of very important practical significance because it can be used to control the required Na_2CO_3 concentration very quickly and accurately during the mixing of Na_2CO_3-activated cementitious materials.

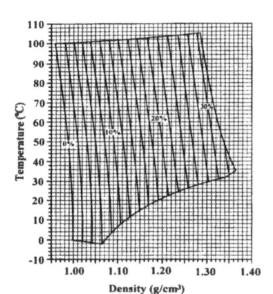

Figure 2.5 Relationships between specific gravity, concentration and temperature of sodium carbonate solutions (General Chemical Industrial Products Inc. 2003).

2.3.4 Application of soda ash in cement and concrete production

Soda ash was used as a chemical admixture for portland cement concrete. When the dosage is low, sodium carbonate can be used as an accelerator of cement hydration, but acts as a retarder when the dosage is high (Odler *et al.* 1976). Soda ash has been proven a good activator for a lot of cementing components, which will be discussed in the following chapters of this book.

2.4 Sodium silicate

2.4.1 Introduction

Sodium silicate is the generic name for a series of compounds with the formula $Na_2O \cdot nSiO_2$. Theoretically, the ratio n can be any number. Sodium silicates with different n have different properties that may have many diversified industrial applications. Commercial liquid sodium silicates have a ratio from 1.60 to 3.85. Sodium silicate liquids outside the range have limited stability and are not practical.

Sodium silicate was first discovered by Van Helmont in 1640 when he combined silica with an excess of alkali and got a liquid in damp places. Johann Nepomuk von Fuchs rediscovered it in 1818 during his experiments. He dissolved silica in caustic soda potash, observed the glass-like properties of the solution, and named it waterglass (Vail 1928). After some investigations, he proposed a variety of applications for the silicates such as glue, cements, paints, detergents, hardening agent for natural and artificial stones, etc. However, it did not become very popular until 1887 when W. Gossage & Sons of Windnes, England exhibited a soap which contained 30% of 20° Baume solution of sodium silicate.

The manufacture of soluble silicates dates back to 1864 when it was introduced by the Philadelphia Quartz Company. At that time, sodium silicate was mainly used for the manufacture of soap. The production volume grew very slowly. Soluble silicates are usually relatively inexpensive. However, their properties are very sensitive to their composition and storage conditions. Many potential applications have failed because necessary precautions were not taken either in the use or in the storing and handling (Wills 1982). Although many researches have been conducted on soluble silicates, some structure changes and property sensitivity of alkali silicate solutions are still not well understood. It has been reported that soluble alkali silicates are the most effective activators for most alkali-activated cementing materials. This section will discuss the production, structure and properties of soluble silicates in great detail.

2.4.2 Production of sodium silicate

The sodium silicate glass is obtained by melting primary sand and sodium carbonate at $1,350-1,450°C$. Then the glass is dissolved in an autoclave at $140-160°C$ under suitable steam pressure. Theoretically, silica and soda ash can be combined in all proportions. The practical range for commercial silicates, however, is determined by their very low solubility at high ratios and instability at normal temperatures at low ratios. Sodium silicate glass components can be designated as $Na_2O \cdot nSiO_2$ – where n is the ratio of the components and fall in the practical range from 0.4 to 4.0 (Occidental Chemical Corporation 1992b, PQ Corporation 1994). Figure 2.6 is the flow chart for the production of sodium silicate.

At the moment, most producers in the world use a spray process to reduce the moisture content in the liquid sodium silicate after the autoclave dissolves to the desired value. In the spray drying process, the silicate species in the solution change only slightly. Recently, a microwave drying method has been developed (Schramm 1999). It is reported that the microwave-dried silicate is dissolved faster than conventional spray-dried silicate powders. A new process has also been developed to manufacture hydrated sodium and potassium silicate (Russian Patent RU 213 4247 1999).

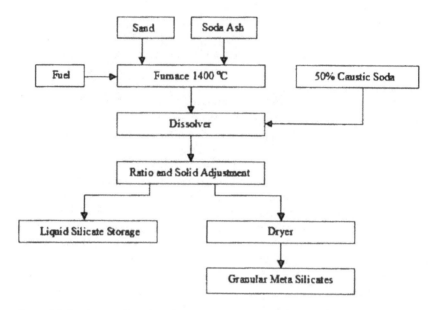

Figure 2.6 Production flow chart for sodium silicate (based on the PQ Corporation 1994).

NaOH is added to a sodium silicate solution with a high modulus to produce a sodium silicate solution with a lower modulus. In many cases, it is not convenient to do it on site. It can also be expected that, for a given modulus and concentration, the adjusted silicate solution may have different species as compared with the one manufactured directly. Korneev and Brykov (2000) proposed a method to produce amorphous hydrated alkali silicate with modulus over 1. In this method, organic solvents mixed with water are used to extract hydrated silicate after the ground silicate clod is dissolved in a dissolver. This method allows production of water-glasses of the optimal modulus and concentration directly.

2.4.3   Structure and properties of solid silicates

2.4.3.1   Physical and thermodynamic properties of solid sodium silicates

Commercial grades of liquid sodium silicate can be produced with mass ratio of SiO_2 to Na_2O from 1.60 to 3.85. This means that both the vitreous and liquid sodium silicates have no definite composition. The structure and properties of vitreous and liquid silicates also vary with their composition. However, for certain compositions, some definite soluble silicate can form. For example, if a solution contains the same molecular proportions of Na_2O

Table 2.3 Physical and thermodynamic properties of anhydrous and hydrous sodium silicates (Vail and Wills 1952)

Formula	Density (g/cm³)	Melting point (°C)	Heat of formation ΔH (Kcal/ g·mol)	Free energy ΔG (Kcal/ g·mol)	Entropy S
$Na_2O \cdot SiO_2$	2.614	1089	−359.8	−338	29
$Na_2O \cdot 2SiO_2$	2.50	874	−576.1	−541.2	39.4
$Na_2O \cdot nSiO_2$			$151.8 - 208.3n$	$-142.6 - 195.6n$	$18 + 11n$
$Na_2O \cdot SiO_2 \cdot 5H_2O$	1.75	72.2	−722	−631.5	77
$Na_2O \cdot SiO_2 \cdot 6H_2O$	1.81	62.9	−792.6	−688.2	87
$Na_2O \cdot SiO_2 \cdot 9H_2O$	1.65	47.9	−1005.1	−803.3	107

and SiO_2 and is concentrated, crystalline sodium metasilicate will form, which can exist in four forms: $Na_2O \cdot SiO_2$; $Na_2O \cdot SiO_2 \cdot 5H_2O$; $Na_2O \cdot SiO_2 \cdot 6H_2O$; and $Na_2O \cdot SiO_2 \cdot 9H_2O$. Another definite soluble silicate is $Na_2O \cdot 2SiO_2$. Table 2.3 summarizes some physical and thermodynamic properties of some anhydrous and hydrous sodium silicates.

2.4.3.2   *Dissolution of solid silicates*

Sodium silicates are usually sold in liquid form with modulus ranging from 1.6 to 3.3 except granular metasilicate. The dissolution of solid sodium silicate is an endothermic reaction. The dissolution rate and solubility of vitreous silicates decreases as the modulus increases. When the modulus of sodium silicate glass is less than two, its dissolution can be regarded as a congruent process. The dissolution experiments indicated that the constitution of the anhydrous Na_2SiO_3 solution became essentially constant within 5 minutes although complete dissolution took over 10 minutes. Trimethylsilylation (TMS) analyses indicate that the constitution of the solution became essentially constant within 5 minutes if sodium hydroxide pellets were added to a 0.1 mol/L $Na_2O \cdot 3.41SiO_2$ solution to give a modulus of 1 (Dent Glasser and Lachowski 1980).

The dissolution of alkali silicate glass with a modulus greater than 2 is a very complicated incongruent process and is still not well understood. Many factors such as temperature, water to solid ratio, modulus, sizes of the glass particles, additives and presence of impurities have effects on the dissolution. When glass particles are in contact with water, the ionic bonds between alkali and oxygen atoms can be broken easily under the action of water, and alkali ions are dissolved into water. Silicate network formers of glass with a high modulus cannot be completely dissolved in water at room temperatures (Korneev and Danilov 1996). The dissolution of alkali ions into water leaves a corroded layer on the surface of glass particles, which differs from either the glass or the solution phase. The hydrolysis of the silicate network can be expressed as follows:

$$\equiv Si \cdot O^- + H_2O \rightarrow \equiv Si \cdot OH + OH^- \tag{2.4}$$

The hydrolysis process results in the formation of the hydroxyl ions and increases pH of the solution, which further attack the network formers as follows:

$$\equiv Si-O-Si \equiv + OH^- \rightarrow \equiv SiOH + \equiv SiO^- \tag{2.5}$$

This reaction destroys some silicate network formers and releases silicate anions into the solution. It can be seen that the dissolution of sodium silicate glass is an incongruent process. The concentration profiles of different species in the corroded layer can be illustrated in Figure 2.7. Figure 2.8 shows how glass particle sizes affect the incongruent dissolution process.

As the modulus of a glass increases, there will be less alkali content in the glass. This means that it will be more difficult to dissolve the glass into water. Results in Table 2.4 indicate that the dissolution of sodium silicate slow down as the modulus and particle size of sodium silicate increases. Also, hydrated sodium silicates are dissolved much faster than anhydrous sodium silicate.

Another way to produce sodium silicate is by dissolving silica in caustic soda solution. Silica with low reactivity, such as quartz, can be used to produce sodium silicate solutions with a modulus below 2.5, and highly reactive silica can be used to produce sodium silicate solutions with high modulus.

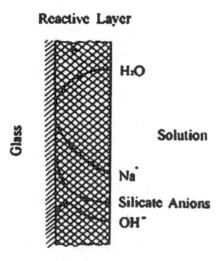

Reactive Layer

Glass

H₂O

Solution

Na⁺

Silicate Anions

OH⁻

Figure 2.7 Profiles of ion concentration in corroded sodium silicate glass surface layer in water (based on Korneev and Danilov 1996).

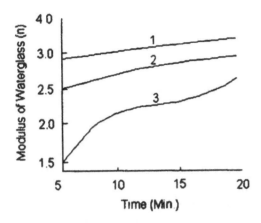

Figure 2.8 Change in modulus (n) of the solution with time while boiling powder of glass $Na_2O_3 \cdot 2SiO_2$ in 3 fold (by Mass) quantity of water powder particle size: 1–0.074 mm, 2–0.150 mm, 3–0.400 mm (based on Vail and Wills 1952).

Table 2.4 Dissolving rates of amorphous potassium and sodium silicates in water (3 parts of water + 1 part of powder) (Weldes and Lange 1969)

Modulus (SiO_2/Me_2O)	Type of silicate	Particle size (μm)	Time required to dissolve certain percentage of the glass powder			T (°C)
			50%	75%	100%	
3.22	Anhydrous sodium soluble silicate	300	60 hours	–	–	
3.22	Hydrated sodium soluble silicate (18.5% H_2O)	150	19 min.	45 min.	–	
2.00	Anhydrous sodium soluble silicate	300	10 hours	70 hours	–	25
2.00	Hydrated sodium soluble silicate (18.5% H_2O)	150	27 sec.	54 sec.	–	
2.50	Anhydrous potassium soluble silicate	300	60 min.	7.5 hours	48 hours	
3.22	Anhydrous sodium soluble silicate	300	15% during 30 min.			
3.22	Hydrated sodium soluble silicate (18.5% H_2O)	150	54 sec.	76 sec.	100 sec.	
2.00	Anhydrous sodium soluble silicate	300	17 min.	1 hour	–	50
2.00	Hydrated sodium soluble silicate (18.5% H_2O)	150	15 sec.	22 sec.	29 sec.	
2.50	Anhydrous potassium soluble silicate	300	12 min.	45 min.	–	

2.4.4 Properties and structure of liquid silicates

2.4.4.1 Concentration, specific gravity and modulus of liquid silicates

The composition of alkali silicate solutions can be expressed by two parameters: one is the modulus of the solution, and the other one is SiO_2 or Me_2O content, or the sum of $SiO_2 + Me_2O$. The sum of $SiO_2 + Me_2O$ can be determined by drying the solution and calculating the residue instead of using a complicated chemical analysis. Me_2O content is measured by acid titration with an indicator in a weakly acidic level. Silica is measured by either a gravimetric method or photocalorimetrically by a reaction with a molybdic acid. It will be easier to replace one of the above analyses by measuring density or refractive index of the solution. It is possible to use one analysis of SiO_2, or Me_2O, or a sum of $(SiO_2 + Me_2O)$ and one property of the solution, to accurately determine the concentration and modulus of the alkali silicate solution with the help of calibration charts or empirical formulae. Shtyrenkov et al. (Korneev and Danilov 1996) proposed an empirical formula for determination of the modulus of a sodium silicate solution n as follows:

$$n = 55.16(\rho - 1)N - 2.28 \tag{2.6}$$

where
ρ = density of sodium soluble silicate solution;
N = normality of alkali in the solution determined by titration.

2.4.4.2 Viscosity

The viscosity of alkali silicate solutions changes very little regardless of the ratio when Na_2O concentration is smaller than about 7%, but increase drastically with ratio if the Na_2O concentration is higher than about 7%. For a given temperature, viscosity of a sodium silicate solution increases as concentration and modulus of the solution increase, as shown in Figure 2.9 (Vail and Wills 1952). For a given modulus, there is a "threshold" concentration, above which the viscosity of the solution starts to increase drastically. The threshold concentration depends upon both the modulus and the nature of cation of the solution. For a given modulus, the threshold concentration for the potassium silicate solutions is lower than that for sodium silicate solutions. For a given solid content, the solution gives the lowest viscosity at a modulus of around 2, as shown in Figure 2.10.

The reasons for high viscosity of the silicate solutions are different, in its nature, from those for high molecular organic polymers (Korneev and Danilov 1996). Thus, an average molecular mass cannot be used to determine the characteristic viscosity of alkali silicate solution like polymers. For

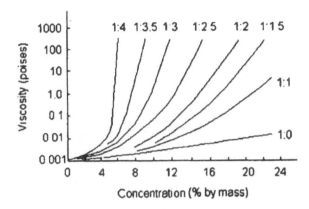

Figure 2.9 Effect of concentration and modulus on viscosity of sodium silicate solution at 20 °C (numbers in the figure are molar ratios) (based on Vail and Wills 1952).

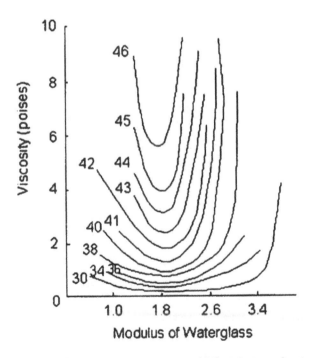

Figure 2.10 Maximum concentration of SiO_2 solutions of various silica modulus with maximum permissible viscosity for their transporting through pipelines at room temperature (based on Vail and Wills 1952).

a given content of alkali (Na_2O), an increase in modulus of the system leads to an increase in viscosity. When the modulus of a solution is greater than 4 but lower than 25, it tends to form a gel, which has a very low viscosity. For a given concentration and modulus, the nature of the silicate anions in the solution has a great effect on viscosity of the solution. Therefore, viscosity can reflect the difference in silicate anions to a great extent, but cannot be used to determine these silicate anions (Christophlienk 1985). An increase in temperature decreases the viscosity of alkali silicate solutions significantly, as shown in Figure 2.11.

2.4.4.3 Structure of liquid sodium silicate

Early studies using cryoscopic, ultra centrifuging, light scattering, ultrafiltration, conductivity, molybdate and TMS methods provided detailed information on mean molecular weights, particle size and degree of condensation of silicate anions in a silicate solution. However, more structural information with respect to the anionic structure of silicate solutions has become available only with the development and refinement of methods such as ^{29}Si-NMR (nuclear magnetic resonance) spectroscopy, paper chromatography, TMS, infrared spectroscopy (IR) and Raman spectroscopy, over the last twenty years. Despite these important advances in analytical techniques, the relationships between structure, properties and silicate species are still not completely understood, because the silicate species in solutions are so variable.

Based on ^{29}Si-NMR analysis, SiO_4 structural units can be classified into seven types: Q^0, Q^1, Q^2_{cy-3}, Q^2, Q^3_{cy-3}, Q^3 and Q^4. The superscripts on the Q represent the number of linkages between the given Si atom and

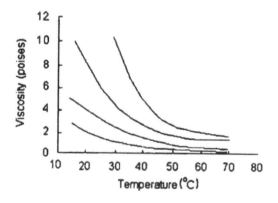

Figure 2.11 Relationship between viscosity of silicate solutions and temperature: $1 - n = 2.4$; 13.7% Na_2O; $2 - n = 2.9$; 10.9% Na_2O; $3 - n = 3.2$; 9.2% Na_2O; $4 - n = 2$; 14.5% Na_2O (based on Vail and Wills 1952).

neighbouring Si atoms by $=Si-O-Si=$ bonds. The symbols Q^2_{cy-3} and Q^3_{cy-3} designate intermediate or branched SiO_4 structural units in cyclo-tristructure (6-membered rings). Figure 2.12 shows the resonance ranges of these individual structural units in silicate solutions. The chemical shifts of the Q^0 to Q^4 structural units in the range from -70 to -110 ppm indicate that each Si atom in the units is always coordinated by four oxygen atoms.

When an industrial sodium silicate solution (6.14 M $Na_{0.61}Si$) is added with NaOH to obtain solutions of constant SiO_2 concentration (4 M) but variable Na_2O/SiO_2 ratios, the distribution of different structural units of the solutions, after storage for several weeks, is shown in Table 2.5.

When sodium silicate solutions with different moduli are adjusted to the same modulus and SiO_2 concentration by adding NaOH and water, almost identical molybdate reaction curves are obtained (Hoebbel and Ebert 1988). This indicates that the mean condensation degrees of these silicate anions are similar. It was found that the recovered low-molecular Si species in an NaOH-adjusted sodium silicate solution (original $n = 3.41$) was about

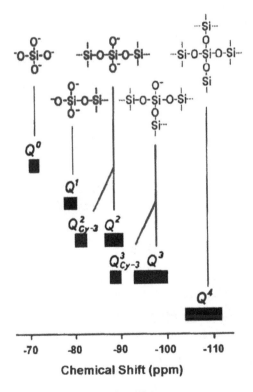

Figure 2.12 Ranges of the ^{29}Si chemical shift of the silicate structural units (based on Hoebbel and Ebert 1988).

Table 2.5 Effect of Na$_2$O/SiO$_2$ ratio on the distribution of different structural units of the silicate solutions (Hoebbel and Ebert 1988)

	SiO$_2$/Na$_2$O ratio	Percentage of structural units (%)					
		Q^0	Q^1	Q^2_{cy-3}	Q^2	Q^3	Q^4
4 M Na$_{0.61}$Si	3.28	1	7	0	33	53	7
4 M Na$_{1.5}$Si		12	25	22	36	6	0
4 M Na$_2$Si	1.0	20	31	28	13	0	0

10% higher than those in a solution made from anhydrous Na$_2$SiO$_3$ (Dent Glasser and Lachowski 1980).

Concentration of sodium silicate solutions also affects the mean condensation degree of silicate anions. If a 6.14 M Na$_{0.61}$Si solution ($R = 3.28$) is diluted with water to SiO$_2$ concentrations from 4 M to 0.1 M and tested by molybdate method after several weeks of storage, the final slopes of the molybdate reaction curve increase with decreased SiO$_2$ concentration in the solutions (Figure 2.13), which indicates a decreased mean condensation degree of the anions. It should be mentioned that the reaction of the silicate solution with molybdic acid, which is equal to an average of about 90% SiO$_2$ conversion, falls to 75% in the 0.1 M sodium silicate solution. This decrease in conversion is due to the formation of high-polymeric, molybdate-inactive anionic species. The occurrence of these high-polymeric

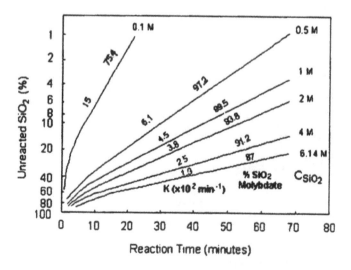

Figure 2.13 Effect of silicate concentration on the molybdate reaction rate of the silicate solutions (based on Hoebbel and Ebert 1988).

anionic species indicated that not only hydrolysis reactions, but also condensation happens when sodium silicate solutions are diluted. Impurities, melting temperature, autoclaving temperature, autoclaving time and storage time also affect the distribution of silicate species.

A decrease in modulus or a decrease in SiO_2 concentration of the solution increases hydrolysis of $=Si–O–Si=$ in accordance with Eqn (2.5), or it decreases the mean condensation degree of the anions. In a very diluted solution with extremely low modulus, it is composed primarily of monosilicate anion.

A slow freezing and thawing process may change the composition of silicate anions. Particularly, a partially frozen or thawed solution may have different modulus and concentration from those of the whole system.

2.4.4.4 pH value and stability of alkali silicate solutions

The pH of an alkali silicate solution is dependent on both its concentration and modulus, as shown in Figure 2.14. Alkali silicate solutions have very high buffer capacity. A sharp change in pH value is observed only after neutralization of the alkali in the solution. pH value is the most important characteristic determining stability of high-modulus silicate solutions, that is, their inclination to the formation of gel or coagulation. The stability of alkali silicate solutions increases steadily as their pH values increase (Figure 2.14).

As reported by Korneev and Danilov (1996), alkali silicate solutions tend to over-saturate. This tendency increases with the increase in modulus. The introduction of crystallization seeds can initiate crystallization of solid phase in the over-saturated solution. The crystallized phase does not have

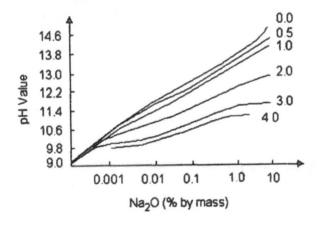

Figure 2.14 Effect of concentration and modulus on pH values of sodium silicate solutions at 20 °C (numbers in the figure are ratios of solutions) (based on Vail and Wills 1952).

a specific composition, but a mixture of several compounds. This crystallization behavior is due to the complexity of anion species in the solution, especially in high-modulus solutions. Solutions with high alkalinity and low modulus contain much more simple anion composition, mainly monomer. In a solution with modulus of approximately 1, the crystallized substance can be a single compound – crystalline alkali meta-silicate hydrate.

2.4.5 Applications of sodium silicates in cement and concrete production

Sodium silicate was widely used as an accelerator for concrete. At the moment, it is still often used as an accelerator for shotcrete. Throughout this book, you can see sodium silicate is the most effective alkaline activator for many cementing systems. With the development of a new technology for producing hydrated silicate powders, the applications of soluble glass in construction have greatly expanded, such as dry glue mixtures, adhesives, oil well cements, special cements for immobilization by cementation of different hazardous, toxic and radioactive wastes, acid resistant concrete, etc. (Korneev and Brykov 2000, Aborin *et al.* 2001).

2.5 Sodium sulphate

2.5.1 Introduction

Sodium sulphate, also known as disodium sulphate (Na_2SO_4), in its natural form is found in two principal minerals: thenardite (anhydrous Na_2SO_4) and mirabilite ($Na_2SO_4 \cdot 10H_2O$). Thenardite, the anhydrous form, was named after the French chemist, Louis Jacques Thenard (1777–1875). Mirabilite, the hydrous form, is commonly called Glauber's salt as it was discovered by German chemist J. R. Glauber (1603–1668).

Thenardite contains 43.68% Na_2O and 56.32% SO_3. It ranges from colourless to white and may be tinted shades of grey or brown. It has a specific gravity of 2.67 and a solubility of 15.9% by mass in water at 20 °C. It commonly occurs in the massive form without visible crystals. Its crystals are frequently tabular pyramids of the orthorhombic system. It has a melting point of 882 °C and decomposition temperature of 1100 °C. Its pH values range from 6.0 to 9.0 for 1% solution.

2.5.2 Resources of sodium sulphate

Sodium sulphate comes from either natural resources or from various manufacturing processes. Natural sodium sulphate is mainly from naturally occurring sodium-sulphate-bearing brines or crystalline evaporite deposits found in alkaline lakes in areas with dry climates and restricted drainage, from sub-surface deposits and brines (Kostick 1993b). The sodium-sulphate-bearing minerals are summarized in Table 2.6. The main sodium sulphate deposits are in China, Mexico, Canada, the United States, the former USSR and Spain. Less important are Argentina, Chile, Iran and Turkey.

Synthetic sodium sulphate is a by-product formed mainly during the production of viscose rayon, hydrochloric acid, sodium dichromate, ascorbic acid, battery acid recycling, boric acid, cellulose, lithium carbonate, rayon, resorcinol, and silica pigments. Most synthetic sodium sulphate can be used for other purposes (McIlveen and Cheek 1994).

Table 2.6 Summary of sodium sulphate-bearing minerals (McIlveen and Cheek 1994)

Mineral	Chemical formula	% Na_2SO_4
Thenardite (anhydrous)	Na_2SO_4	100
Hanksite	$9Na_2SO_4 \cdot 2Na_2CO_3 \cdot KCl$	81.7
D'ansite	$9Na_2SO_4 \cdot MgSO_4 \cdot 3NaCl$	81.2
Lecontite	$(Na, NH_4, K)_2SO_4 \cdot 2H_2O$	<79.8
Vanthoffite	$3Na_2SO_4 \cdot MgSO_4$	78
Hectorfloresite	$4Na_2SO_4 \cdot CaClO_3$	74.1
Sulphohalite	$2Na_2SO_4 \cdot NaCl \cdot NaF$	73.9
Burketite	$Na_6(SO_4)_2(CO_3)$	72.8
Eugsterite (Fritzsche's salt)	$2Na_2SO_4 \cdot CaSO_4 \cdot 2H_2O$	62.3
Darapskite	$NaNO_3 \cdot Na_2SO_4 \cdot H_2O$	58.0
Hydro-glauberite	$5Na_2SO_4 \cdot 3CaSO_4 \cdot 6H_2O$	57.9
Glauberite	$Na_2SO_4 \cdot CaSO_4$	51.1
Loeweite	$MgSO_4 \cdot Na_2SO_4 \cdot 2.5H_2O$	46.2
Ferrinatrite	$3Na_2SO_4 \cdot Fe_2(SO_4)_3 \cdot 6H_2O$	45.6
Mirabilite (Glauber's salt)	$Na_2SO_4 \cdot 10H_2O$	44.1
Bleodite (Astrakanite)	$MgSO_4 \cdot Na_2SO_4 \cdot H_2O$	42.5
Kroehnkite	$CuSO_4 \cdot Na_2SO_4 \cdot 2H_2O$	42.1
Nickelbloedite	$Na_2Ni(SO_4)_2 \cdot 4H_2O$	40.3
Sideronatrite	$Na_2Fe(SO_4)_2 \cdot (OH) \cdot 3H_2O$	38.9
Caracolite	$Pb(OH)Cl \cdot Na_2SO_4$	35.4
Palmierite	$(K, Na)_3Na(SO_4)_2$	<31.9
Tychite	$2MgCO_3 \cdot 2Na_2CO_3 \cdot NaSO_4$	27.2
Aphthitalite (glaserite)	$(K, Na)_3 \cdot Na(SO_4)_2$	21–38
Tamarugite	$Na_2SO_4 \cdot Al_2(SO_4)_3 \cdot 12H_2O$	20.3
Natrochalcite	$Cu_4(OH)_2(SO_4)_3 \cdot Na_2SO_4 \cdot 2H_2O$	18.8
Almeriite	$Na_2SO_4 \cdot Al_2(SO_4)_3 \cdot 5Al(OH)_3 \cdot H_2O$	15.9
Mendozite (soda alum)	$Na_2SO_4 \cdot Al_2(SO_4)_3 \cdot 24H_2O$	15.5
Natrojarosite	$Na_2Fe_6(OH)_{12}(SO_4)_4$	14.7
Noselite	$3Na_2Al_2Si_2O_8 \cdot Na_2SO_4$	14.3
Slavikite	$(Na, K)_2SO_4 \cdot Fe_{10}(OH)_6(SO_4)_{12} \cdot 63H_2O$	<4.6

2.5.3 Na₂SO₄ – H₂O phase diagram

Figure 2.15 is the $Na_2SO_4 - H_2O$ phase diagram. It can be seen that the maximum solubility is only about 30% even when it is heated. In addition to mirabilite ($Na_2SO_4 \cdot 10H_2O$), another hydrate, $Na_2SO_4 \cdot 7H_2O$, also exists. However, $Na_2SO_4 \cdot 7H_2O$ is unstable and has not been found in natural environments.

Mirabilite contains 55.9% of crystallized water and is opaque to colourless. It has a specific gravity of 1.48 and forms as efflorescent, needle-like monoclinic crystals, but generally is found in the massive form. Mirabilite has a very low melting point of 32.4 °C, slightly higher than room temperature.

Mirabilite was once an item of commerce. Because of its low melting point along with efflorescence when exposed to ambient conditions, sodium sulphate anhydrous is preferred for most applications. However, small quantities of mirabilite are still sold for some special uses.

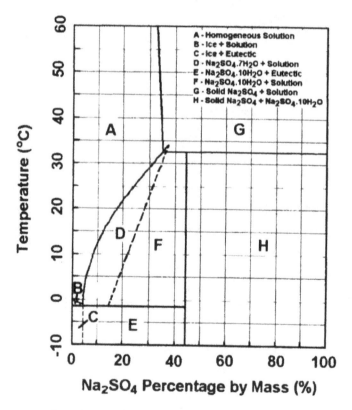

Figure 2.15 $Na_2SO_4 - H_2O$ phase diagram.

2.5.4 Applications of sodium sulphate in cement and concrete production

Many researches have confirmed that sodium sulphate can be a very effective alkaline activator for portland cement and lime-based cementing materials. The introduction of sulphates in cementing systems usually promotes the formation of ettringite at early and later ages. Detailed description of sodium sulphate-activated cementing systems can be found in Chapters 9 and 10.

2.6 Summary

This chapter has discussed the production, properties and applications of the four most commonly used activators: sodium hydroxide, sodium carbonate, sodium sulphate and sodium silicate. Sodium carbonate and sodium sulphate can be either from natural resources or manufacturing processes, while sodium hydroxide and sodium silicate are only from manufacturing processes. These alkalis have very different properties and should be handled differently during applications.

Chapter 3

Cementing components

3.1 Introduction

Chapter 2 discussed the most commonly used alkaline activators. This chapter will discuss cementing components that are commonly used in alkali-activated cements and concretes, which include granulated blast furnace slag, granulated phosphorus slag, steel slag, coal fly ash, volcanic glasses, zeolite, metakaolin, silica fume and non-ferrous slags.

3.2 Blast furnace slag

3.2.1 Production of blast furnace slag

In the production of iron, iron ore, iron scrap and fluxes (limestone and/or dolomite) are charged into a blast furnace along with coke for fuel. Figure 3.1 is an illustration of a blast furnace. Iron ore is either hematite (Fe_2O_3) or magnetite (Fe_3O_4) and the iron content ranges from 50 to 70%. This iron-rich ore can be charged directly into a blast furnace without any further processing. Iron ore that contains lower iron content must be processed or beneficiated to increase its iron content. Once these materials are charged into the furnace top, they go through numerous chemical and physical reactions while descending to the bottom of the furnace.

The coke descends to the bottom of the furnace to the level where the preheated air or hot blast enters the blast furnace. The hot air reacts with the coke to produce carbon monoxide, which then reduces iron oxides in the iron ores to pure iron and produces carbon dioxide, which leaks out of the furnace at the top. The limestone descends in the blast furnace and is decomposed into calcium oxide and additional carbon dioxide. CaO becomes the blast furnace flux for the slag and removes sulphur and other impurities. The slag is also formed from any remaining silica (SiO_2), alumina (Al_2O_3), magnesia (MgO) or lime (CaO) that enter with the iron ore or coke. The molten slag appears above the pig iron at the bottom since it is has a lower density. Its temperature is close to that of the molten pig iron,

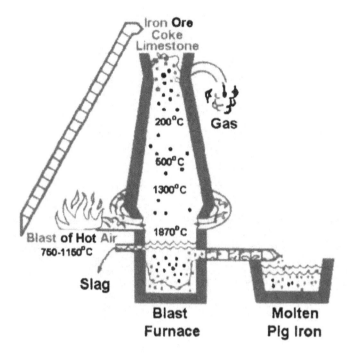

Iron **Ore**
Coke
Limestone

200°C

Gas

500°C

1300°C

1870°C

Blast **of Hot** Air
750-1150°C

Slag

**Blast
Furnace**

**Molten
Pig Iron**

Figure 3.1 Schematic representation of a blast furnace.

which ranges between 1400 and 1600 °C. The slag rises to the surface and is tapped off from time to time.

The higher the CaO/SiO_2 ratio of the slag is, the more effective is the desulphurization. Blast furnace operators may blend different types of limestone to produce the desired slag chemistry and to create optimum slag properties for operation, such as a low melting point and a high fluidity.

3.2.2 Chemical composition of blast furnace slag

The composition of blast furnace slag varies with the type of iron made and the type of ore used, which can be represented in a $CaO-SiO_2-Al_2O_3-MgO$ quaternary diagram. Osborn *et al.* (1954) conducted a large-scale study on practical furnace operation and determined the optimum slag composition for furnace operation in a $CaO-MgO-Al_2O_3-SiO_2$ system, as shown in Table 3.1. To give consistent iron production, the raw material burden needs to be carefully controlled and the range of chemical compositions of slag should be fairly narrow for a specific ore and furnace operation. However, the chemical composition of slag varies with ores and furnace operation, and may vary within a wide range. Table 3.2 lists chemical analyses of blast

Table 3.1 Optimal chemical composition of slags (% by mass) (Osborn et al. 1954)

No.	Al_2O_3	SiO_2	CaO	MgO
1	5.0	36.0	43.0	16.0
2	10.0	32.0	44.0	14.0
3	15.0	28.5	44.0	12.5
4	20.0	24.0	45.0	11.0
5	25.0	19.0	48.0	8.0
6	25.0	12.0	57.0	6.0
7	30.0	9.0	56.0	5.0
8	35.0	7.0	54.0	4.0

Table 3.2 Chemical composition of blast furnace slag from several countries in the world (% by mass)

SiO_2	Al_2O_3	Fe_2O_3	CaO	MgO	Na_2O	K_2O	S	TiO_2	MnO	Origin	Reference
35.04	13.91	0.29	39.43	6.13	0.34	0.39	0.44	0.42	0.43	Australia	Collins and Sanjayan (2001a)
35.3	9.9	0.6	34.7	14.6	0.3	0.4	1.0	0.5		Canada	Shi (1992)
36.23	9.76	1.99	39.4	10.5				0.7		China	Shi et al. (1991b)
36	9	1.3	41	8			1.1	0.9	0.85	Finland	Gjorv (1989)
35.4	12.9	0.3	41.8	6.8	0.26	0.38	1.0	1.65	0.42	Japan	Sato et al. (1986)
35	13.5	2.3	36.5	7.5			0.6	2.0	1.25	Norway	Gjorv (1989)
35.3	9.4	1.1	39.7	10.03	0.98		1.16	0.72	0.98	Sweden	Byfors et al. (1989)
34.9	7.12	1.02	42.87	10.30	0.24	0.50	1.16	0.39		USA	Hogan and Rose (1986)
34.2	11.3	1.17	41.6	8.21	0.26	0.4	0.48	0.77	0.25	UK	Osborne and Singh (1995)

furnace slag from several countries in the world. It seems that these slags have very similar SiO_2 and CaO contents, but have an obvious difference in Al_2O_3, MgO and TiO_2 contents.

3.2.3 Cooling of blast furnace slag

3.2.3.1 Slow cooling

A slow cooling of slag melts leads to a stable solid, which consists of crystalline Ca–Al–Mg silicates. Melilite – a solid solution of gehlenite C_2AS and akermanite C_3MS_2 – is the most common mineral. Of course, the mineral composition of a slowly cooled slag is determined by its chemical composition, as indicated in the $CaO–MgO–Al_2O_3–SiO_2$ phase diagram in Figure 3.2. These minerals, which are often identified in slowly cooled slag, are summarized in Table 3.3. The only crystalline compound in slowly

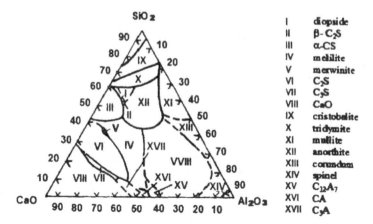

Figure 3.2 Crystallized phases in CaO–MgO–Al₂O₃–SiO₂ system.

Table 3.3 Minerals in air-cooled blast furnace slag (Smolczyk 1980)

Mineral name	Chemical formula	Abbreviation
Melilite (solid solution of gehlenite and akemanite)	$2CaO \cdot Al_2O_3 \cdot SiO_2 + 2CaO \cdot MgO \cdot 2SiO_2$	$C_2AS + C_2MS_2$
Merwinite	$3CaO \cdot MgO \cdot 2SiO_2$	C_3MS_2
Dicalcium silicate	$2CaO \cdot SiO_2$	C_2S
Rankinite	$3CaO \cdot 2SiO_2$	C_3S_2
Wollastonite	$CaO \cdot SiO_2$	CS
Diopside	$CaO \cdot MgO \cdot 2SiO_2$	CMS_2
Monticellite	$CaO \cdot MgO \cdot SiO_2$	CMS
Spinel	$MgO \cdot Al_2O_3$	MA
Magnesium silicate	$2MgO \cdot SiO_2$	M_2S
Sulphide	CaS, MnS, FeS	
Others	FeO, Fe_2O_3	

cooled slag, which possesses cementitious property, is β-C₂S. The sulphur in slowly cooled slag is mostly present as oldhamite (CaS). Thus, slowly cooled crystalline slag has little or no cementing property, but has mechanical properties similar to basalt and can be used as aggregates (Regourd 1986).

3.2.3.2 Fast cooling

As stated above, slowly cooled slag has little or no cementitious properties. The molten slag must be quickly cooled in order to enhance its cementitious property. The granulation of slags started in 1853 and is carried out in practice by a variety of methods. In some cases, the granulation of slags is done in central plants. The molten slag arising from pig iron production is transported in ladles to these plants and is poured into concrete pits filled with water. It is reported that the glass content of the slag granulated in this way may reach only 60–70% (Mass and Peters 1978). Concrete pits may be constructed near the blast furnace and kept filled with water.

Nowadays, molten slag is granulated directly by high-pressure water (about 0.6 MPa) jets at the furnace as it leaves the spout. To prevent pollution of the atmosphere by H_2S and SO_2, water is sprayed from special nozzles onto those parts where vapour and fumes are produced, and the system is totally enclosed with a venting stack. One tonne of slag consumes about three cubic meters of water. The water for granulation of the slag can be recirculated and the necessary pressure can be generated by a pump. Granulated slag contains about 30% water.

A pelletizer was developed in Canada and now is being widely used (Margesson and Englang 1971). In this process, molten slag is first cooled with water and then flung into the air by a rotary drum with a speed of 300 rpm. The water consumption is only about one cubic meter per ton of slag and the residual moisture content is only about 10% or even less.

The major advantage of using pelletized slag for cement production is its low moisture content resulting in less energy for drying. The other advantage is that the grinding of pelletized slag could consume up to 15% less energy than the grinding of granulated slag to obtain the same fineness (Cotaworth 1980). A later investigation indicated that the grindability of blast furnace slag appeared to be inversely related to glass content (Hogan 1983). Because water-granulated blast furnace slag has a higher glass content than pelletized slag, this may explain why pelletized slag can be more easily ground than water-granulated slag.

3.2.4 Structure of vitreous blast furnace slag

3.2.4.1 Vitreous structure theory

The structure of a glassy slag can be described using a two-dimensional framework of SiO_4 tetrahedra, represented schematically in Figure 3.3 (Bregg and Klaringboul 1967). According to the network theory proposed by Zachariasen (1932), the components of a glass can be classified into three groups: (1) network formers, (2) network modifiers and (3) intermediates.

Network formers are characterized by small ionic radii, the highest possible ionic valencies and are surrounded by four oxygen atoms. Together with oxygen atoms, they form a more or less disordered three-dimensional network through tetrahedra. The bond energies between these network formers and oxygen atoms are usually higher than 335 KJ/mol. Si and P are typical network formers in vitreous blast furnace slag. The higher the content of the network formers, the higher is the condensation degree of the glass.

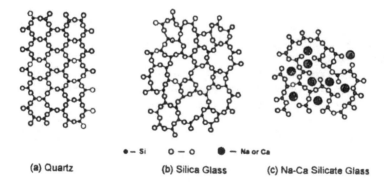

(a) Quartz   (b) Silica Glass   (c) Na-Ca Silicate Glass

● – Si   O – O   ● – Na or Ca

Figure 3.3 Two-dimensional representation of crystalline and vitreous structure (based on Din 1979).

Network modifiers have coordination number of 6 or 8, and have large ionic radii. The presence of network modifiers disorders and depolymerizes the network. The bond energies between the network modifiers and oxygen atoms are usually less than 210 KJ/mol. Na, K and Ca are typical network modifiers in vitreous blast furnace slag.

Intermediates can act as both the network formers and modifiers. The amphoteric metals Al and Mg are typical intermediates in vitreous blast furnace slag. Their coordination number is 4 when they act as network formers, and 6 as network modifiers. Their bond energies with oxygen atoms range from 210 to 335 KJ/mol. From the bond energies of network formers and network modifiers, it can be anticipated that the more the network formers are, the less reactive is the glass.

The condensation degree of network formers determines the number of non-bridge oxygen atoms and structural units within the vitreous structure (Govorov 1976). The condensation degree of SiO_4 tetrahedra in a vitreous material can be calculated as follows:

$$f_{Si} = \frac{SiO_2}{Me_2O + MeO + 3Me_2O_3 + 2MeO_2 + 5Me_2O_5} \tag{3.1}$$

Based on the condensation degree, slag can be classified into the following four types:

1 Orthosilicate-type slag – with a condensation degree from 0.25 to 0.286, and consisting of isolated SiO_4 tetrahedra isolated by Ca^{2+} cations.
2 Melilite-type slag – with a condensation degree from 0.286 to 0.333 and consisting of SiO_4 tetrahedra partially linked with each other to form diorthogroups $(Si_2O_7)^{6-}$ or linked with AlO_4 tetrahedra.

3   Wollastonite-type slag – with a condensation degree from 0.286 to 0.333, and consisting of small circular or chains of SiO_4 tetrahedra; when $(SiO_2 + 2/3Al_2O_3) < 50\%$ by mol, formation of separated tetrahedra and diorthogroups; when $(SiO_2 + 2/3Al_2O_3) > 50\%$ by mol, restricted space groupings.

4   Anorthite-type slag – with a condensation degree from 0.286 to 0.333, and consisting of a three-dimensional framework consisting of SiO_4 and Al-O tetrahedra with Ca^{2+} cations in cavities.

The role of MgO in blast furnace slags is dependent upon the CaO and Al_2O_3 content (Coale et al. 1973, Cheron and Landinois 1968, Taylor 1964). It was found that the slag exhibited the highest reactivity when the MO_6/MO_4 ratio is 0.35 (Satarin 1976). However, Yuan et al. (1987) felt that the optimum ratio was not a fixed value, but varied with the nature of the slag. They proposed a parameter α, or average percentage of ionic bonding, to evaluate the hydraulic reactivity of a glassy slag:

$$\alpha = \frac{\sum\limits_i^n [1 - \exp(-\tfrac{1}{4}\overline{x_o - x_m^2})].M_i}{\sum\limits_i^n M_i} \times 100 \qquad (3.2)$$

where

x_o = valence of oxygen atoms;
x_m = valence of network modifiers;
M_i = mass percentage of oxide i; and
n = number of oxides

Experimental results indicated that the higher the α is, the higher the hydraulic reactivity of the slag (Yuan et al. 1987). Although there are some controversies about the effect of the glass structure on the hydraulic reactivity of slag, it is generally accepted that the more disordered the slag is, the more reactive is the slag.

Chen and Yang (1989) found that granulated slag contains about 40 to 50% SiO_4^{4-}, 10 to 15% $Si_2O_7^{6-}$, 2 to 5% $Si_3O_{10}^{8-}$ and 1 to 5% $Si_4O_{12}^{8-}$. They also found that there was a very good relationship between the hydraulic reactivity and content of low-molecular-weight polymers of slags.

3.2.5 Measurement of hydraulic reactivity

So far, granulated blast furnace slag is mainly used as a partial replacement of portland cement in concrete. An extensive review on standards for fly ash and slag (Swamy 1993) indicated that some countries have specifications on both slag and portland slag cement. However, some countries have only performance standards for portland slag cement, but not on the hydraulic reactivity of slag. There are two different ASTM methods to determine the hydraulic reactivity of slag – ASTM C 595 (2003) and ASTM C 1073 (2003). ASTM C 595 requires blending of slag with cement to compare the strength of the blended cement with portland cement and ASTM C 1073 requires mixing of ground blast furnace slag with NaOH solution to measure the strength of the hardened slag mortars.

In ASTM C 595, the slag activity test with portland cement is the same as ASTM C 311 (2003) except for the use of a test mortar with 75 g of portland cement and an amount of slag determined by 175 g × density of slag/density of cement.

ASTM C 989 (2003) specifies that slag activity shall be evaluated by determining the compressive strength of both portland cement mortars and corresponding mortars made with the same mass of 50–50 mass combination of slag and portland cement. The slag activity index at 7 and 28 days can be calculated as follows:

$$\text{Slag activity index} = \frac{SP}{P} \times 100 \tag{3.3}$$

where

SP = average compressive strength of slag-reference cement mortar cubes at designed ages, MPa and

P = average compressive strength of reference cement mortar cubes at designed ages, MPa.

In ASTM C 1073 (2003), hydraulic activity of ground slag with alkali was specified. 100% ground slag is used as cement and specimens are prepared in accordance with ASTM C 109 (2003), except for the mixing water, which will be replaced by the volume of 20% NaOH solution equal to the volume of water, which would give a W/C ratio of 0.45 by mass. Immediately upon completion of moulding, the moulds are placed into a container with 50 ml water in order to assure 100% relative humidity during curing cycle. Then, the containers with moulds are placed into a curing chamber maintained at $55 \pm 2\,°C$. At the end of 23 ± 0.25 hours of curing, the moulds are taken from the container. The specimens are taken from moulds and are stored in room air another one hour for strength test. The strength is regarded as the hydraulic reactivity of the slag. The reason for developing this test method was that the chemical composition of cement used in ASTM C 595 and C 989 could have a significant effect on the strength of the blended cement (Frigione 1986, Mantel 1994). Although this method does not use cement to evaluate the hydraulic reactivity of the slag, different slags may show different sensitivity to NaOH (Shi and Day 1996c). Thus, this method is not good for evaluation of hydraulic reactivity of slags from different sources.

3.2.6 Factors affecting hydraulic reactivity of granulated blast furnace slag

3.2.6.1 Introduction

Many factors can affect the hydraulic reactivity of a slag. For slag itself, the factors include chemical composition, glass content and fineness (Botvinkin 1955). Externally, the factors include curing temperature and the material to be used with the slag for testing. This section mainly discusses how chemical composition, glass content and fineness affect the hydraulic reactivity of ground granulated blast furnace slag.

3.2.6.2 Chemical composition

As discussed above, the chemical composition determines the structure of a vitreous slag and plays an important role in determining its hydraulic properties. Table 3.4 lists different proposed chemical hydraulic indexes or hydraulic moduli based on chemical composition of granulated blast furnace slag for use as a partial replacement of portland cement. These hydraulic indexes can be classified into three types. Type I considers only the main component SiO_2 in the slag. K_3 is used as a quality criterion for blast furnace slag in many countries. For example, it requires $K_3 > 1.0$ in

Table 3.4 Hydraulic indexes of slag based on chemical composition (Smolczyk 1978, 1980, Shi 1987)

Type	Hydraulic index		
I	$K_1 = 100 - SiO_2$		$K_2 = \dfrac{100 - SiO_2}{SiO_2}$
	$K_3 = \dfrac{CaO + MgO + Al_2O_3}{SiO_2}$		$K_4 = \dfrac{CaO + MgO + Al_2O_3 - 10}{SiO_2}$
	$K_5 = \dfrac{CaO + 1.4MgO + 0.6Al_2O_3}{SiO_2}$		$K_3 = CaO + 0.5MgO + Al_2O_3 - 2.0SiO_2$
	$K_7 = \dfrac{6CaO + 3Al_2O_3}{7SiO_2 + 4MgO}$		
II	$K_8 = \dfrac{CaO + 0.5MgO + CaS}{SiO_2 + MnO}$		$K_9 = \dfrac{CaO + 0.5MgO + Al_2O_3}{SiO_2 + FeO + (MnO)^2}$
	$K_{10} = \dfrac{CaO + MgO + Al_2O_3 + BaO}{SiO_2 + MnO}$		$K_{11} = \dfrac{CaO + MgO + Al_2O_3}{SiO_2 + MnO + TiO_2}$
III	$K_{12} = \dfrac{CaO + MgO + 0.3Al_2O_3}{SiO_2 + 0.7Al_2O_3}$		$K_{13} = \dfrac{CaO + MgO}{SiO_2 + 0.5Al_2O_3}$

European and Canadian standards, $K_3 > 1.4$ in Japanese standard. K_7 was proposed based on the 28-day strength of an MgO-rich slag.

The second type of hydraulic index includes the effects of some minor components. K_{11} is used in Chinese standards and is required for a value of more than 1.2. Type III indicates that Al_2O_3 has an adverse effect on the hydraulic index. K_{12} was proposed based on 28-day strengths and K_{13} based on strengths after 28 days.

The presence of some minor components can also affect the hydraulic reactivity of slag. It was found that the presence of S^{2-}, within the range of its solubility limit (up to 2–2.5% S^{2-}), had a favourable effect on the hydraulic activity of blast furnace slag (Shkolnick 1986). The further increase of S^{2-} content did not show any difference. This may be because S^{2-} causes disorder of the network. Ti exists in six-coordination when TiO_2 content is less than 4% and in four-coordination when TiO_2 content is higher than 4% (Smolczyk 1980). Thus, Ti does not have a significant effect on hydraulic reactivity of the slag when TiO_2 content is less than 4%. The Chinese national standard GB203 (1992) limits the TiO_2 content to less than 10%.

The selectivity of activator is actually an indication of compatibility between chemical composition of the slag and activator(s) (Shi and Day 1996c). In order to examine how activator and the chemical composition of slag affect the strength of alkali-activated slag cement, eighteen vitreous materials from $CaO-Al_2O_3-SiO_2$ system, eight from $CaO-MgO-Al_2O_3-SiO_2$ system, and three from $CaO-Al_2O_3-SiO_2-TiO_2$ system have been synthesized (Glukhovsky and Raksha 1979). The CaO content varied between 30 and 50%, Al_2O_3 between 5 and 20%, MgO between 5 and 20% and TiO_2 between 5 and 15%. The iso-strength curves of alkali-activated slag cements made with materials from the $CaO-Al_2O_3-SiO_2$ system and different activators are shown in Figure 3.4.

Regardless of the curing conditions and the nature of activators, the cements made with synthetic slags consisting of 15–20% Al_2O_3 and 40–50% CaO show the highest strength. As CaO and Al_2O_3 contents in the synthetic slag decrease, the strength of corresponding cements decreases as well. Among the three activators, sodium metasilicate-activated slag cement gives the highest strength. It was also noticed that a replacement of 5–20% CaO with MgO in the glass does not essentially change the strength of the cements. However, the replacement of 5–15% SiO_2 in the glass composition with TiO_2 decreases strength. The decrease in strength is dependent upon the alkaline activator used. For example, with a 15% replacement of SiO_2 with TiO_2 in the synthetic slag, the strength of the cement is only 75% of the reference (Ti-free synthetic slag), when soda is used as an alkaline activator, 81% when sodium hydroxide is used, and 86% when sodium metasilicate is used. However, the effect of TiO_2 on the strength of alkali-activated-slag cements is much less than that on portland slag cement or lime-slag-gypsum cements.

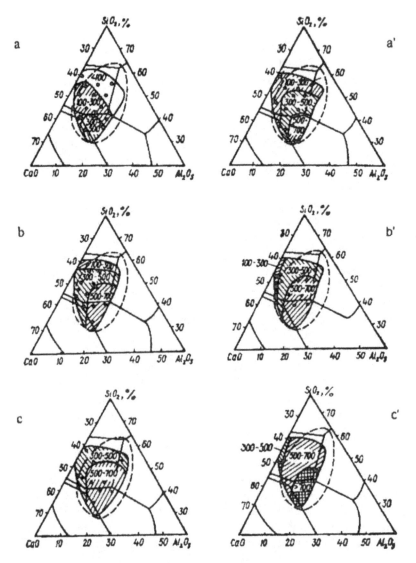

(*a* and *a'* – Soda solution with ρ=1180 kg/m³; *b* and *b'* – Sodium Hydroxide solution with ρ=1180 kg/m³; and *c* and *c'* – Sodium Metasilicate solution with ρ=1300 kg/m³);
a, *b*, and *c* – cements cured under normal conditions;
a', *b'*, and *c'* – cements cured under a regime: 2+6+2 hrs, T=368±5K.

Figure 3.4 Iso-strength curves of alkali-activated slag cements (Glukhovsky and Raksha 1979).

3.2.6.3 Glass content

As stated above, the hydraulic reactivity of granulated blast furnace slag is associated with its disordered glassy structure. Therefore, the glass content should play an important role in determining the hydraulic properties of a slag. However, results from different researchers are very controversial. Schwiete and Dolbor (1963) quenched hematite slags in different ways and found that the three-day strength correlated linearly with the glass content of slags. Demoulian *et al.* (1980) found that the strength of cement increases with glass content up to 95% and then decreases as glass content further increases. In some earlier studies, it was reported slags containing 40% glass content showed higher hydraulic reactivity than slags containing 80% glass (Budnikov and Gorshkov 1965, Botvinkin 1955). Demoulian *et al.* (1980) found that a glass content of less than 70% is acceptable, and Schwiete and Dolbor (1963) even felt that 30–40% glass content was also acceptable. Coale *et al.* (1973) suggested that the minimum glass content of quenched slag should be 85% for the assessment of slag as a potential cement replacement. Smolczyk (1980) stated that the presence of finely distributed crystals would have little detrimental effect on strength because these finely distributed crystals could remain partially or completely encapsulated in reactive glass even when ground to cement fineness. Other researchers felt that the crystals, such as merwinite, were mostly encapsulated in glass and their presence made the glass particles more reactive (Demoulian *et al.* 1980, Frearson and Uren 1986). Glukhovsky and Krivenko (1981) suggested that hydraulic activity of slag (especially in the presence of an alkali) would depend upon a combination of physical, chemical and structural peculiarities of both the vitreous and crystalline phases in the slag. Most standards in the world do not specify the glass content except the British Standard that requires a minimum glass content of 90% (Mehta 1986).

3.2.6.4 Grinding

Prolonged grinding increases not only the surface area of a material, but also the number of imperfections or active centres which exist at the edges, corners, projections and places where the interatomic distances are abnormal or are embedded with foreign atoms. These centres are in a higher energy state than in the normal structure. The more such centres are, the higher the reactivity (Gregg 1961, Dave 1981). In one study, it was observed that percussive dry grinding could cause obvious crystal distortion of kaolinite (Millers and Oulton 1970). In another study, it was noticed that impaction and friction milling of high alumina cement altered its crystallinity and notably modified its hydraulic behaviour (Scian *et al.* 1991). Quartz can be a pozzolanic material when it is finely ground (Alexander 1960). Generally speaking, the hydraulic reactivity of slag increases as its fineness increases

(Nakamura *et al.* 1986; Sato *et al.* 1986, Osbaeck 1989). Several studies have consistently found that the strength of a slag cement linearly increases with the Blaine fineness of the slag used either as a lime slag cement, or portland slag cement, or alkali-activated slag cement (Osbaeck 1989, Shi and Li 1989b, Wang *et al.* 1994).

3.3   Granulated phosphorus slag

3.3.1   *Production of granulated phosphorus slag*

Elemental phosphorus is produced by smelting a mixture of phosphate rock, silica and coke in an electric furnace. Phosphate ores usually consist of small particles, and the raw material must be agglomerated and hardened before smelting. High-grade lump phosphate may be used without agglomeration. Coke combines with oxygen in the phosphate and releases phosphorus vapour. The chemical reaction can be expressed as follows:

$$Ca_3(PO_4)_2 + 5C + 3xSiO_2 \rightarrow P_2 \uparrow + 5CO + 3(CaO \cdot xSiO_2) \qquad (3.4)$$

The x, or the SiO_2/CaO ratio of phosphorus slag usually ranges from 0.8 to 1.2. Figure 3.5 is an illustration of a phosphorus production process. Phosphate can be reduced to form elemental phosphorus without addition of any silica rock. However, the addition of silica rock makes the melt in the furnace fluid and the phosphate can be more readily reduced (Barber 1975). Iron compounds present in those raw materials are reduced in furnaces to elemental iron, which combines with phosphorus to form a metallic by-product – ferrophosphorus. Phosphorus recovery decreases as

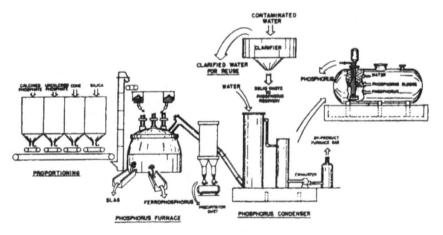

Figure 3.5 Schematic illustration of phosphorus production (Barber 1975).

ferrophosphorus forms. Thus, iron is undesirable in the furnace. Other metals present in those raw materials can also be reduced and the metals mix with the ferrophosphorus. Phosphate minerals always contain some impurities such as fluorapatite, phosphatized limestone, sand and clay. Those unreduced compounds come out in the slag. The production of one ton of elemental phosphorus generates approximately ten tons of slag.

3.3.2 Chemical and mineral compositions

Phosphorus slag is composed mainly of SiO_2 and CaO. The minor components in phosphorus slag, which depend on the nature of phosphate ores used, are 2.5–5% Al_2O_3, 0.2–2.5% Fe_2O_3, 0.5–3% MgO, 1–5% P_2O_5 and 0–2.5% F. Table 3.5 lists the chemical composition of several phosphorus slags produced in the world.

Molten phosphorus slag can be cooled either in air or through water-quenching. If phosphorus slag is air-cooled, it consists mainly of crystalline compounds such as $CaO \cdot SiO_2$ and $3CaO \cdot 2SiO_2 \cdot CaF_2$ can also be detected if CaO/SiO_2 ratio of the slag is at the high end. If it is water-quenched, it mainly consists of a vitreous structure with a glass content of up to 98% due to the high viscosity of the molten slag.

3.3.3 Hydraulic reactivity of phosphorus slag

Air-cooled phosphorus slag does not exhibit any cementitious property and can only be crushed for uses as ballast or aggregate. Granulated phosphorus slag has a vitreous structure similar to that of granulated blast furnace slag. Granulated phosphorus slag is a latent cementitious material but less reactive than granulated blast furnace slag at early age due to its lower Al_2O_3 content and the presence of P_2O_5 and F. The results obtained by Shi *et al.* (1989b) confirmed this. However, if a portion of siliceous materials used as flux is replaced with bauxite to give Al_2O_3 content similar to that in blast furnace slag, the hydraulic reactivity of the granulated phosphorus slag will

Table 3.5 Chemical composition of phosphorus slag from several countries in the world (% by mass)

SiO_2	Al_2O_3	Fe_2O_3	CaO	MgO	P_2O_5	F	Origin	Reference
38.2	3.9	0.4	49.0	1.6	1.6	2.9	Nanjing, China	Shi et al. (1991a)
41.1	4.1	2.5	44.8	2.8	2.4	2.7	Yunnan, China	Wu (1984)
42.9	2.1	0.2	47.2	2.0	1.8	2.5	Germany	Wu (1984)
43.0	3.4	3.2	45.0		3.0	2.7	Former USSR	Wu (1984)

be comparable with that of granulated blast furnace slag (Sun *et al.* 1984). The hydraulic index of phosphorus slag is defined as follows:

$$K_{14} = \frac{CaO + MgO + Al_2O_3}{SiO_2 + P_2O_5} \tag{3.5}$$

In the Chinese and the former USSR standards for granulated phosphorus slag for use as a cement replacement, they require a minimum K_{14} of 1.2 and a maximum P_2O_5 content of 2.5%. P_2O_5 in the slag can exist in the form of both orthophosphates and condensed phosphates. Orthophosphates are very soluble and increase the time of setting of portland cement. Condensed phosphates act as the network formers and decrease the reactivity of granulated phosphorus slag because P—O bonds have higher bonding energy than Si—O or Al—O bonds. Generally speaking, the replacement of portland cement with granulated phosphorus slag increases the setting time of the cement (Sun *et al.* 1984, Wu 1984). No effects from F in the phosphorus slag were observed on the properties of the cement (Sun *et al.* 1984). The replacement of gypsum with Na_2SO_4 can activate the potential activity of phosphorus slag and increase the early strength of portland phosphorus slag cement very significantly (Shi *et al.* 1989b). The latent cementitious properties of granulated phosphorus slag can be very effectively activated using alkalis (RST 5024-83 1983, Shi and Li 1989b). The Na_2SiO_3-activated phosphorus slag mortars exhibit a compressive strength of 80 MPa at 28 days. The other interesting fact is that the presence of soluble phosphates does not affect the setting and strength development of alkali-activated granulated phosphorus slag cements (Shi and Li 1989b). Alkali activation of phosphorus slag will be discussed in the following chapters.

3.4 Steel slag

3.4.1 Production of steel slag

Steel slag is a by-product from either the conversion of iron to steel in a basic oxygen furnace (BOF), or the melting of scrap to make steel in an electric arc furnace (EAF). In the basic oxygen process, hot liquid blast furnace metal, scrap and fluxes, which consist of lime (CaO) and dolomitic lime, are charged to a furnace. A lance is lowered into the converter and high-pressure oxygen is injected. The oxygen combines with and removes the impurities in the charge (Figure 3.6a). These impurities consist of carbon as gaseous carbon monoxide, and silicon, manganese, phosphorus and some iron as liquid oxides, which combine with lime and dolomitic lime to form steel slag. At the end of the refining operation, the liquid steel is tapped (poured)

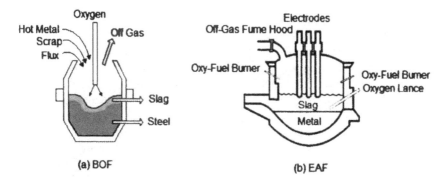

Figure 3.6 Schematic illustration of a BOF and an EAF (Shi 2004b).

into a ladle while steel slag is retained in the vessel and subsequently tapped into a separate slag pot.

Unlike the basic oxygen process, the EAF does not use hot metal, but uses "cold" steel scraps, which would otherwise be unsightly and environmentally damaging. It can be charged with limited amounts of iron scrap, pig iron and direct reduced iron. The EAF is a kettle-shaped structure with a removable lid, as shown in Figure 3.6b. The three graphite electrodes that heat the furnace pass through the lid. An electric current is passed through the electrodes to form an arc. The heat generated by this arc melts the scrap. During the melting process, other metals (ferro-alloys) are added to the steel to give it the required chemical composition. Also oxygen is blown into the electric arc furnace to purify the steel. After samples have been taken to check the chemical composition of the steel, the EAF is tilted to allow the slag, which is floating on the surface of the molten steel, to be poured off. The EAF is then tilted in the other direction and the molten steel poured into a ladle.

After being tapped from the furnace, the molten steel is transferred to a ladle for further refining to remove additional impurities still contained within the steel. This operation is called ladle refining because it is completed within the transfer ladle. During ladle refining, additional steel slags are generated by again adding fluxes to the ladle to melt. These slags are combined with any carryover of furnace slag and assist in absorbing deoxidation products (inclusions), heat insulation and protection of ladle refractories. The slags produced at this stage of steel-making are generally referred to as ladle slags. Because the ladle refining stage usually involves comparatively high flux additions, the properties of the slag from ladle refining stage are quite different from those of the slag from BOF.

3.4.2 Cooling of steel slag

There are several methods for cooling molten steel slag: natural air-cooling, water-spray, water-quenching, air-quenching and shallow box chilling. Natural air-cooled steel slag is cooled in air naturally after pouring into a pit or on ground. It consists mainly of big lumps and some powder. The powder results from the conversion of β-C_2S to γ-C_2S at around 675 °C. Since β-C_2S and γ-C_2S have different crystal structure and density, the conversion of β-C_2S to γ-C_2S is accompanied by an increase in volume of nearly 10% and results in the shattering of the crystals into dust.

In the water-spray cooling process, molten steel slag is poured into a pit and cooled in air. Water is sprayed onto the surface of the steel slag after the molten steel slag is solidified so the solidified steel slag will break into pieces itself due to the temperature differences, which will make the slag handling and metal recovery much easier. It may also prevent the conversion of β-C_2S to γ-C_2S if the water spray takes place at higher temperatures.

Water-quenching is widely used for molten blast furnace slag but rarely used for steel slag since molten steel slag has a higher viscosity than molten blast furnace slag. Thus, water can be trapped in steel slag easily and cause explosion. Several steel plants used the water-quenching process and the Qiangtao steel plant in Qingtao, China is still using the water-quenching method to process steel slag. Water-quenched steel slag consists of particles from 3 to 5 mm (Zhu *et al.* 1989).

Air-quenching process was recently developed by Chongqing Iron and Steel Co. Ltd, Chongqing, China, and is being used in several steel plants in China (Ye and Liao 1999). Basically, the molten steel slag in slag cart is poured into a slot, underneath which there is an air nozzle. Compressed air is blown onto slag stream with a pressure between 0.35 to 0.6 MPa. The compressed air should be turned on before pouring the molten slag. The slag is blown into small particles with size ranging from 3 to 5 mm and fall into a pond filled with water. This process is very simple and has no secondary environmental pollution. However, most metals are oxidized during the air-quenching process. Air-quenched steel slag is much more difficult to be ground than air-cooled or water-sprayed steel slag.

The shallow box chilling was developed in Japan. Montgomery and Wang (1991) described the process in detail. Molten steel slag of around 100 mm thickness is poured into a pan and cooled for four minutes. Then, water will be sprayed onto the slag for approximately 20 minutes dropping the temperature of the steel slag down to around 500 °C. After the initial water spray, the slag will be transported to a spraying station for further water spraying for four minutes to lower the temperature to about 200 °C. Finally the slag is placed into a water pool for further cooling and magnetic separation. The steel slag cooled in this manner has a particle size of 30–50 mm and low free lime content (2–4%). Because water is sprayed onto a relatively thin steel slag layer, the risk of explosion due to steam generation and entrapment is avoided.

3.4.3 Chemical composition of steel slag

Chemical composition of steel slag is highly variable and changes from batch to batch even in one plant depending on raw materials, type of steel made, furnace conditions, etc. Table 3.6 lists the chemical composition ranges for different types of steel slag. Steel slag from EAF for the production of carbon steels is very similar to that from BOF. However, the slag from EAF for the production of alloy or stainless steels is quite different. It has a lower FeO content and a very high content of Cr, which leads to classifying the slag as a hazardous waste in US and Canada.

Chemical composition of ladle slag is significantly different from that of steel furnace slag in that the former has a low FeO content. Some steel-making operations use Al for further refining purposes. In these cases, the ladle slag has a high Al_2O_3 content. In other operations, they use CaF_2 for further refining purposes. Then the ladle slag consists mainly of CaO and SiO_2.

3.4.4 Mineral composition of steel slag

Since the chemical composition of steel slag varies significantly from source to source, it can be expected that the mineralogical composition of steel slag can be very different from source to source. Reported minerals in steel slag include olivine, merwinite, C_3S, β-C_2S, γ-C_2S, C_4AF, C_2F, RO phase (CaO-FeO-MnO-MgO solid solution), free-CaO and free-MgO (Shi 2004b). Table 3.7 summarizes reported minerals in different steel slags.

Regardless of operations, the mainly mineral composition of ladle slag is γ-C_2S, which is converted from β-C_2S during cooling. The conversion of

Table 3.6 Chemical composition range of steel slags (%) (Shi 2004b)

Components	BOF	EAF (carbon steel)	EAF (alloy/stainless)	Ladle
SiO_2	8–20	9–20	24–32	2–35
Al_2O_3	1–6	2–9	3.0–7.5	5–35
FeO	10–35	15–30	1–6	0.1–15
CaO	30–55	35–60	39–45	30–60
MgO	5–15	5–15	8–15	1–10
MnO	2–8	3–8	0.4–2	0–5
TiO_2	0.4–2	N/A	N/A	N/A
S	0.05–0.15	0.08–0.2	0.1–0.3	0.1–1
P	0.2–2	0.01–0.25	0.01–0.07	0.1–0.4
Cr	0.1–0.5	0.1–1	0.1–20	0–0.5

Table 3.7 Minerals in air-cooled steel slags (Shi 2004b)

Mineral name	Chemical formula	Abbreviation
Merwinite	$3CaO \cdot MgO \cdot 2SiO_2$	C_3MS_2
Tricalcium silicate	$3CaO \cdot SiO_2$	C_3S
Dicalcium silicate	$2CaO \cdot SiO_2$	$\beta\text{-}C_2S, \gamma\text{-}C_2S$
Rankinite	$3CaO \cdot 2SiO_2$	C_3S_2
Wollastonite	$CaO \cdot SiO_2$	CS
Diopside	$CaO \cdot MgO \cdot 2SiO_2$	CMS_2
Monticellite	$CaO \cdot MgO \cdot SiO_2$	CMS
Calcium aluminate	$CaO \cdot Al_2O_3$	CA
Calcium ferrite	$CaO \cdot Fe_2O_3$	CF
Magnesium silicate	$2MgO \cdot SiO_2$	M_2S
Sulphide	CaS, MnS, FeS	
RO Phase	$FeO\text{–}MnO\text{–}CaO\text{–}MgO$	RO
Lime	CaO	
Periclase	MgO	
Others	FeO, Fe_2O_3	

$\beta\text{-}C_2S$ to $\gamma\text{-}C_2S$ is accompanied by an increase in volume of nearly 10% and results in the shattering of the crystals into dust because of their different crystal structures and densities. Thus, ladle slag is sometimes called falling slag.

The RO phase consists primarily of FeO, which has a cubic structure similar to MnO, CaO and MgO in the solid solution. This phase, along with the ferrite phase, is the last to solidify and locates in between silicate grains.

The free lime in steel slag comes from two sources: residual free lime from the raw material and precipitated lime from the molten slag. Wachsmuth et al. (1980) found that there is a relationship between residual free lime, precipitated lime and total free lime content, which is plotted in Figure 3.7. When the total free lime content in steel slag is less than 4%, it mainly comes from precipitation of lime from molten slag. When the total free lime content is more than 4%, the precipitated lime does not change much with total free lime content and the free lime is mainly attributed to the residual lime. The precipitated lime content is about 2% at a total free lime content of 4%, and increases to around 2.8% as the total free lime content increases to 12%. Thus, the volume soundness of steel slag is mainly determined by the content of residual lime (Geiseler 1996).

MgO in steel slag comes from dolomite, which is used as a flux, and MgO refractory, which is used as the lining of the steel furnace. A high content of MgO is often detected because refractory material comes into the slag, which also causes soundness problems of steel slag. Steel slag is used as asphalt concrete aggregate in many countries (Ye and Burstrom

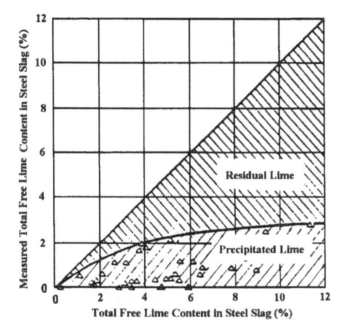

Figure 3.7 Relationship between residual lime, precipitated lime and total lime content in steel slag (Wachsmuth *et al.* 1980).

1995, Geiseler 1996). The Ministry of Transportation of Ontario, Canada, banned the use of steel slag in asphalt concrete on provincial highways in 1993 because of the soundness problem.

3.4.5 Cementing properties of steel slag

The presence of C_3S, C_2S, C_4AF and C_2F contributes to the cementitious property of steel furnace slag. Table 3.8 summarizes the relationship between basicity, main mineral phase and hydraulic reactivity of steel slag. It can be seen that the reactivity of steel slag increases with its basicity. However, free-CaO content also increases with the increase of the basicity of steel slag. The C_3S content in steel slag is much lower than that in portland cement. Thus, steel slag can be regarded as a weak portland cement clinker (Tang 1973).

A speciality cement, steel and iron slag cement, which is composed mainly of steel furnace slag, blast furnace slag, cement clinker and gypsum, has been commercially marketed in China for more than 20 years (Sun and Yuan 1983, Wang and Lin 1983). The strength of the steel slag cement also depends on the basicity of the steel slag used, as shown in Figure 3.8. Steel

Table 3.8 Reactivity, basicity and mineral compositions of steel slag (Tang 1973)

Hydraulic reactivity	Types of steel slag	Basicity		Major mineral phases
		CaO/SiO_2	$CaO/(SiO_2 + P_2O_5)$	
Low	Olivine	0.9–1.5	0.9–1.4	Olivine, RO phase and Merwinite
	Merwinite		1.4–1.6	Merwinite, C_2S and RO phase
Medium	Dicalcium silicate	1.5–2.7	1.6–2.4	C_2S and RO phase
High	Tricalcium silicate	>2.7	>2.4	C_3S, C_2S, C_4AF, C_2F and RO phase

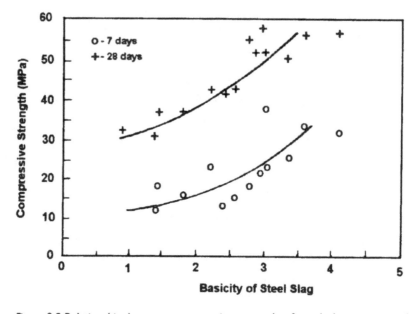

Figure 3.8 Relationship between compressive strength of steel slag cement and basicity of steel slag (based on Chinese Academy of Building Materials 1985).

slag cement can be used for general construction uses, especially suitable for mass concrete and pavement applications due to its special features.

The Chinese national standard GB 13590-92 (1992) specifies the composition, properties, testing, storage and applications for steel and iron slag cement. Another standard YB/T 022-92 (1992), which was developed and administrated by the Ministry of Metallurgical Industry of China, specifies the composition, quality, testing and storage of steel slag that can be used for the production of steel and iron slag cement.

Steel furnace slag itself can display very good cementing properties under the action of a proper alkaline activator(s). Several studies (Petropavlovsky 1987, Li and Wu 1992, Shi *et al.* 1993) have confirmed that the use of alkaline activator can increase the strength, especially early strengths, and other properties of the steel furnace slag cement. Usually, some other materials such as blast furnace slag or fly ash should be used together with steel slag in order to eliminate the soundness problem. Alkali-activated steel slag–blast furnace slag cement can show very high strength and corrosion resistance (Petropavlovsky 1987, Bin *et al.* 1989, 1992, Shi 1999), which is discussed in detail in Chapter 10.

Ladle slag consists mainly of γ-C_2S and does not display obvious cementitious properties under normal hydration condition. However, it shows significant cementing properties under the activation of alkalis. Its cementitious property increases with the fineness of the slag (Shi 2002).

3.5 Pozzolans

3.5.1 Definition of pozzolans

The term "pozzolan" comes from the US simplification of "pozzolana" which derived from the location "Pozzuoli, Italy". Here the Romans found a reactive silica-based material of volcanic origin which they called "pulvis puteolanus" (Everett 1967). Today, both the terms "pozzolan" and "pozzolana" are used.

According to ASTM C 618 (2003), a pozzolan is defined as a "siliceous or siliceous and aluminous material which in itself possesses little or no cementitious value but will, in finely divided form and in the presence of moisture, chemically react with calcium hydroxides at ordinary temperatures to form compounds possessing cementitious properties". Typical examples are volcanic glasses, tuff, fly ashes and silica fumes.

It is estimated that about 5% of the solid surface of the earth is covered by volcanic rocks or effusives (Lorenz 1985). At the same time, a large number of industrial by-products, such as fly ash, slag, silica fumes, are produced every year.

3.5.2 Classification of pozzolans

There have been several classification systems proposed for pozzolans. The most common classification divides pozzolans into two groups – natural and artificial pozzolans, as shown in Figure 3.9. Natural pozzolans refer to

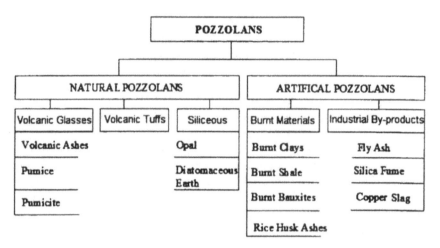

Figure 3.9 Classification of pozzolans (Shi 2001).

materials that show pozzolanic reactivity without any processing. They can be sub-classified into three categories:

1   volcanic glasses: derived from volcanic rocks in which amorphous glass is produced from fusion. Volcanic ashes, pumice and pumicite are typical examples.
2   volcanic tuffs (zeolites): altered volcanic glass formed by the action of groundwaters on volcanic glass under high temperature. Many factors, such as the characteristics of the volcanic glass, groundwater properties, temperature and pressure, affect the zeolitization process.
3   siliceous: usually formed from the precipitation of silica from solution or from the remains of organisms. The common materials are diatomaceous earth, diatomaceous stone, opal and chert.

Artificial pozzolans are divided into two categories according to their origin – "industrial by-products" and "burnt materials". Blast furnace slag, fly ash, silica fume, copper slag and nickel slag are typical industrial by-products of the iron industry, power generating plant, copper and nickel production respectively. "Burnt materials" refer to those that have pozzolanic reactivity only after calcination: burnt clay, burnt shale, burnt rice husks and burnt bauxite are examples.

3.5.3   Chemical composition of pozzolans

Regardless of the sources or nature, pozzolans consist mainly of SiO_2 and Al_2O_3, as summarized in previous publications (Day 1992, Shi 1992). The total content of SiO_2 and Al_2O_3 is usually above 70%. The other oxides in pozzolans include Fe_2O_3, CaO, MgO, Na_2O, K_2O, etc. The total content of K_2O and Na_2O can be more than 10% in some zeolites.

3.5.4 Evaluation of pozzolanic reactivity of pozzolans

Since Vicat proposed the first method – a lime absorption test in 1937, many methods have been developed to evaluate pozzolanic reactivity of pozzolanic materials. Several publications (Moran and Gilliland 1950, Day 1992, Shi 2001) have summarized these methods, as shown in Table 3.9.

Lime absorption, solubility of a pozzolan or the electrical conductivity change of a solution due to the dissolution of a pozzolan may correlate well with the performance tests of one or more pozzolans, but not of all pozzolans. The performance of a hardened paste depends not only on the

Table 3.9 Summary of methods for evaluation of pozzolanic reactivity of pozzolans (Shi 2001)

Method		Evaluation criteria
Lime absorption		amount of absorbed lime in a pozzolan/saturated $Ca(OH)_2$ solution at different ages
Setting time		setting time of 1:4 lime pozzolan pastes by Vicat test
Solubility		
in saturated $Ca(OH)_2$ solution		decrease of Ca^{2+} in solution due to the addition of a pozzolan
in alkali		dissolved SiO_2 or $SiO_2 + R_2O_3$ of pozzolan in alkaline solution*
in acid		dissolved $SiO_2 + R_2O_3$ of pozzolan in acid
in alkali then in acid		dissolved $SiO_2 + R_2O_3$ of pozzolan after treatment in acid then in alkali solution
Electrical conductivity		change of electrical conductivity in saturated $Ca(OH)_2$ solution or HF (HF + HNO_3) solution within certain time after the addition of a pozzolan
Mechanical strength	pozzolan + portland cement	tensile strength difference of mortars cured at 18° and 50 °C strength ratio of portland pozzolan cement to pure portland cement mortars
	pozzolan + lime	strength of lime–pozzolan mixtures cured under controlled conditions at a specified age

* R – Al or Fe.

reaction rate and degree, but also on the nature of reaction products. Performance standards are becoming prominent because more diverse materials and combinations of materials are being used to produce satisfactory products. In practice, people are primarily concerned about the performance of the materials used.

Malquori (1960) stated that an evaluation of pozzolanic materials for purpose of their partial replacement for portland cement must be based on two factors: (1) the mechanical strength of mortars and concretes made with a portland pozzolan cement mixture, and (2) the reduction of free calcium hydroxide in the hardened pozzolanic cement. Takemoto and Uchikawa (1980) also suggested that the quality of natural or artificial pozzolanic cements should be evaluated by strength tests. Many standards now use the compressive or tensile strength of mortars. The mortars are prepared with a specified ratio of pozzolan to portland cement or pozzolan to lime, and curing conditions are closely controlled.

3.5.5 Volcanic glasses

3.5.5.1 Volcanic ashes

Volcanic ashes are formed during explosive eruptions by shattering of solid rocks and violent separation of magma (molten rock) into tiny pieces. Explosive eruptions are generated when ground water is heated by magma and abruptly converted to steam and also when magma reaches the surface so that volcanic gases dissolved in the molten rock expand and escape (explode) into the air extremely rapidly. Volcanic ash is composed of fragments of rock, minerals and glass that are less than 2 mm (0.08 inch) in diameter. Figure 3.10 shows particle size distribution of two volcanic ashes from Bolivia and Guatemala. Although these two ashes are from different locations, they have similar particle size distribution and consist mainly of particles ranging between 20 and 300 μm.

Figure 3.11 shows the raw volcanic ash particles from Bolivia before and after grinding. The raw volcanic particles are very irregular and pitted. Many tiny particles are stacked in the pits on the surface of pozzolan particles. After grinding, most pits are destroyed and more small irregular particles are produced. Some pits with tiny particles can still be seen on the surface of coarse particles even after grinding.

3.5.5.2 Pumice and pumicite

During an explosive eruption, volcanic gases dissolved in the liquid portion of magma also expand rapidly to create a foam or froth; the liquid part of the froth quickly solidifies to glass above ground. Pumice is actually a kind of glass and not a mixture of minerals. It is colourless or light grey and has

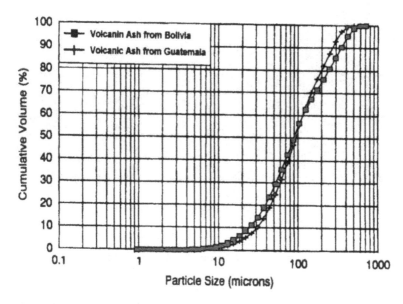

Figure 3.10 Particle size distribution of volcanic ashes from Guatemala and Bolivia (Shi 1992).

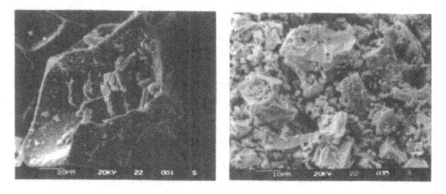

Figure 3.11 Volcanic ashes before and after grinding (Shi 1992, Shi and Day 1995b).

the general appearance of rock froth. The viscosity of the lava, the quantity of water vapour and gas, and the rate of cooling together determine the fineness of the vesicular substance. Large amounts of gas result in a finer-grained variety known as pumicite. The chemical composition is that of granite. Coarser-grained rock, with fewer and larger air spaces, is called scoria; it is usually associated with dark-colored igneous rocks of diorite or gabbro composition.

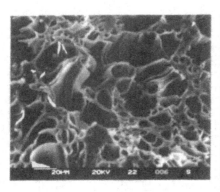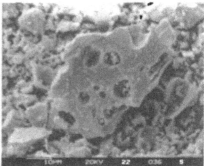

Figure 3.12 SEM pictures of pumice before and after grinding in a ball mill (Shi 1992, Shi and Day 1995b).

The main use for pumice and pumicite is lightweight aggregate for concrete blocks and assorted building products. Other major applications for pumice and pumicite include abrasive, absorbent, filter aid, etc. In 2000, the total consumption of pumice and pumicite was 697,000 metric tons at an average of $24.27/ton (Bolen, 2000).

Figure 3.12 shows the SEM pictures of a pumice sample from Guatemala before and after grinding (Shi and Day 1995b). The honeycomb-like structure can be observed with the help of SEM under low magnification. The channels inside the pumice range from several to $100\,\mu m$, and the majority of the particles are less than $20\,\mu m$ after grinding, very small irregular particles without pores form, and some big particles with pores can still be observed. Since pumice consists mainly of vitreous phase, ground pumice can be a good pozzolanic material (Shi and Day 1995b).

3.5.5.3 Pozzolanic reactivity of volcanic glasses

Pozzolanic reactivity of volcanic glasses results from their glassy structure. As discussed in Section 3.2.6, any factor that affects the glassy structure will influence the pozzolanic reactivity of a volcanic glass. The most significant factors include chemical composition, glass content, fineness, curing temperature and use of chemical activators. The detailed discussion of how the pozzolanic reactivity of volcanic glasses is affected by different factors can be found in the literature review (Shi 2001).

3.5.6 Zeolites

Zeolites are hydrated aluminosilicates with symmetrically stacked alumina and silica tetrahedra which result in an open and stable three-dimensional

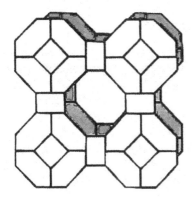

Figure 3.13 Illustration of structural unit of zeolites.

honeycomb framework of consistent diameter interconnecting channels (Figure 3.13). These channels allow the easy movement of the resident ions and molecules into and out of the structure. The diameter of these open channels is what differentiates each species of natural zeolites and is what gives rise to their unique properties. Within these channels are positively charged cations such as sodium, potassium, barium and calcium and even relatively large molecules and cation groups such as water, ammonia, carbonate ions and nitrate ions, attached and held by the structure's negative charge. Zeolites are characterized by their ability to lose and absorb water without damage to their crystal structures. Over 150 types of zeolite structures have been identified.

Natural zeolites are formed over millions of years by nature from the alteration of volcanic ash in alkaline waters under high pressure and then crystallized. They have pozzolanic property and are widely used as a cement replacement in China (Guo and Liang 1980, Kasai *et al.* 1992, Feng 1993).

Feng (1993) discussed the pozzolanic reactivity of different types of zeolite and the properties of cement and concrete containing ground zeolite as a cement replacement. Zeolite may show much higher lime absorption and strength with lime than calcined clay. The pozzolanic reaction between zeolite and $Ca(OH)_2$ was discussed in detail by Guo and Liang (1980). When lime is present, stacked alumina and silica tetrahedral are destructed under the action of OH^-. Destructed alumina and silica tetrahedral react with Ca^{2+} in the solution and form C–S–H and C–A–S–H. A calcination of a zeolite between 600 and 900 °C can improve its pozzolanic reactivity since the heating destructs the alumina and silica tetrahedral and makes the material more vulnerable to OH^- attack.

3.5.7 Coal fly ash

3.5.7.1 Production and characteristics of fly ashes

Fly ashes are by-products of power generation at coal-fired power plants, collected by the fabric filters and/or the electrostatic precipitators that remove the particulates (solid particles) from the smoke, making the smoke cleaner and less harmful to the environment. Fly ashes are heterogeneous fine powders consisting mostly of rounded or spherical glassy particles of SiO_2, Al_2O_3, Fe_2O_3 and CaO. The composition of fly ash depends on the coal used, but also on the various substances injected into the coal or gas stream to reduce gaseous pollutants or to improve efficiency of particulate collectors. When limestone and dolomite are used for desulphurization of the exit gases, CaO and MgO contents in fly ash will be increased. Conditioning agents such as sulphur trioxide, sodium carbonate and bicarbonate, sodium sulphate, phosphorus, magnesium oxide, water, ammonia and triethylamine are often used to improve the collection efficiency. There are also irregular or angular particles including both unburned coal remnants and mineral particles. According to ASTM C 618 (2003), fly ash belongs to Class F if the $(SiO_2 + Al_2O_3 + Fe_2O_3) > 70\%$, and belongs to Class C if $70\% > (SiO_2 + Al_2O_3 + Fe_2O_3) > 50\%$. Both fly ashes consist mainly of spherical particles. No difference in their shape and size could be discerned. Class F fly ash particles have a clean surface (Figure 3.14a), while there are deposits of various condensates, such as alkalis and sulphates, on the surface of Class C fly ash particles, as shown in Figure 3.14b.

The most abundant phase in fly ashes is glass. Crystalline compounds usually account for from 5 to 50% and include quartz, mullite, hematite, spinel, magnetite, melilite, gehlenite, kalsilite, calcium sulphate, alkali sulphate (Hemmings and Berry, 1988). High calcium fly ash may contain appreciable

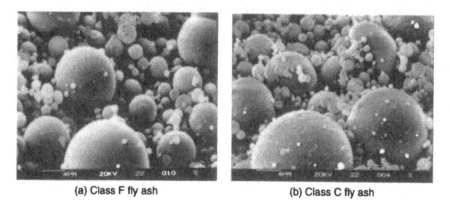

(a) Class F fly ash   (b) Class C fly ash

Figure 3.14 SEM pictures of classes F and C fly ash particles (Shi 1996a).

amounts of free CaO, C_3A, C_2S, $CaSO_4$, MgO and $4CaO \cdot 3Al_2O_3 \cdot SO_3$. The X-ray diffraction (XRD) technique is very useful in identifying these crystallized substances in fly ashes. A broad diffraction halo, which is attributed to the glassy phase, always appears on the XRD patterns of fly ashes. The position of the diffraction halo is related to the lime content in the fly ash (Diamond 1983, McCarthy *et al.* 1988).

Most of the fly ash particles are solid spheres and some are hollow cenospheres. Also present are plerospheres, which are spheres containing smaller spheres. The presence of internal voids in the glassy ash particles (produced during cooling) and voids in the agglomerated particles causes wide variations in the particle density. The variation in composition, especially iron and carbon contents, also causes differences in density (Minnick *et al.* 1971). The average density of fly ash without grinding varies between 1900 to 2800 kg/m³. The bulky density varies greatly with packing and changes from 540 to 860 kg/m³ without close compaction.

3.5.7.2 Pozzolanic and cementitious properties of fly ashes

Class F fly ashes usually have a low CaO content and have very little or no cementitious properties. The pozzolanic reactivity of Class F fly ashes is mainly determined by the characteristics of the aluminosilicate glass and particle sizes of the fly ashes. The factors affecting the pozzolanic reactivity of fly ashes are the same as those affecting volcanic glasses.

Class C fly ashes contain, in addition to glass phase, appreciable amount of crystalline phases such as free CaO, C_3A, C_2S, $CaSO_4$, MgO and $4CaO \cdot 3Al_2O_3 \cdot SO_3$ and have cementing properties (McCarthy *et al.* 1984). The hydration behaviour of C_3A and C_2S in fly ash is the same as that in portland cement (Ghosh and Pratt 1981). Many people often contribute the cementing properties of Class C fly ashes to its crystalline phases. Actually, glass phases with high lime content can also show very high hydration reactivity and give self-hardening cementing properties (McDowell 1986).

3.5.8 Metakaolin

3.5.8.1 Production of metakaolin

Metakaolin is obtained by heating kaolin-containing clays expressed as follows:

$$\underset{\text{kaolin}}{Al_2(Si_2O_5)(OH)_4} \xrightarrow{560-580\,^{\circ}C} \underset{\text{metakaolin}}{Al_2O_3 \cdot 2SiO_2} + 2H_2O \qquad (3.6)$$

Researches have indicated that the metakaolin produced under heating temperature of 600 to 900 °C shows the highest pozzolanic reactivity (Gao

Table 3.10 Effect of calcination on the structural changes of clays (Shi 2001)

Step	Temperature (°C)	Structure change
Step I	<180	Loss of surface and adsorbed water
Step II	180–500	Dehydroxylation of the clay structure
Step III	600–800	Rupture of bonds and collapse of clay structure
Step IV	900–1000	Formation of new high temperature phases

et al. 1989). The changes in the structure of clay during calcination are summarized in Table 3.10. If the temperature is above 900°C, crystalline mullite ($Al_6Si_2O_{13}$) or spinel ($MgAl_2O_4$) and amorphous silica will form, and the reactivity of metakaolin decreases.

3.5.8.2 Pozzolanic reactivity of metakaolin

Metakaolin can react with lime and produce both C–S–H and hydrated gehlenite (Jambor 1963):

$$Al_2O_3 \cdot 2SiO_2 + 3Ca(OH)_2 + nH_2O \longrightarrow C\text{-}S\text{-}H + C_2ASH_8 \qquad (3.7)$$

Metakaolin has a particle size range from 0.5 to 20 μm, and is a highly reactive pozzolan. A partial replacement of portland cement with metakaolin can increase strength development, reduce permeability and improve durability of concrete (Caldarone *et al.* 1994, Hooton *et al.* 1997, Ashbridge *et al.* 1996). The addition of Na_2SO_4 significantly accelerates the pozzolanic reaction between metakaolin and lime (Shi *et al.* 1999). Metakaolin can react with alkalis and form cementing materials, which will be discussed in detail in Chapters 10 and 11 (De Silva and Glasser 1991, Davidovits 1994).

3.5.9 Condensed silica fume

Condensed silica fume, also known by other names, such as volatilized silica, microsilica or simply silica fume, is a by-product of the manufacture of silicon or of various silicon alloys by reducing quartz to silicon in an induction arc furnace at temperatures up to 2000°C. Gasified SiO_2 at high temperatures condenses in the low-temperature zone to tiny spherical particles consisting of non-crystalline silica.

Chemical composition of silica fume depends not only upon the raw materials used, but also upon the quality of electrodes and the purity of silicon product. Generally speaking, the impurities in condensed silica fume decrease as the amount of silicon increases in the final products. The by-products from the silicon metal and the ferrosilicon alloy industries, producing alloys with 75% or higher silicon content, contain 85 to 95% non-crystalline silica; the by-product from the production of ferro-silicon alloy with 50% silicon contains a much lower silica content. Minor components in silica fume are 0.1–0.5% Al_2O_3, 0.1–5% Fe_2O_3, 2–5% carbon, 0.1–0.2% S, less than 0.12% CaO, less than 0.1% TiO_2, less than 0.07% P_2O_5 and less than 1% alkalis (Uchikawa 1986).

The material removed by filtering the outgoing gases in bag filters has a spherical shape and consists mainly of amorphous phase. Its particle has an average diameter in the order of 0.1 μm, as shown in Figure 3.15, and surface areas in the range 20 to 25 m^2/g. Silica fume dissolves rapidly in $Ca(OH)_2$ saturated solution and is highly pozzolanic. The use of silica fume in concrete increases the water requirement of the concrete appreciably unless water-reducing admixtures are used. Because of its small particles and highly reactivity, silica fume is often used as a component of densely-packed cementing materials, including alkali-activated cements, with high strength and low permeability. The book by Malhotra *et al.* (1987) contains more detailed information about silica fume and silica fume concrete.

3.5.10 Non-ferrous slag

Non-ferrous slag usually refers to the slag from the production of lead, zinc, nickel and copper. One half of the total non-ferrous slag is from nickel production, one third from copper production, and the rest from zinc production. They consist mainly of silica, iron and magnesium oxides. In some

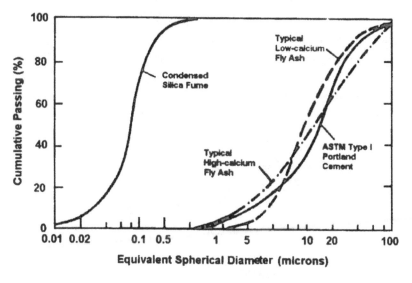

Figure 3.15 Particle size distribution of silica fume, portland cement and fly ashes (based on Mehta 1986).

cases, they may contain a significant amount of CaO and/or MgO. Minor components in non-ferrous slag include TiO_2, Cr_2O_3, MnO, Na_2O, K_2O, etc. and can be up to 25%. They contain no more than 0.5% copper, 0.15% nickel, 0.03% cobalt, 1–1.5% zinc and 0.2% lead.

Granulated non-ferrous slag consists mainly of vitreous $(Mg, Fe)SiO_3$ (90–95%) rich in $FeSiO_3$ and a small amount of crystalline minerals usually less than 6%. These crystalline minerals include pyroxene, clinoferrosilite, sulphides (mainly pyrrhotine $Fe_{1-x}S$), periclase (MgO), magnetite (FeO + Fe_2O_3) and chromium spinel $(Fe^{2+}, Mg)x(Fe^{3+}, Cr, Al)_2O_3$.

The hydraulic activity of the granulated non-ferrous slag can be evaluated using a coefficient of quality as follows (TU 67-648-84):

$$K = \frac{CaO + MgO + Al_2O_3 + Fe\,O_3 + 1/2FeO}{SiO_2 + 1/2FeO} \tag{3.8}$$

Based on the coefficient of quality and chemical composition, non-ferrous slag can be classified into three types as shown in Table 3.11.

Since non-ferrous slag usually has a low content of CaO and MgO, they exhibit pozzolanic properties (Baragano and Rey 1980, Roper and Auld 1983, Douglas and Mainwaring 1985). As CaO content in non-ferrous slag increases, they can exhibit cementitious properties (Krivenko *et al.* 1984, Deja and Malolepszy 1989, 1994). It is reported that alkali-activated non-ferrous slag cements (a combination of Type II lead, nickel and zinc slags) exhibited compressive strength from 40 to 80 MPa in both normal and steam curing conditions (Krivenko *et al.* 1984).

3.5.11 Use of pozzolans in alkali-activated cement and concrete

Pozzolan does not exhibit cementing property when it is mixed with water only. However, alkali-activated pozzolanic materials, such as coal fly ash, metakaolin and ground plate glass can give high strength. Many researches have used pozzolan to replace ground granulated blast furnace slag or

Table 3.11 Classification of non-ferrous slags

Item	Criteria for different types of slag		
	I	II	III
Coefficient of Quality, K	>1	0.7–1	<0.7
Major oxide content (mass %)			
SiO_2	26–32	33–52	33–52
CaO + MgO	>20	>17	>8
FeO	<30	<30	<35

ground steel slag to modify the properties or to reduce the cost of the alkali-activated cements and concrete. These materials are discussed in detail in Chapters 10 and 11.

3.6 Summary

This chapter has discussed the production and characteristics of granulated blast furnace slag, granulated phosphorus slag, steel slag, coal fly ash, volcanic glasses, zeolite, metakaolin, silica fume and non-ferrous slag. Granulated blast furnace slag, granulated phosphorus slag and steel slag have cementitious property, the rest are pozzolanic materials. These materials are used as the major cementing components of alkali-activated cements and concretes alone or as a combination of two or even more. However, their physical and chemical characteristics may vary significantly from source to source, which affects the consistency of alkali-activated cements and concretes.

Chapter 4

Hydration and microstructure of alkali-activated slag cements

4.1 Introduction

In cement chemistry, hydration refers to the chemical reactions between cement and mixing water. In this book, hydration refers to the chemical reaction between ground slag and mixing water in the presence of activator(s). For alkali-activated slag cements, the alkaline activator(s) can be added in three ways: (1) dissolved in mixing water, (2) interground with slag and (3) blended with the ground slag before mixing with water. When ground slag is mixed with water or a solution containing activator(s), slag particles will disperse in the water, which results in a paste.

Although extensive research has been conducted on the hydration of portland cement, the hydration mechanism is still not fully understood. The hydration of alkali-activated slag cements is more complicated than that of portland cement since the characteristics of slag may vary significantly from source to source. The nature and dosage of activator(s) also have a great effect on the hydration mechanisms, products and microstructure of alkali-activated slag cements. An adequate understanding of the hydration chemistry and microstructure of alkali-activated slag cements is necessary for a full appreciation of the properties of cements and concretes. This chapter discusses the hydration and microstructure of alkali-activated slag cements and how different factors affect the hydration, products and microstructure of hardened pastes.

4.2 Heat of hydration

4.2.1 Portland cement

Calorimetry has played an important role in understanding the early hydration chemistry of portland cement. The early hydration of portland cement can be divided into five periods based on heat evolution curves, as schematically represented in Figure 4.1: (I) initial (pre-induction) period;

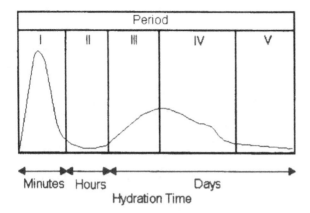

Figure 4.1 Schematic representation of hydration periods of portland cement (Shi and Day 1995d).

(II) induction period; (III) acceleration (post-induction) period; (IV) deceleration period; and (V) diffusion period. The peak during the pre-induction period is generally attributed to the rapid dissolution of alkali sulphates and aluminates, initial hydration of C_3S and formation of AFt (Jawed *et al.* 1983). Bensted (1987) believes that the peak can also be attributed partly to the combined effects of heat of wetting of the cement, the hydration of free lime and partly to the hydration of $CaSO_4 \cdot 0.5H_2O$ to $CaSO_4 \cdot 2H_2O$:

$$CaO + H_2O \rightarrow Ca(OH)_2 + 15.58 \, kJ/mol \qquad (4.1)$$

$$CaSO_4 \cdot 0.5H_2O + 1 \cdot 5H_2O \rightarrow CaSO_4 \cdot 2H_2O + 4.10 \, kJ/mol \qquad (4.2)$$

$CaSO_4 \cdot 0.5H_2O$ originates from the decomposition of $CaSO_4 \cdot 2H_2O$ during grinding. The heat evolution curve then declines to an induction period lasting a few hours. This is followed by an acceleration peak, the major peak, which results from the hydration of C_3S to C–S–H, the principal cementing reaction. Usually, the magnitude of the second peak is lower than that of the first peak. A third peak which is attributed to the transformation of ettringite ($C_3A \cdot 3CaSO_4 \cdot 32H_2O$) to monosulphate ($C_3A \cdot CaSO_4 \cdot 13H_2O$) can be observed. In period IV, hydration products and microstructure continue to develop, which are controlled by moderate chemical reactions and diffusion processes. Period V is a gradual densification period and is controlled by very slow diffusion processes.

4.2.2 Effect of the nature of activators on heat evolution

Many studies have investigated the heat evolution characteristics of alkali-activated slag cements. It is agreed that the characteristics of the slag, and the nature and dosage of activators play a critical role in determining the heat evolution characteristics of alkali-activated slag cements. Shi and Day (1995d) examined the heat evolution curves of alkali-activated slag cements activated by different activators belonging to groups 1, 2, 3 and 6 as described in Chapter 2, and divided the curves into three classes as shown in Figure 4.2.

For Class I, one peak occurs during the first few minutes and no peaks appear thereafter. Examples include the mixture of ground blast furnace slag and water or Na_2HPO_4 solution at both 25 and 50 °C. They exhibited only a very small initial heat evolution peak, no other peaks can be recorded during the 72-hour testing period, and did not set and harden (Shi and Day 1995d). These mixtures may set depending on the characteristics of the slag.

When ground granulated slag particles contacts water, the Si–O, Al–O and Ca–O bonds on their surface break under the polarization effect of OH^- (Royak et al. 1978, Teoreanu 1991). These dissolved species exist in water in the form of $(H_2SiO_4)^{2-}$, $(H_3SiO_4)^-$, $(H_4AlO_4)^-$ and Ca^{2+}. Because Ca–O bond is much weaker than Si–O and Al–O bonds, the concentration of Ca^{2+} in the water is much higher than $(H_2SiO_4)^{2-}$, $(H_3SiO_4)^-$ and $(H_4AlO_4)^-$, and an Si–Al-rich layer forms quickly on the surface of slag particles (Rajaokarivony-Andriambololona et al. 1990). The Si–Al-rich layer may adsorb some H^+ in the water, which results in an increase of OH^- concentration or pH of the solution. However, this concentration of OH^- still could not break enough Si–O and Al–O bonds for the formation of a significant amount of C–S–H, C–A–H or C–A–S–H. Although elevating temperature can enhance the polarization effect (Teoreanu 1991), no accelerated hydration peak can be detected even at 50 °C. The only very small initial peak is attributed to the wetting and dissolution of slag grains and adsorption of some ions onto the surface of slag grains. No obvious hydration indication could be observed under ESEM after 24 hours of hydration in water (Jiang et al. 1997), and only a small amount of C–S–H forms even at 150 days (Rajaokarivony-Andriambololona et al. 1990).

Class II heat evolution curve is the same as that of portland cement: only one initial peak appears before the induction period and one accelerated hydration peak appears after the induction period. The initial peak is attributed to the wetting and dissolution of slag particles and the second peak is due to the accelerated hydration of the slag. The hydration of NaOH-activated slag cement is a typical example (Shi and Day 1995d). However, the nature and dosage of activator and hydration temperature have a very significant effect on the position and magnitude of these two peaks.

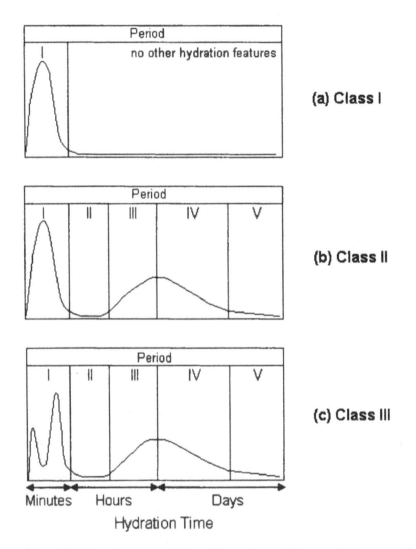

Figure 4.2 Schematic representation of three classes of heat evolution curves for alkali-activated slag cements (Shi and Day 1995d).

The activators in this class have high pH values. When their solutions are mixed with slag, OH^- in the solution can break not only Ca–O bonds, but also a significant number of Si–O and Al–O bonds. Because $Ca(OH)_2$ has a much higher solubility than C–S–H, C–A–H and C–A–S–H, $Ca(OH)_2$ cannot precipitate from the solution. A very thin layer consisting of low Ca/Si ratio C–S–H, C–A–H and C–A–S–H, which has a very low solubility, precipitates very quickly through the solution. This can be observed under an SEM in NaOH-activated slag hydrated for 20 minutes at 23 °C (Shi 1997). It seems the precipitation of C–S–H and C–A–H does not lead to a noticeable induction period, which may be determined by the nature of these hydration products.

For Class III curve, two peaks, one initial and one additional initial peak, appear before the induction period and one accelerated hydration peak appears after the induction period. The initial peak could be lower or higher than the additional initial peak depending on the nature of activators and hydration temperature. In this study, this type of hydration includes the slag activated by Na_2SiO_3, and Na_2CO_3 at 25 °C, Na_3PO_4 at 50 °C and NaF at both 25 and 50 °C. The initial peak can be attributed to the wetting and dissolution of slag particles. The additional initial peak can be attributed to the reaction between dissolved Ca^{2+} from slag particles and anions or anion groups from activators (Shi and Day 1995d, 1996b). This reaction and corresponding reaction products play an important role in determining the setting time and strength of the cement pastes (Fernandez-Jimenez and Puertas 2003). The formation of C–S–H due to Ca from slag and silicate anions from the activator was confirmed by NMR analysis (Brough and Atkinson 2002).

Because the initial and the additional initial peaks appear very close, they may merge into one peak depending on the activator dosage, slag activity and hydration temperature. The merger was observed from the hydration of Na_2SiO_3 and Na_2CO_3 activated slag at 50 °C (Shi and Day 1995d, 1996b).

A combination of two or more activators will change the heat evolution characteristics and total heat evolution of the cement, depending on the nature and portion of each activator. The use of more NaOH promotes and strengthens the accelerated hydration peak, while the use of more Na_2CO_3 retards and decreases the accelerated hydration peak (Fernandez-Jimenez and Puertas 2003). Chemical admixtures can also affect the heat evolution of alkali-activated slag cements. It was found that the addition of a vinyl co-polymer-based superplasticizer significantly retarded, while a polyacrylate co-polymer based superplasticizer slightly accelerated the heat evolution of waterglass-activated slag cement (Puertas *et al.* 2003).

4.2.3   Effect of activator dosage on heat of hydration

Generally speaking, the hydration of alkali-activated slag cements is accelerated and the cumulative heat of hydration is increased as the activator dosage is increased. However, the effect is also dependent on the nature of the slag and activators used (Shi and Day 1996b, Krizan and Zivanovic 2002, Cincotto *et al.* 2003). The effect of activator dosage on cumulative heat of hydration of alkali-slag cement at 25 and 50 °C is illustrated in Figures 4.3 to 4.6. At 25 °C, increasing NaOH dosage shows an increase in

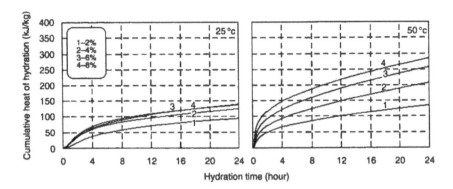

Figure 4.3 Effect of NaOH dosage on cumulative heat of hydration at 25 and 50 °C (Shi and Day 1996b).

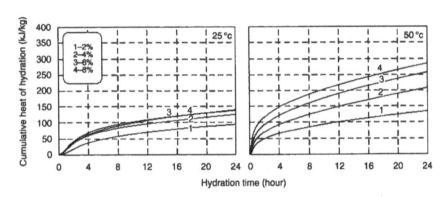

Figure 4.4 Effect of Na_2CO_3 dosage on cumulative heat of hydration at 25 and 50 °C (Shi and Day 1996b).

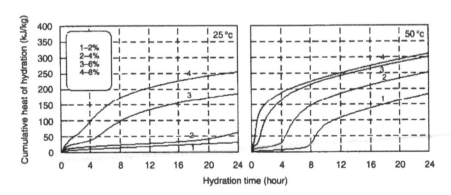

Figure 4.5 Effect of Na_2SiO_3 dosage on cumulative heat of hydration at 25 and 50 °C (Shi and Day 1996b).

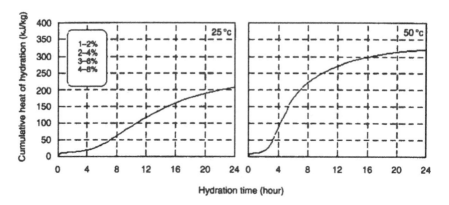

Figure 4.6 Cumulative heat of hydration of portland cement at 25 and 50 °C (Shi and Day 1996b).

cumulative heat of hydration above the dosage of 2%; at 50 °C, the effect is more marked. For the Na_2CO_3 case, increased dosage above 2% shows a significant reduction in total heat after 8 to 12 hours. At 50 °C, the opposite effect is observed. The cumulative heat of hydration of Na_2SiO_3 activated slag increases with Na_2SiO_3 dosage, especially at 25 °C (Figure 4.3). The effect of activator degree on heat evolution of alkali-activated slag cements were well confirmed by Krizan and Zivanovic (2002) later.

One can compare the cumulative heat of hydration of portland cement at 25 and 50 °C (Figure 4.6) with that of alkali-activated slag cement. During the first 24 hours, NaOH or Na_2CO_3 activated slag releases less heat than portland cement at a given iso-thermal temperature for all activator dosages. Compared with portland cement the slag activated with 2 or 4% Na_2SiO_3 releases less heat, 6% Na_2SiO_3 releases heat similar at both 25 and 50 °C; and 8% Na_2SiO_3 releases more heat at 25 °C and similar heat at 50 °C.

It has been reported that alkali-slag cements have lower heat evolution rate and total heat evolution than portland cement and can be regarded as low-heat cements (Glukhovsky *et al.* 1980). The results described above indicated that both heat evolution rate and total heat evolution of the same cement could be higher than those of portland cement. The heat evolution rate of Na_2CO_3 activated slag and the total heat evolution of NaF activated slag at 50 °C surpass those of portland cement respectively.

For waterglass, the effect of activator dosage on hydration also varies with its modulus. It was found that for a given Na_2O dosage, the hydration of slag was retarded and decreased as the modulus of waterglass (Krizan and Zivanovic 2002). This is more obvious when the dosage is less than 6% Na_2O by the mass of the slag.

4.2.4 Effect of water-to-slag ratio on heat evolution rate

It is well known that water-to-cement ratio has a great effect on the properties of hardened cement and concrete. Actually, an increase in water to
cement accelerates the hydration of portland cement (Bensted 1983). The
effect of water-to-cement or slag ratio on the heat evolution rate of portland
cement and alkali-activated slag cements with fixed amount of activators
(4% by Na_2O) is illustrated in Figure 4.7. As water-to-cement or slag ratio
is increased from 0.45 to 0.6 as shown in Figure 4.7, it does not show a
significant effect on the induction period and the general trend of the heat
evolution of portland cement, but it lowers the accelerated hydration peak
slightly. It can be seen that water-to-solid ratio does not affect the heat evolution rate thereafter. The effect of water-to-solid ratio on NaOH-activated
slag is similar to the effect on portland cement (Figure 4.7a). For Na_2CO_3-
activated slag, increasing water-to-solid ratio from 0.45 to 0.60 changes
the accelerated hydration peak into a more diffuse peak. Conversely, the

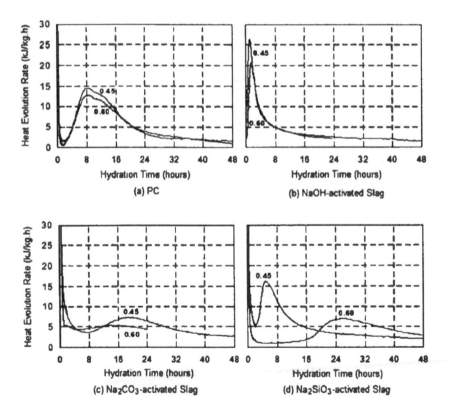

Figure 4.7 Effect of water-to-cement or water-to-slag ratio on heat evolution rate of
portland and alkali-activated slag cements containing 4% Na_2O at 25 °C (Shi
and Day 1996b).

increase of water-to-slag ratio has a very marked effect on the heat evolution of Na_2SiO_3 activated slag. No obvious induction period could be observed for the water-to-slag ratio of 0.45, but an induction period of about 15 hours appears by increasing the water-to-slag ratio from 0.45 to 0.60. Also, the increase of water-to-slag ratio changes the pronounced sharp accelerated peak into a very diffuse peak.

The principal heat evolution characteristics of the alkali-slag cements activated by the various alkaline activators are summarized in Table 4.1. Accelerated hydration peak refers to the peak at the end of induction period, which usually corresponds to the second peak in the course of hydration of portland cement. However, it is the third peak for some activated slag because the reaction between slag and activators produces a second peak.

4.2.5   Effect of initial pH of activator solution

Some publications stated that the activation of slag depends on the initial pH of activator solutions and the early hydration of slag is accelerated as the pH of activator solution increases (Yuan *et al.* 1987, Roy *et al.* 1992, Song *et al.* 2000). For a given amount of Na_2O, NaOH solution has higher pH than any other sodium-containing compounds, NaOH-activated slag shows the shortest induction period and the highest accelerated hydration peak. Song *et al.* (2000) suggested that the required minimum pH value for NaOH activator is 11.5. Although the pH of Na_2SiO_3 solution is higher than that of Na_2CO_3 activated slags, the former shows a longer induction period and a lower accelerated hydration peak than the latter. Also, NaF solution has a pH of only about 10, but it can still activate the hydration of slag. A previous work indicated that NaF activated ground phosphorus slag gave a high strength at room and high temperatures (Shi *et al.* 1989a). It seems that hydration characteristics of alkali-activated slag rely more upon the nature of anion or anion group of the activator than upon the initial pH of the solution, as stated before based on strength tests (Shi *et al.* 1989a).

4.2.6   Effect of temperature

Generally speaking, the rate of a chemical reaction will be doubled with a temperature increase of every 10 °C. Of course, it can be expected that an increase in hydration temperature will accelerate the hydration and heat evolution of alkali-activated slag cements significantly.

Results in Figures 4.3 to 4.5 have indicated that an increase in hydration temperature can have a significant effect on the heat evolution of alkali-activated slag cements. Fernandez-Jimenez and Puertas (1997a) conducted a detailed study on the effect of hydration temperatures and activator dosage on heat evolution rates and hydration kinetics. Figure 4.8 shows the heat

Table 4.1 Hydration characteristics of alkali-activated slag cements activated by various activators (activator dosage - 4% Na_2O by mass of slag) (Shi and Day 1995d)

No.	Activator	Initial pH of the solution at 23°C	Hydration temperature (°C)	Height of the 1st peak (kJ/kg.h)	Starting time of accelerated hydration peak (h)	Height of the accelerated hydration peak (kJ/kg.h)	Total heat during first 24 hours (kJ/kg)
1	PC*	5.2 (water)	25	63.16	1.99	14.57	207.89
			50	92.80	1.11	48.41	320.50
2	NaOH	13.87	25	17.94	0.25	21.60	125.37
			50	39.51	0.11	58.21	207.56
3	Na_2SiO_3	12.89	25	31.04	17.75**	7.25**	63.13
			50	74.56	2.17	47.90	252.90
4	Na_2CO_3	12.16	25	27.45	5.25	6.30	114.89
			50	112.31	0.52	56.71	304.90
5	Na_3PO_4	12.61	50	61.83	14.58**	9.17**	199.92
6	Na_2HPO_4	9.18	25	40.15	N/A	N/A	45.44
			50	81.96	N/A	N/A	64.39
7	NaF	10.17	25	18.06	N/A	N/A	52.13
			50	40.75	5.9**	27.74**	331.85

* 100% ASTM Type I portland cement used as a control.
** Refer to the third peak.

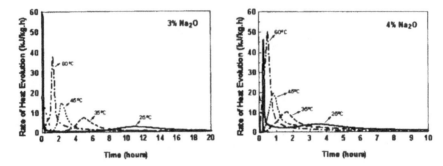

Figure 4.8 Effect of temperature on heat evolution rate of sodium silicate-activated slag cements (Fernandez-Jimenez and Puertas 1997a).

Table 4.2 Effect of temperature on characteristics of second heat evolution peak (Fernandez-Jimenez and Puertas 1997a)

Hydration temperature (°C)	Concentration of sodium silicate			
	3% by mass of Na$_2$O		4% by mass of Na$_2$O	
	Peak value (kJ/kgh)	Time of appearance (h)	Peak value (kJ/kgh)	Time of appearance (h)
25	2.68	11.13	3.96	3.45
35	7.18	4.87	10.34	1.54
45	14.71	2.37	20.12	0.87
60	38.40	1.27	50.45	0.5

evolution curves of alkali-activated slag cements containing 3 and 4% Na$_2$O (by mass) at 25, 35, 45 and 60 °C. The characteristics of the accelerated hydration peaks are summarized in Table 4.2. For a given Na$_2$O dosage, the accelerated hydration peak is intensified and appears earlier as hydration temperature increased. For a given hydration temperature, the accelerated hydration peak is intensified and appears earlier as activator dosage increased.

4.2.7 Apparent hydration activation energy of alkali-activated slag cements

Several researchers have confirmed that the total heat evolution of cement with time can be described by the semi-empirical equation (Knudsen 1980,

Roy and Idorn 1982, 1985, Shi *et al.* 1991a, Zhou *et al.* 1993, Fernandez-Jimenez and Puertas 1997b):

$$P = P_\infty \cdot \frac{K_T \cdot (t - t_o)}{1 + K_T \cdot (t - t_o)} \qquad (4.3)$$

where
    P = total heat evolution at time t (kJ/kg);
    P_∞ = total heat evolution at $t = \infty$ (kJ/kg);
    K_T = heat evolution rate constant at curing temperature T (hours^{-1});
    t = actual hydration time at temperature T (hours);
    t_o = the time at the end of the induction period (hours).

The P_∞ and K_T can be obtained either from the non-linear regression of the total heat evolution curves, or from a linear regression of the plot between $1/P$ versus $1/(t - t_o)$. Knudsen found that the heat evolution rate constant K_T equals to the reciprocal of t_{50}, which it is time to achieve 50% of P_∞.

The kinetic parameters and apparent activation energy of alkali-activated slag cements obtained by different researchers are summarized in Table 4.3. It seems that the apparent activation energy varies with the activator. For a given Na_2O dosage, NaOH-activated slag cement shows lower apparent hydration activation energy than Na_2SiO_3-activated slag cement. On the other hand, the apparent activation energy decreases as the activator dosage increases. Usherov-Marshak *et al.* (1998) reported similar results.

4.3 Degree of hydration of slag in alkali-activated slag cement

4.3.1 Solvent selective dissolution of hydrated slag

Since slag and the main hydration products of alkali-activated slag cement are amorphous, XRD cannot be used to quantitatively monitor the hydration process. However, Kondo and Ohsawa (1968) developed a salicyclic acid/methanol/acetone based selective dissolution method to determine the amount of hydrated slag in hardened cement pastes. In the method, 0.5 g sample is mixed with 2.5 g of salicyclic acid, 35 ml of acetone and 15 ml of methanol in a beaker for one day at room temperature. Then, it is filtered and the residue is washed with methanol, dried and heated at 850 °C for 10 minutes. The mass of dried residue is the amount of unhydrated slag. The original slag sample has to be treated in the same way. Later, Demoulian *et al.* (1980) established an ethylene diamine tetraacetic acid (EDTA) based selective dissolution method. Luke and Glasser (1987)

Table 4.3 Kinetic parameters and apparent activation energy obtained from different alkali-activated slag cements

No	Slag	Activator	Water-to-slag (cement) ratio	Temperature (°C)	T_{50} (h)	P_∞ (kJ/kg)	Activation energy (kJ/mol)	Reference
1	Blast furnace slag	50% portland Cement	0.4	27	83.61	125	49.1	Roy and Idorn (1982)
2	Phosphorus slag	3% NaOH (by Na_2O)	0.4	38	34.57	108	38.89	Shi et al. (1991a)
				60	9.86	107		
				46	5.46	196		
3	Phosphorus slag	3% $Na_2O \cdot SiO_2$ (by Na_2O)	0.4	58	3.21	201	64.62	Shi et al. (1989a)
				46	23.25	367		
4	Blast furnace slag (600 m^2/kg)	7.5% $Na_2O \cdot SiO_2$ (by Na_2O)	0.7	58	9.61	402	53.63	Zhou et al. (1993)
				20	22.75	390		
				40	11.31	289		
				60	3.28	261		
				25		131		
5	Blast furnace slag (440 m^2/kg)	4% $Na_2O \cdot SiO_2$ (by Na_2O)	0.4	35		117	57.6	Fernandez-Jimenez and Puertas (1997a)
				45		112		
				60		61		

compared those methods and indicated that the EDTA method was better than the salicyclic acid/methanol/acetone method. However, the salicyclic acid/methanol/acetone method has been used by several researchers (Moranville-Regourd 1998) with success and is still being commonly used.

Zhou *et al.* (1993) determined the amount of hydrated slag in the alkali-activated slag cement pastes activated by 10 different sodium silicate solutions using the salicyclic acid/methanol/acetone method. The slag hydrates very fast during the first 24 hours, then slowly thereafter. At seven days, the hydration degree of slag varies from 15 to 35%. The degree of hydration of slag varies very significantly depending on the nature of activator used. For a given concentration of SiO_2 in the solution, the pH of the solution and the hydration degree of slag increase with the increase of Na_2O concentration or with the decrease of modulus of the sodium silicate solution. Shen *et al.* (1996) measured the degree of hydration of sodium silicate-activated slag in which slag had Blaine finenesses of 600, 800 and 1000 m^2/kg. They found that the hydration degree was only 38 to 44% at three days and increased by approximately 3% from 3 to 28 days regardless of the fineness of the slag. The thickness of the hydrated layer on slag particles is independent of particle size, but the hydration degree of the slag particles increases with the increase of total surface area of slag particles (Sato *et al.* 1986). In another study (Zhong and Yang 1993), it was found the hydration degree of the slag was about 25% at 28 days when the slag had a Blaine fineness of 340 m^2/kg. It can be concluded that the hydration of slag depends on the nature of activator and the fineness of the slag.

The hydration degree of slag in alkali-activated slag cement is much lower than that of a typical ASTM Type I portland cement at 3, 7 and 28 days which are around 50, 70 and 80% (Hansen, 1986). It seems that the degree of hydration of slag in alkali-activated slag cement is much lower than that of portland cement.

4.3.2 Recrystallization method

Belitsky *et al.* (1993a,b) noticed that two characteristic peaks corresponding to melilite and merwinite appeared on XRD patterns of a granulated blast furnace slag after heating at 900 °C for 15 minutes, but the peak intensity melilite increased and that of merwinite decreased with time when the slag hydrated in the presence of an alkaline activator. They used XRD to quantitatively estimate the amount of melilite and merwinite in the slag and alkali-activated slag cement pastes after heating, and then calculated the hydration degree of slag in the cements by dividing the amount of melilite and merwinite decreased in cement pastes due to hydration to their total amount in the slag. The calculated hydration degrees reached 30 to 50% during the first few hours of hydration, and then increased slowly with time. Those initial hydration degrees seem too high compared with the results discussed above. The nature and dosage of activators also affect the change of intensity of those characteristic peaks.

4.4 Non-evaporable water

Water in hydrated cement pastes can be arbitrarily classified into evaporable water and non-evaporable water. Non-evaporable water refers to the water that cannot be removed under certain drying conditions. The conditions for several drying methods can be found in the book authored by Mindess and Young (1981). The non-evaporable water in hardened cement pastes after oven-drying at 105 °C can be regarded as chemically combined water and is often used as an indication of the degree of hydration of cement. Usually, the weight loss during calcination at 700 to 900 °C after oven-drying is regarded as the amount of non-evaporable water in the hardened paste. The non-evaporable water for completely hydrated portland cement pastes is approximately 22 g per 100 g of a typical portland cement (Powers 1958).

The measurement of non-evaporable water has also been used to follow the hydration of alkali-activated slag cements. Table 4.4 shows the effect of water-to-slag ratio on non-evaporable water of alkali-activated granulated phosphorus slag cements (Shi and Li 1989b). For NaOH-activated slag, no marked effect can be observed from one to seven days. However, the non-evaporable water increases with water-to-slag ratio from 14 to 30 days. Increasing water-to-slag ratio decreases the non-evaporable water content of Na_2SiO_3-activated phosphorus slag significantly up to 14 days. No noticeable effect can be observed from 14 to 30 days. The decrease of non-evaporable water content at early ages can be attributed to the dilution effect of alkaline activators. The later hydration of slag is controlled by the diffusion of water, so the non-evaporable water increases with the increase of water-to-slag ratio after seven days (see Table 4.4).

Table 4.4 Effect of water-to-slag ratio on non-evaporable water (Shi and Li 1989b)

Activator	Water-to-slag ratio	Non-evaporable water (%)				
		1 d	3 d	7 d	14 d	30 d
3% NaOH (by Na_2O)	0.29	4.18	7.44	9.81	9.56	11.52
	0.33	4.24	7.84	9.77	9.79	11.89
	0.37	3.94	7.50	9.64	10.44	12.53
	0.40	3.69	7.34	9.43	10.68	13.82
3% $Na_2O \cdot SiO_2$ (by Na_2O)	0.29	3.03	7.39	9.75	10.61	12.29
	0.33	2.32	5.88	9.79	11.19	12.97
	0.37	1.94	4.94	8.38	10.80	12.53
	0.40	1.70	3.99	8.40	11.49	13.00

In the same study (Shi and Li 1989b), it was found that, for a given water-to-slag ratio and Na_2O dosage, the non-evaporable water of alkali-activated phosphorus slag cement increased with the increase of the modulus of the waterglass (from 0 to 1.5) during the first 12 hours. No observable effect could be observed thereafter. Another study (Wu 1999) used a wider range of modulus for sodium silicate, but a fixed amount of sodium silicates. It was found, at 28 days, that the non-evaporable water content increased with the modulus of sodium silicates from 0.4 to around 1.2, reached the maximum at the ratio of around 1.5, then decreased with the increase of the modulus. Belitsky et al. (1993b) also found that sodium silicate with a ratio of 1.5 gave higher non-evaporable water content than the ratios of 1 and 2. However, they obtained non-evaporable water from 6 to 9% after 28 days of hydration, which was lower than the values reported by other studies.

Compaction was applied to produce high strength alkali-activated slag cement paste using a low water-to-slag ratio of 0.12 (Xu 1988, Xu et al. 1993). The non-evaporable water content of the cement pastes decreased as the compaction pressure was increased from 0 to 100 MPa, especially after seven days. At pressures in excess of 100 MPa, however, the non-evaporable water content is almost constant and independent of compaction pressure.

4.5   Silicate polymerization

It was Lentz (1969) who first used the trimethylsilylation (TMS) technique to measure the silicate polymerization in the hydration of cement. Tamas, Sarkar and Roy (1976) modified the technique using trimethylchlorosilane and dimethylformamide in addition to hexamethyl disiloxane, which gives more realistic values for evaluation of polymerized silicates. Studies on fully hydrated cement pastes, hydrated at room temperatures by gas-liquid and gel-permeation chromatography, have indicated that they contain 9–11% monomer, 22–30% dimer, 1–2% linear trimer and 44–51% polymer.

Chen and Yang (1989) measured the content of low-molecular-weight polymers, which include SiO_4^{4-}, $Si_2O_7^{6-}$, $Si_3O_{10}^{8-}$ and $Si_4O_{12}^{8-}$, in eight granulated blast furnace slags from different steel companies in China using the TMS method and found that granulated slag contains about 40 to 50% SiO_4^{4-}, 10 to 15% $Si_2O_7^{6-}$, 2 to 5% $Si_3O_{10}^{8-}$ and 1 to 5% $Si_4O_{12}^{8-}$.

Several studies have been conducted to measure the silicate polymerization in hydration of alkali-activated slag cement. Yu and Wang (1990) determined the silicate polymerization in NaOH-activated slag cements and found that the percentage of monomer is almost constant, but dimer and trimer increase and polymer decreases with time from one to seven days. From 7 to 28 days, monomer and trimer decrease, and polymer increases with time. This indicates the hydration of NaOH-activated slag is

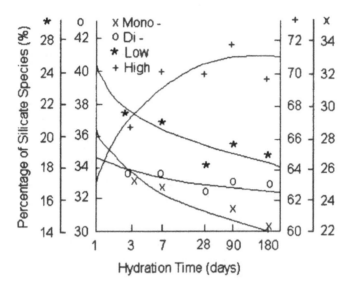

Figure 4.9 Silicate polymerization in sodium silicate-activated slag cement paste (based on Zhong and Yang 1993).

a depolymerization process from one to seven days and a polymerization process from 7 to 28 days.

Figure 4.9 shows the changes of silicate species in hardened sodium silicate-activated slag pastes (Zhong and Yang 1993). It can be seen that the contents of monomer, dimer, low-molecular-weight polymer decreased with time, but high-molecular-weight polymer increased with time. Thus, the hydration of waterglass-activated slag can be regarded as a polymerization process. This has been confirmed by other studies (Lu 1989, Yu and Wang 1990). In addition to that, the polymerized silicates from the activator sodium silicate may play two more roles: (1) bridging slag particles and (2) filling voids between slag particles. This may also explain why sodium silicate activated-slag cement pastes show higher strength than slag cement pastes activated by other activators.

4.6 Kinetics of hydration

Several models have been proposed to describe the hydration kinetics of portland cement. All of them assume that cement particles are spherical although they are actually very irregular. Jander proposed an equation for solid-state sintering controlled by diffusion, which can satisfactorily fit kinetic data with the following general form (Jawed *et al.* 1983):

$$(1 - \sqrt[3]{1 - \alpha})^N = Kt \tag{4.4}$$

where
  α = reaction degree;
  K = reaction constant;
  t = reaction time; and
  N = reaction grade.

Kondo *et al.* (1976) classified a reaction process based on the reaction grade N in the above equation:

1 If the reaction is controlled by reactions happening on the surface of grains, or by the dissolution of reactants or by the precipitation of reaction products, then $N \leq 1$.
2 If the reaction is controlled by the diffusion of reactants through a layer of porous reaction products, then $1 < N \leq 2$.
3 If the total reaction is controlled by the diffusion of reactants through a layer of dense reaction products, then $N > 2$.

Ginstling and Brounstein (1950) proposed an equation when the thickness of the diffusion layer on cement particles is considered:

$$[1 - (1 - \alpha)^{\frac{1}{3}}] - \frac{2}{3}[1 - (1 - \alpha)^{\frac{1}{3}}]^3 = Kt \tag{4.5}$$

According to Ginstling and Brounshtein (1950), the Jander's Equation is valid when the ratio of the thickness of the hydration product to the initial radius of the particle is less than 0.5 and the Ginstling-Brounstein Equation is valid when the ratio is greater than 0.5.

Taplin (1962) considered the differences between outer and inner hydration products and derived an equation as follows:

$$\left(\frac{1}{D_i}\right)\left[1 - \left(1 - \frac{2}{3}\alpha\right) - (1 - \alpha)^{\frac{2}{3}}\right]$$
$$-\left(\frac{1}{D_0} - \frac{1}{D}\right)\left[1 + \frac{2\gamma\alpha}{3} - (1 + \gamma\alpha)^{\frac{2}{3}}\right] = Kt \tag{4.6}$$

where D_i and D_0 are the diffusion coefficients of inner and outer C–S–H respectively, and γ is the volume ratio of outer to inner C–S–H.

Fernandez-Jimenez and Puertas (1997b) estimated the degree of reaction of sodium silicate-activated slag cement using the division of total heat $Q_{(t)}$ at time t by the total or maximum heat Q_{max} released when hydration is completed:

$$\alpha = \frac{Q_{(t)}}{Q_{max}} \tag{4.7}$$

$$Q_{(t)} = Q_{max} \cdot \frac{K_T \cdot (t - t_o)}{1 + K_T \cdot (t - t_o)} \tag{4.8}$$

where t_0 is the time when accelerated hydration starts and K_T is the reaction constant.

The analyses based on results at 25, 35, 45 and 60°C indicated that the Jander Equation could describe the reaction kinetics better than the Ginstling-Brounstein Equation (Fernandez-Jimenez and Puertas 1997b). The reaction grades at all those temperatures varied between 1.80 and 2.03. This means that the hydration of the sodium silicate-activated slag cement is controlled by the diffusion of reactants through a layer of porous reaction products.

The nature of activators also has a significant effect on the hydration kinetics of the slag. It was noticed that the reaction grade N was only 1.2 when NaOH was used as an activator and was increased to 2.6 when Na_2CO_3 was used as an activator (Fernandez-Jimenez, Puertas and Arteaga, 1998). This means that the hydration of NaOH-activated slag cement is almost a first-order reaction process and the hydration of Na_2CO_3-activated slag cement is a diffusion controlled process.

4.7 Hydration products

4.7.1 Introduction

Since the chemical composition of portland cement does not change significantly from source to source, it is generally accepted that fully hydrated portland cement consists of 70% calcium silicate hydrate (C–S–H), 20% calcium hydroxide, 7% calcium sulphoaluminate hydrates and 3% unhydrated cement particles. C–S–H is not a well-defined compound and its C/S ratio varies with the composition of the cement and hydration conditions. Several models have been proposed to describe C–S–H. Details on these models can be found in references (Mindess and Young 1981; Ramachandran et al. 1981, Hearn et al. 1994, Popovics 1998).

Many researchers have investigated and reported the hydration products of alkali-activated slag cements (Glukhovsky et al. 1974; Teoreanu et al. 1980, Krivenko 1986, Kutti 1992, Wang and Scrivener 1995, Jiang et al. 1997). It is generally agreed that its main hydration product is C–S–H. There is no doubt that the minor hydration products of alkali-activated slag cement will change with the nature of the slag and activator. This section will discuss the hydration products and their characteristics of alkali-activated slag cement from published work.

4.7.2 Effect of chemical composition of blast-furnace slag on hydration products

As discussed in Chapter 3, the chemical composition of the slag varies with the type of iron being made and the type of ore being used. There is no doubt that the chemical composition of blast furnace slag has a significant effect on hydration process, hydration product and properties of hardened alkali-activated slag cements. In many cases, the MgO content of blast furnace slag is low and the slag can be described by the $CaO-SiO_2-Al_2O_3$ system. The phase diagram of $CaO-Al_2O_3-SiO_2-H_2O$ system, as shown in Figure 4.10, indicates that five different products such as C–S–H, $Ca(OH)_2$, C_4AH_{13}, C_2ASH_8 and CS_2H could appear in this system, while calcium hydroxide and gehlenite hydrate cannot co-exist at equilibrium (Dron 1974). In reality, equilibrium is attained only locally on a scale of microns or less due to difficulty of ionic transport within the pastes (Lachowski *et al.* 1980). It must also be recognized that some calcium hydroxide may be made essentially non-reactive by the formation of an impermeable coating of C–S–H (Malquori 1960). Thus, it is not surprising that calcium hydroxide and gehlenite hydrate have been detected at the same time (Shi and Day 1995a).

It can also be expected that hydration products vary with the chemical composition of the slag in the presence of an activator. Granulated phosphorus slag consists mainly of CaO and SiO_2, and its Al_2O_3 and MgO contents are usually very low. XRD analyses and SEM observations indicate that C–S–H is the only hydration product of activated ground granulated phosphorus slag when NaOH is used as an activator (Shi 1987, Shi *et al.* 1989a). However, C_2ASH_8 and C_4AH_{13} have also been detected in NaOH-activated blast-furnace slag in addition to C–S–H (Regourd 1980). C_2ASH_8 or gehlenite hydrate is an AFm phase which has an interlayer

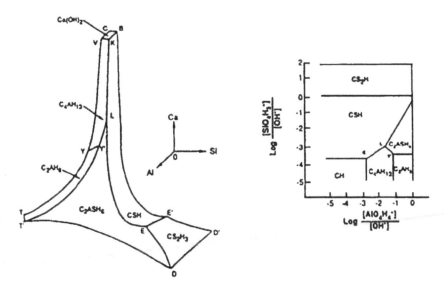

Figure 4.10 Phase diagram in $CaO-Al_2O_3-SiO_2-H_2O$ system (Dron 1974).

aluminosilicate anion (Taylor 1990). M_4AH_{13}, instead of C_4AH_{13}, occurs if the MgO content of the slag is high. Since the strength of hardened cement pastes depends on initial porosity, degree of hydration and space filling, and the intrinsic strength of the reaction products (Helmuth 1987), the chemical composition of slag will have a direct influence on strength development of cements.

4.7.3 Effect of activators on hydration products

Voinvitch and Dron (1976) proposed stoichiometric equations for the hydration of a slag possessing a chemical composition of C_5S_3A in different activation media:

(1) In water

$$C_5S_3A + 12H_2O \rightarrow \frac{1}{3}C_4AH_{13} + \frac{7}{3}C\text{-}S\text{-}H + \frac{2}{3}C_2ASH_8 \qquad (4.9)$$

(2) In lime solution

$$C_5S_3A + 2CaO + 16H_2O \rightarrow C_4AH_{13} + 3C\text{-}S\text{-}H \qquad (4.10)$$

(3) Activated by calcium sulphate

$$C_5S_3A + 2CaSO_4 + \frac{76}{3}H_2O \rightarrow 3C\text{-}S\text{-}H + \frac{2}{3}C_3A \cdot 3CaSO_4 \cdot 32H_2O$$

$$+ \frac{2}{3}Al(OH)_3 \qquad (4.11)$$

(4) Activated by soda-sulphate

$$C_5S_3A + 4CaSO_4 + 2NaOH + 34H_2O \rightarrow 3C\text{-}S\text{-}H + C_3A \cdot 3CaSO_4 \cdot 32H_2O$$

$$+ Na_2SO_4 \qquad (4.12)$$

It can be seen that activators can have a great effect on hydration products. It was found that C-S-H, C_4AH_{13} and C_2ASH_8 are products of NaOH and lime-activated slag, and ettringite forms when sulphates are present (Regourd 1980). X-ray diffraction patterns of alkali-activated slag and ASTM Type III portland cement pastes hydrated for 540 days at 23°C are shown in Figure 4.11. Hydration products of alkali-activated slag are different from those of portland cement pastes. The main hydration product of the three alkali-activated slag cement pastes is C-S-H, but minor hydration products vary with the nature of activators. C_4AH_{13} appears in $Na_2O \cdot SiO_2$-activated slag, $C_3A \cdot CaCO_3 \cdot 12H_2O$ in Na_2CO_3-activated slag

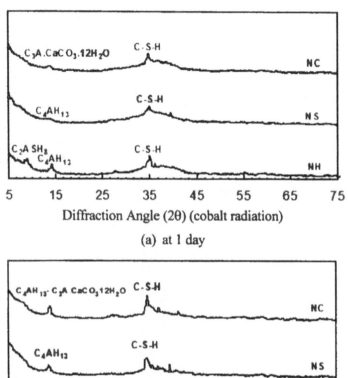

Figure 4.11 XRD patterns of alkali-activated slag cement pastes hydrated at 23 °C (Shi and Day 1996a).

and C_4AH_{13} and C_2ASH_8 in NaOH-activated slag (Schilling *et al.* 1994a, Shi and Day 1996a). The diffraction intensities of C_4AH_{13} and C_2ASH_8 increase as the NaOH concentration and hydration time increase (Schilling *et al.* 1994a). Minor difference could be noticed between NaOH- and KOH-activated slags (Rajaokarivony-Andriambololona *et al.* 1990). Change in hydration products due to the presence of different activators will change the hydration process and microstructure formation, which in turn affect the strength development of the cement.

Several studies (Raksha 1975, Wang and Scrivener 1993, 1995) examined NaOH- and waterglass-activated slag cement pastes using XRD. As hydration time increased, the diffuse peak attributed to the $CaO-SiO_2-Al_2O_3$ vitreous structure decreased, but a peak belonging to C–S–H intensified. The crystallinity of C–S–H in NaOH-activated slag cement is obviously higher than that in waterglass-activated slag cement. This is also confirmed by an NMR study (Wang and Scrivener 2003). A crystalline compound $Mg_6Al_2CO_3(OH)_{16}.4H_2O$ (hydrotalcite) also appears in NaOH-activated slag cement.

Glukhovsky et al. (1980) reported a series of crystalline compounds such as hydrogarnet, zeolitic phases such as analcite, natrolite and gismondite, and mica phases including nepheline, paragonite, etc. from the hydration of alkali-activated slag cements based on XRD analyses. However, many researchers (Shi 1987, Yu and Wang 1990, Hong et al. 1993, Wang and Scrivener 1995) could not identify those crystalline phases in hardened alkali-activated slag cement pastes cured at room temperatures, even after 15 years of curing (Malolepszy 1993). An increase in curing temperature under atmospheric pressure definitely increases the crystallinity of C–S–H regardless of the nature of activators used (Wang and Scrivener 2003).

4.7.4 Hydration products under autoclave curing

Under autoclave conditions, the hydration process and products are very different from those under atmospheric pressure. Crystalline compounds such as xonotlite, tobermorite and sodium-zeolite (NAS_2H_3), could be identified in hydrothermal treated alkali-activated slag cement pastes (Shi et al. 1991b, Malolepszy 1993). Shi et al. (1991b) examined the hydration products of portland cement, alkali-activated blast furnace slag cement and alkali-activated phosphorus slag cement pastes after one day, five days and 15 days of hydration at 150 °C. For portland cement paste, crystalline $Ca(OH)_2$ and some unidentified hydration products were detected at one day. These unidentified hydration products disappeared, and only $Ca(OH)_2$ and $C_2SH(A)$ could be detected at 5 and 15 days. From the beginning to the end, no $C_3A(F)H_6$ was detected; this is in agreement with the results of Kalousek et al. (1949). The main hydration products for alkali-activated blast furnace slag cement paste were C–S–H(B) and xonotlite. $3CaO \cdot MgO \cdot 2SiO_2$ in the slag could be detected in alkali-activated blast furnace slag cement paste at one day, but disappeared after five days. Only C–S–H(B) and tobermorite could be detected in alkali-activated phosphorus slag cement paste.

When Sugama and Brothers (2004) used a 20% (by mass) sodium silicate solution (SiO_2/Na_2O mol ratio of 3.22) as the alkali activator and autoclaved sodium silicate-activated slag cements up to 200 °C, they found that C–S–H and tobermorite phases are the main hydration products. As the temperature was further increased to 300 °C, well-formed tobermorite and xonotlite crystals were identified. It also resulted in an undesirable porous microstructure, causing the retrogression of strength and enhancing water permeability. Jiang (1997) examined the hydration products of NaOH-activated slag hydrated at 25 and 700 °C. At 25 °C, only C–S–H and hydrotalcite were detected. However, when the temperature was increased to 700 °C, a series of well-crystallized products such as mica, nephelite, melilite and sodium calcium silicate formed.

4.8 Microstructure development

4.8.1 Microstructure at atmospheric pressure

No obvious hydration indication could be noticed on the surface of slag particles after 24 hours in pure water (Jiang *et al.* 1997). However, the surface of all slag particles was etched and coagulated products precipitated at some locations after 20 minutes in NaOH solution, as shown in Figure 4.12a (Shi

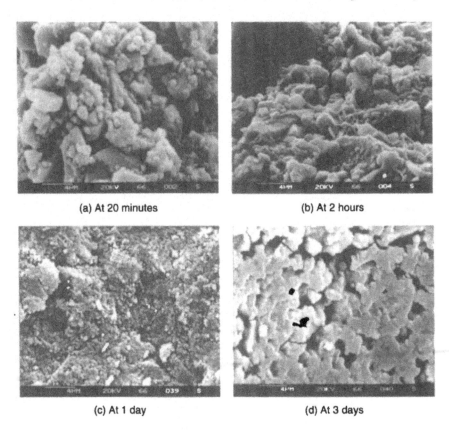

(a) At 20 minutes   (b) At 2 hours

(c) At 1 day   (d) At 3 days

Figure 4.12 Microstructure development of NaOH-activated slag pastes (Shi 1997).

1997). As time proceeded to two hours, most voids between slag particles were filled with coagulated products. Hexagonal plates, which were embedded in the coagulated products, appeared (Figure 4.12b). However, the configuration of slag particles could still be identified. At one day, a uniform hardened structure was observed on fracture surface. At this time, although structure looked loose (Figure 4.12c), it was impossible to identify the configuration of slag particles. Features of hydration products could only be notified in air voids. At three days, the fracture surface of the activated slag paste appeared to be similar to but denser than that at one day. Features of hydration products in air voids are shown in Figure 4.12d. In addition to coagulated particles, some hexagonal plates were also observed. The C–S–H here has a foil-like morphology compared with that in a conventional portland cement paste, which has a high Ca/Si ratio and fiber-like morphology (Song *et al*. 2000).

Microstructure development of Na_2CO_3-activated slag pastes is shown in Figure 4.13. At 20 minutes, most slag particles were intact except for

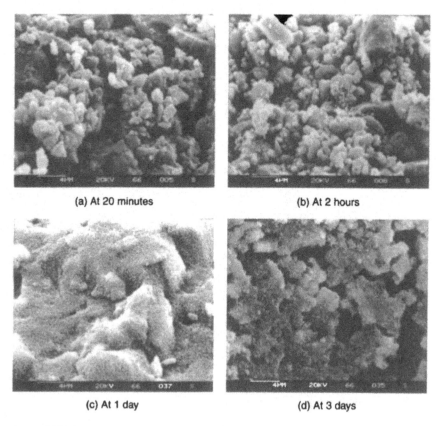

(a) At 20 minutes   (b) At 2 hours

(c) At 1 day   (d) At 3 days

Figure 4.13 Microstructure development of Na_2CO_3-activated slag pastes (Shi 1997).

precipitated gel-like products on some slag particles (Figure 4.13a). At two hours, all slag particle surfaces were etched and more gel-like products filled voids between slag particles (Figure 4.13b). As time proceeded to one day, a foil-like product was observed, which had really different morphology from that either at early ages or in NaOH-activated slag. The foil-like product covers most spaces between slag particles (Figure 4.13c). The configuration of slag particles was only identified at occasional locations. At three days, the fracture surface of hardened pastes showed a dense structure and the morphology of products could only be identified in voids (Figure 4.13d). No well-crystallized products were observed up to three days.

Figure 4.14 shows the microstructure development of Na_2SiO_3-activated slag pastes. During the first two hours, the microstructure characteristics of Na_2SiO_3-activated slag pastes (Figure 4.14a and 4.14b) were very similar

(a) At 20 minutes   (b) At 2 hours

(c) At 3 days

Figure 4.14 Microstructure development of Na_2SiO_3-activated slag cement pastes (Shi 1997).

to those of Na_2CO_3-activated slag pastes. However, at one day, most slag particles were covered by a layer of coagulated products but still separated. The morphology of C–S–H was very similar to that observed at two hours. At three days, the fracture surface appeared to be denser that that of Na_2CO_3-activated slag pastes.

Wang and Scrivener (1993, 1995) used SEM and backscattered electron (BSE) images to observe the morphologies of NaOH- and sodium silicate-activated slag cement pastes. They also observed tiny platy hydration products in NaOH-activated slag cement pastes since day one. The crystallized hydration products grew with time as indicated by XRD analysis. Well-crystallized hydrogehlenite could be clearly observed in a 12-year-old alkali-activated slag cement concrete (Malolepszy and Deja 1988).

The morphology of early hydration product of sodium silicate-activated slag cement pastes observed by Wang and Scrivener (1993, 1995) was similar to those as shown in Figures 4.11 to 4.13. The interesting fact is that unhydrated slag particles could still be clearly observed in NaOH- and sodium silicate-activated slag pastes after one year of hydration, which are well illustrated in BSE images. A clear ring with darker colour than bulk pastes could also be observed around unhydrated particles (Figures 4.15a and b). Different steam curing systems under atmospheric pressure did not show a noticeable effect on the morphology of hydration products at early ages (Malolepszy and Deja 1988).

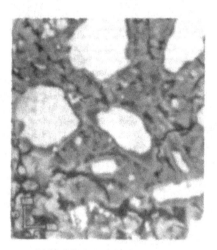
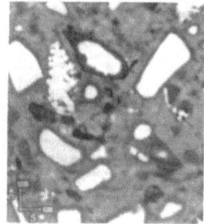

(a) NaOH-activated slag cement   (b) $Na_2O \cdot SiO_2$-activated slag

Figure 4.15 BSE images of alkali-activated slag cement pastes after one year of hydration (Wang and Scrivener 1995).

4.8.2 Microstructure of alkali-activated slag cement pastes under autoclave conditions

As discussed above, crystalline hydration products such as tobermorite and xonotlite can be detected by XRD in alkali-activated phosphorus slag and alkali-activated blast furnace slag cement pastes under autoclave conditions. SEM observation confirmed that crystalline fibrous xonotlite could be observed at the edge of air voids in autoclaved alkali-activated blast furnace slag cement pastes. The longer the curing time is, the larger the quantity and the crystal size of xonotlite are (Shi *et al.* 1991b). The microstructure of alkali-activated phosphorus slag cement paste was similar to that of alkali-activated blast furnace slag cement paste except that instead of fibrous xonotlite, thin plate tobermorite could be observed at the edge of air voids in the pastes. The longer the curing time is, the more and the larger the tobermorite crystals are. Figure 4.16 shows SEM pictures of tobermorite in Na_2CO_3-activated phosphorus slag cement pastes autoclaved at 175 °C. Malolepszy and Deja (1988) also observed fibrous tobermorite in a paste cured for 90 days at room temperature, followed by an autoclaving curing.

Sugama and Brothers (2004) used 20 % (by mass) sodium silicate solution (SiO_2/Na_2O mol ratio of 3.22) as the alkali activator, and autoclaved the sodium silicate-activated blast furnace slag cements at temperatures up to 200 °C. They found that the paste displayed an outstanding compressive strength of more than 80 MPa, and C–S–H and tobermorite phases were the detected hydration products. As the curing temperature

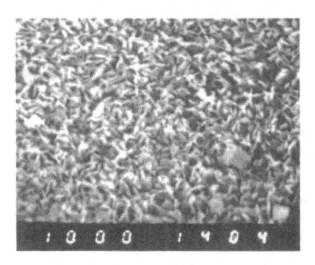

Figure 4.16 Tobermorite in Na_2CO_3-activated phosphorus slag cement cured at 185 °C (Tang 1986).

was increased to 300°C, well-crystallized tobermorite and xonotlite were detected. At the same time, a more porous microstructure with lower strength was obtained. Thus, it is important to understand how the temperature will affect hydration products and structural characteristics of the cements.

4.8.3 Microanalysis

Wang *et al.* (Wang and Scrivener 1993, 1995, Wang 2000) analyzed the chemical composition of hydration product using Energy Dispersive X-ray Microbeam Analysis (EDS) and found that Al, Mg and Na were detected in all analyses and fairly distributed throughout the paste. They compared the composition of platy crystals, rims and bulk pastes. Detailed analyses on those platy crystals in NaOH-activated slag cement pastes confirmed that those crystals were C_4AH_{13}. Figure 4.17 shows plots of Mg/Ca vs Al/Ca ratios for the rims and bulk pastes in waterglass and NaOH-activated slag pastes hydrated for one year. There is a clear linear relationship between the two ratios with the points lying on a line with Mg/Al ratio of 2.2. This relationship is consistent with the view that both the rims and bulk pastes consist of C–S–H with small crystals of hydrotalcite-type phase dispersed throughout the C–S–H as suggested by the X-ray examination. Rims around unhydrated slag particles contained a much higher proportion of the hydrotalcite-type phase than the bulk paste, which accounts for its darker appearance in the BSE images.

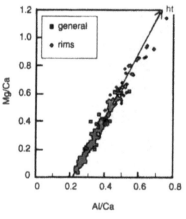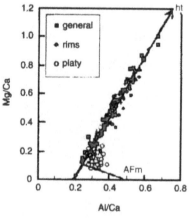

(a) Microanalysis of waterglass-activated slag hydrated for one year

(b) Microanalysis of NaOH-activated slag hydrated for one year

Figure 4.17 Microanalysis of products in waterglass- and NaOH-activated slag cement pastes hydrated for one year (Wang and Scrivener 1995).

When the straight line is extrapolated to Mg/Ca = 0, an Al/Ca ratio of 0.22 is obtained. This means that the Al existed either in solid solution within the C–S–H structure or in an AFm form finely intermixed with it. Richardson and Groves (1992) observed that Al could replace the bridging Si in the dreierketten C–S–H structure in portland slag cement pastes cured at elevated temperature, and felt that the majority of this aluminium may exist in solid solution of C–S–H structure after one year of hydration.

In Figure 4.17b, Mg/Ca ratios are plotted against Al/Ca ratios for those plates, rims and bulk pastes in the NaOH-activated slag cement pastes after one year of hydration, respectively. The relationship is very similar to that from the waterglass-activated slag cement pastes. It also confirmed that that paste consists of C–S–H and hydrotalcite-type phase. When extrapolating to Mg/Ca = 0, an Al/Ca ratio of 0.16 is obtained at one day and 0.20 at one year. Aluminium can be accommodated in bridging sites in the silicate chains and exists in AFm-type layers intermixed with the C–S–H type layers as described in Taylor's C–S–H gel model (Richardson et al. 1993, Taylor 1993).

If the analyses from all points are corrected for the aluminium associated with the hydrotalcite-type phase, a new graph can be plotted to show the relationship between the aluminium and calcium in the C–S–H and AFm phases (Wang and Scrivener 1995). This indicates that the Al/C ratio is always around 0.5, confirming the identification of the plates as AFm and indicating a similar relationship for the calcium and aluminium within the 'C–S–H gel'.

Richardson et al. (1994) found that, as with portland blast furnace slag cement (Richardson and Groves 1992), the microstructural features in KOH-activated slag cement can be classified into outer product, formed in the originally water- or solution-filled spaces, and inner product, formed within the boundaries of the original anhydrous grains.

On the ^{27}Al NMR spectra (Figure 4.18), only one broad peak appears on the spectrum for slag, which reflects the presence of Al in a range of distorted tetrahedral environments within the glassy phase of the slag. On hydration, this peak becomes narrow and lower. This corresponds to a reduced amount of glassy phase mixed with a phase with Al present in 4-fold co-ordination with a small range of chemical shifts, i.e. in a better environment than the glass (Richardson et al. 1994, Fernández-Jiménez et al. 2003). Al exists in 4-fold co-ordination in the C–S–H phase. The octahedral peak can be attributed to C_4AH_{13}, AFm layers which exist as distinct phases (Schilling et al. 1994b), to the Mg, Al hydroxide present in the inner product (Richardson et al. 1993), or to its substitution for Ca in the central CaO layer of a dimer-rich-tobermorite-like structure present in the NaOH-activated slag cement (Stade and Muller 1987, Grutzeck, 1997, Jiang et al. 1997).

Figure 4.19 shows the ^{29}Si MAS NMR Spectra of unhydrated slag and NaOH-activated slag cement (Jiang 1997). The unhydrated slag has

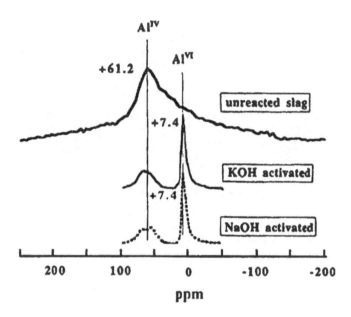

Figure 4.18 ^{27}Al NMR spectra of anhydrous slag, KOH (5 M)- and NaOH-activated slag cement (Jiang 1997).

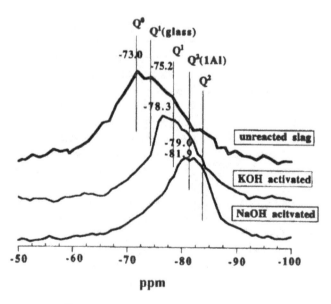

Figure 4.19 ^{29}Si NMR spectra of anhydrous slag and eight-year-old KOH (5 M)-activated slag (Jiang 1997).

peaks centred at -73.0 and -75.2 ppm. The broad peak (about 16 ppm at half-maximum) centred around -73 ppm is similar to that reported by Kirkpatrick (1988) for an akermanite glass ($Ca_2MgSi_2O_7$) in both position and peak width since the composition of the slag is similar to that of akermanite. In virtually all silicates, Si occupies a tetrahedral site. Silicate minerals can be classified according to the degree of polymerization of the tetrahedra. The Q^n notation represents the number of bridging oxygens per tetrahedron – thus Q^0 (isolated), Q^1 (dimer and end groups), Q^2 (chains), Q^3 (sheets) and Q^4 (framework) (Putnis, 1992). The peak of unreacted slag at about -73 ppm can be attributed to monomeric silicate anions Q^0, and peak at -75.2 ppm can be attributed to dimeric silicate anions Q^1 (glass). The hydrated KOH- and NaOH-activated slag cements yield a peak near -78.3 ppm and $-79.0, -81.9$ respectively. After hydration, a shift to a more shielded (more negative) position is seen and the peak is much narrower, indicating a greater degree of short-range order. Two features can be observed from these ^{29}Si patterns: (1) the structural modification around the Si nuclei during hydration has been made, as reflected by the shift in peak position, (2) extensive reaction is observed, with the broad glass peak greatly reduced in the pattern of the reacted materials. The KOH-activated sample shows a strong peak at -78.3 ppm, which can be assigned to dimeric Q^1 silicate species. The overlapping peaks from alkali-activated slag cement show the highest intensity from -79 ppm to -84 ppm. Peaks in this range have a higher degree of polymerization, which can be attributed to Q^1, Q^2 (1A1) and Q^2 silicate species, with Q^2 (1A1) being a chain midmember site which shares one oxygen with an aluminum-substituted tetrahedron (Schilling *et al.* 1994b). NMR study indicated that Al incorporated into C–S–H even at early ages (Wang 2000b, Fernández-Jiménez *et al.* 2003, Wang and Scrivener 2003). This incorporation, even at early ages, is associated with a longer chain length of the C–S–H than in C_3S or portland cement systems.

4.9   Pore solution chemistry

The pore solution chemistry of alkali-activated cements and concrete is related to the nature and dosage of activators used at early ages, and is related to the characteristics of hydration products at later ages. Song and Jennings (1999) investigated how the initial alkali concentrations affect the pore solution chemistry of NaOH-activated blast furnace slag. They concluded that the solubilities of Si, Ca, Al and Mg are strong functions of the pH. An increase in pH increases the concentration of Si and Al, but reduces that of Ca and Mg in the pore solution. Thus, it can be expected that initial pH can have an effect on the nature of C–S–H by affecting the chemistry of the pore solution.

Jiang *et al.* (1997) measured Na and Ca concentrations in the pore solution of NaOH-activated slag cement with 2.5 M NaOH solution at

$W/S = 0.4$ from one hour to seven days. They found that the Na concentration decreased approximately 10%, but Ca concentration remained almost unchanged. The chemical analyses of extruded pore solution of NaOH-activated and waterglass-activated slag cement pastes are shown in Figure 4.20 (Puertas *et al.* 2004). Although the slag has a high content of Mg, it is under detection limit in the extruded pore solution. The Ca^{2+} concentrations are also very low for both pastes. This is in agreement with a previous publication (Jiang *et al.* 1997). The Si concentration of waterglass-activated paste is very high (2243 mmol) at three hours, due to the introduction of Si from the activator. It decreased drastically with time and showed a concentration of 324 mmol at 24 hours. In contrast, the Si concentration of NaOH-activated paste is always much lower than that of the waterglass-activated slag paste.

The Na concentrations of the solutions from the two pastes decreased slowly with time. This is also in agreement with previous results. Calculations based on solubility products indicate all solutions are over-saturated for grossular (C_3AS_3) and gehlenite hydrate (C_2ASH_8) and tricalcium aluminate hydrate (C_3AH_6). However, the solution from NaOH-activated pastes is not over-saturated with these products last two phases at seven days.

As for NaOH-activated pastes, at the third hour, the solution is also over-saturated in $Ca(OH)_2$, C–S–H (Ca/Si = 1.8) gel, C–S–H (Ca/Si = 1) gel and C_4AH_{13}. At one day, the solution was not over-saturated with $Ca(OH)_2$, and at two days, the solution was not over saturated in any C–S–H gel. At seven days, the solution was only over-saturated in brucite and mainly in hydrotalcite. As for waterglass-activated pastes, the solution was over-saturated in gibbsite, C–S–H (Ca/Si = 1.8) gel, C–S–H (C/S = 1.1) gel, C–S–H (Ca/Si = 0.8) gel and sodium silicoaluminate hydrate. At one day, it was not over-saturated with gibbsite and C–S–H (Ca/Si = 0.8) gel. At two days, the solution was over-saturated in C–S–H (Ca/Si = 1.8) gel

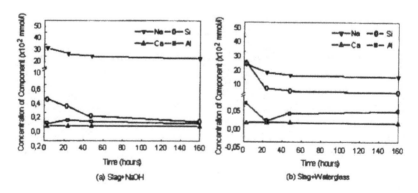

Figure 4.20 Pore solution chemistry of NaOH- and warterglass-activated slag cement pastes (Puertas *et al.* 2004).

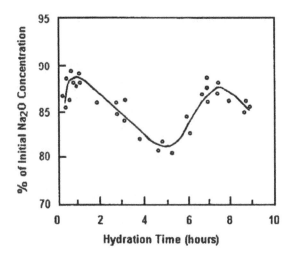

Figure 4.21 Wave character of the change of Na$_2$O concentration in the liquid phase of hardened alkali-activated slag cement [based on Krivenko 1992a].

and C–S–H (Ca/Si = 1) gel. At seven days, the solution was over-saturated in brucite, hydrotalcite and C–S–H (Ca/Si = 1.1) gel.

Hong *et al.* (1993) monitored the concentration of Na$_2$O in the liquid phase of 0.3 N NaOH and Na$_2$SiO$_3$ activated-slag cements with a water-to-slag ratio of 20 from 3 hours to 28 days. It was found that the Na$_2$O concentration in the liquid did not change with time in Na$_2$SiO$_3$-activated slag cements, but decreased slightly with time in the NaOH-activated slag cement. In another study (Krivenko 1992), it was found that 80 to 90% of initial alkali could be detected in the liquid phase of the hardened alkali-activated slag cement (Figure 4.21). The author proposed that the alkaline cation acts as a destruction catalyst, based on the wave character of the change of Na$_2$O concentration in the liquid phase of hardened alkali-activated slag cement, during the destruction of glassy slag structure and the formation of hydrates (Krivenko 1992a). Puertas *et al.* (2004) found that Na concentration of waterglass-activated slag is always lower than that of NaOH-activated slag, while Hong *et al.* (1993) got an opposite trend.

4.10   C–S–H in alkali-activated slag cement paste

C–S–H is an amorphous colloidal microporous material and the main cementitious compound in most hardened cementitious materials. To a large extent, the performance of a hardened cement paste depends upon

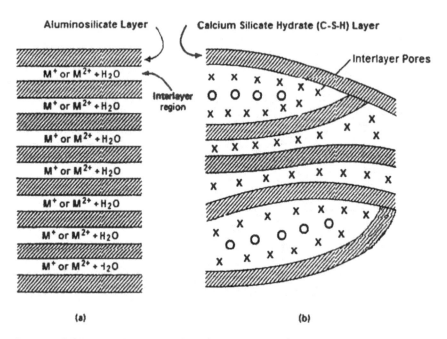

Figure 4.22 Schematic illustration of (a) clay and (b) C–S–H structure (based on Mindess and Young 1981).

C–S–H. The structure of C–S–H is regular within a particular sheet and resembles clay structures (Figure 4.22); however, there is no long-range order (Mindess and Young 1981). C–S–H has a very high surface area of about 100 to 700 m^2/g, depending on measurement techniques.

Investigations of the ternary $CaO–SiO_2–H_2O$ system indicated that solubility curve for C–S–H tended to group along two curves (Jennings 1986). It seems that there are two types of C–S–H present in the $CaO–SiO_2–H_2O$ system. The C–S–H curve exhibiting the low solubility was interpreted as that for C–S–H (I). Figure 4.23 is the relationship between Ca/Si ratio of C–S–H and the equilibrium pH (Beaudoin and Brown, 1992). C–S–H (I) exists at pH of 10 and C–S–H (II) at pH of 11.9. At the same time, C–S–H (I) has a higher buffer capacity than C–S–H (II), which is beneficial for the acidic resistance of the cements.

The average detected Ca/Si ratios of C–S–H gel in alkali-activated slag cement pastes are listed in Table 4.5. It can be seen that the Ca/Si ratio of C–S–H in NaOH-activated slag cement is very close to that of the slag. However, the Ca/Si ratio of the C–S–H in waterglass-activated slag cement is lower than that of the slag due to the introduction of additional SiO_2 from the waterglass. It increases with time due to the further hydration of the slag.

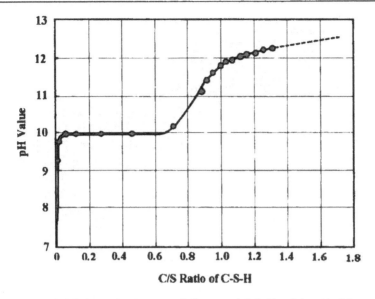

Figure 4.23 Relationship between Ca/Si ratio of C–S–H and the pH of the co-existing solution phase (based on Beaudoin and Brown 1992).

Table 4.5 Average detected C/S ratios of C–S–H in alkali-activated slag cement (Wang and Scrivener 1993)

Material	Hydration time (day)	Ca/Si ratio
Slag	N/A	1.27
NaOH-activated slag cement	1	1.30
$Na_2O \cdot SiO_2$-activated slag cement	1	0.94
	5	1.02
$Na_2O \cdot 1.5SiO_2$-activated slag cement	1	0.88
	5	0.98

Richardson and Groves (1992) noticed that the outer C–S–H is coarse, foil-like and has a semi-crystalline, layered C–S–H related to 1.4 nm tobermorite. The inner product region containing C–S–H Mg and Al-rich precipitates are oriented towards the interface of inner and outer products. It can be seen (Table 4.6) that the outer C–S–H has a lower Ca/Si ratio than

Table 4.6 Comparison of inner and outer products in eight-year-old KOH-activated slag cement pastes (Richardson *et al.* 1994)

Product	Ca/Si (C–S–H)	Al/Ca (C–S–H)	K/Ca	Number
Outer	0.993 ± 0.064	0.195 ± 0.018	0.125 ± 0.024	33
Inner	1.178 ± 0.072	0.155 ± 0.015	0.090 ± 0.016	34

the inner product. On the other hand, the former showed a higher Al/Ca or K/Ca ratio than the latter.

4.11   Selectivity of chemical activators

In addition to the factors affecting the strength development of portland cement, the nature and dosage of activators play a crucial role in determining hydration and strength development of alkali-activated slag. As stated above, effects of activators may be different for slags from different sources (Krivenko 1986).

Malolepszy (1986) tested the strength of different slags under the action of various activators and found that Na_2CO_3 is especially suitable for slags rich in C_2MS, and NaOH is a good activator for slags rich in C_2AS. Under most circumstances, waterglass is the most effective activator; while, the optimal modulus n of the waterglass also varies with the nature of the slag (Bin 1988, Shi and Li 1989a, b).

Similar phenomena were observed in blast furnace slag/fly ash mixtures. Smith and Osborne (1977), and Bijen and Waltje (1989) found NaOH to be the best activator for fly-ash/slag mixtures; waterglass did not show any activation effect on such mixtures. In contrast, Dai and Cheng (1988) found that waterglass was much better than NaOH for fly-ash/slag mixtures. Lu (1989) developed alkali-activated fly-ash/slag concretes with a strength as high as 76 MPa at 28 days with waterglass as the activator.

Shi and Day (1996c) analysed the effects of activators and slag on hydration products and strength of alkali-activated slag cements. They concluded that the selectivity of activators results from the differences in chemical composition of slags, which results in the formation of different hydration products and different strength.

4.12   State of alkalis in alkali-activated slag cement pastes

4.12.1   Introduction

Since alkalis play a critical role in determining the properties and structure of hardened alkali-activated slag cement pastes and concretes, the state and function of alkalis during the hydration and hardening of alkali-activated slag cements have been the focus of many studies. Ilyukhin *et al.* (1979) found that the presence of Na even in a quantity of 0.6% Na_2O by mass starts to affect the stability of silicates and hydrosilicates as well as their hydration rate. This section discusses the interaction between alkalis and

C–S–H, leachability of alkalis from C–S–H and the state of alkalis in hardened alkali-activated slag cement pastes.

4.12.2 Interactions between alkalis and C–S–H

Stade (1989) suggested that the incorporation of alkalis into C–S–H, from room to elevated temperatures, follows the following three mechanisms:

(1) Neutralization of acidic Si–OH groups by Na$^+$ or K$^+$

$$C-S-H + M^+ \rightarrow M-C-S-H \tag{4.13}$$

(2) Ion exchange of Na/K for Ca

$$C-S-H + M^+ \rightarrow M-C-S-H + Ca^{2+} \tag{4.14}$$

This process would increase with increasing C/S ratio
(3) Cleavage of Si–O–Si bonds by Na$^+$ or K$^+$
With attachment of Na$^+$ or K$^+$ to Si–O, represented by:

$$Si-O-Si + 2M^+ \rightarrow 2Si-O-M \tag{4.15}$$

The first mechanism happens mainly to C–S–H with a low Ca/Si ratio because it contains a high concentration of acidic Si–OH group. The second mechanism is applied mainly to C–S–H with a high Ca/Si ratio. The third mechanism depends on the surface area of C–S–H. Due to the complexity of the system, it is also suggested that multiple mechanisms may be operating to uptake alkaline ions at any one time.

It is agreed that the incorporation of alkalis into C–S–H increases as the Ca/Si ratio of C–S–H decreases. It was found (Hong and Glasser 1999) that there is almost a linear relationship between the log scale of the Na/Ca and the Ca/Si ratios of the C–S–H, ranging from 0.85 to 1.8, as shown in Figure 4.24. This means that the incorporation of alkalis into C–S–H decreases swiftly as the Ca/Si ratio of C–S–H increases. The alkali incorporated into C–S–H is almost linearly proportional to the initial concentration of the alkali in the solution. Kalousek (1944) found that the limit of Na incorporated into the N–C–S–H system is $0.25Na_2O \cdot CaO \cdot SiO_2 \cdot xH_2O$. Taylor (1987) found that the Na/Ca atom ratio is about 0.01 in hydrated portland cement pastes, since the Ca/Si ratio of C–S–H in hydrated portland cement paste usually ranges from 1.2 to 2.3 depending on the cement composition, water-to-cement ratio, hydration

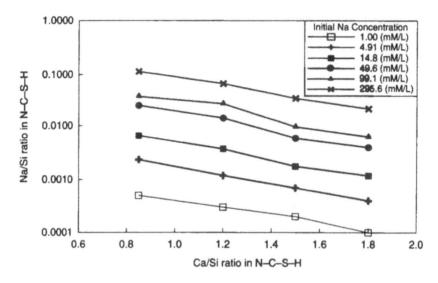

Figure 4.24 Effect of Ca/Si ratio and initial alkali concentrations on Na/Si ratio in C–S–H phase (data from Hong and Glasser 1999).

temperature, age, etc. (Odler 1998). The incorporation of impurities into a crystalline structure is usually governed by crystallochemical restraints that limit the number of possible replacement mechanisms and their stoichiometry, while C–S–H is a poorly crystallized version of the mineral tobermorite and the main restriction for the incorporation of alkali into the structure of C–S–H is probably the condition of electroneutrality (Ramachandran *et al.* 1981).

Curing temperature can have a significant effect on the alkali retention capacity of C–S–H. For a given Ca/Si ratio, C–S–H formed at elevated temperatures has higher crystallinity and lower surface area, and will bind smaller amounts of alkalis than that formed at lower temperature. ^{29}Si NMR spectroscopy indicated that there is little difference between Na-free and Na-containing C–S–H gel with a high Ca/Si ratio; however, some differences were found in C–S–H gels with a low Ca/Si ratio. Q_2/Q_1 ratio decreases as the Na content increases (Atkins *et al.* 1991). This indicates that the presence of Na decreases the Si chain lengths in C–S–H.

Taylor concluded from his analysis of pore solution data that sodium was significantly better bonded into hydrated cement paste than potassium (Taylor 1987). Experimental results obtained by Stade (1989), and Hong and Glasser (1999) indicated that there is no significant difference in uptake of Na and K by C–S–H.

4.12.2.1   Solubility of C–S–H and leachability of alkalis from C–S–H

The presence of alkalis significantly increases the solubility of Si and decreases the solubility of Ca due to the common ion effect. However, it does not change the K_{sp} of C–S–H. The solubility curve of N–C–S–H is very different from that of C–S–H (Atkins *et al.* 1991), but is closer to that of C–S–H after leaching. Analysis of liquid and solid phases indicated that the Na/Ca molar ratio in leachates is much higher than that in the solid phase, which means that Na in preferably leached from the solid phase. The Na/Ca molar ratio in the leachate consistently increases with decreasing Ca/Si ratio of N–C–S–H gel. This is attributed to the fact that gels with lower Ca/Si ratio are relatively enriched in Na.

Several studies have measured the leachable alkalis in alkali-activated slag cement pastes in order to understand the activation mechanism of slag cements. Belitsky *et al.* (1993b) measured non-leachable Na_2O by immersing two grams of 1–2 mm size crushed hardened $Na_2O \cdot nSiO_2$-activated slag cement pastes in 100 ml of distilled water for seven days. The total alkali content was determined by dissolving the pastes in 1 N HCl solution. Experimental results indicated that the non-leachable Na_2O varied only from 0.17 to 0.37% by mass of slag, depending on the age, molar ratio n and the dosage of $Na_2O \cdot nSiO_2$. Eight batches of hardened alkali-activated slag cement pastes were prepared using four types of slag – gehlenite, akermanite, diopside and melilite, and two types of activators – NaOH and Na_2CO_3 and cured at 80, 200 and 250°C for 28, 90, 360 and 720 days (Malolepszy 1993). The cement pastes were ground in an agate mill and the ground samples were washed with distilled water until they reached pH = 7. The measured soluble alkali content varied from 16% to 46% and changed slightly with the type of activator and the composition of slag, but decreased significantly as the curing temperature increased. In another study, Deja (2002a) measured the relative soluble alkali content in samples from reinforced alkali-activated slag concrete floor slabs and wall panels used in a storehouse built in Kraków in 1974, as shown in Table 4.7. The results in the table indicate that the soluble alkali content does not change from 28 days to 27 years.

Wang (2000a) measured Na in NaOH and sodium silicate-activated slag cement pastes with two different techniques. The concentrations of the NaOH and sodium silicate ($Na_2O \cdot SiO_2$) were 4 M and 2 M. The solution-to-slag ratio for both activators was 0.25 ml/g. First, he washed

Table 4.7 Relative free alkali contents in alkali-activated concrete at different ages (Deja 2002a)

Age	28 days	90 days	27 years
Percentage of free alkalis (%)	35.2	29.2	38.1

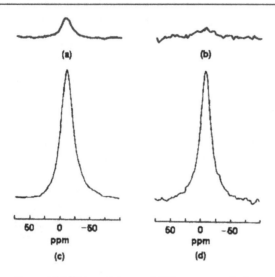

Figure 4.25 ^{23}Na solid-state NMR spectra of alkali-activated slag cement pastes at 80°C for six months: (a) NaOH-activated slag after 3 washings, (b) sodium silicate-activated slag after 10 washings, (c) NaOH-activated slag before washing, (d) sodium silicate-activated slag before washing (Wang 2000a).

3 grams of these ground hardened cement pastes with 15 ml of distilled water for 10 times, then measured the Na content in these washed powders. The amounts of sodium left (percentage of the total sample by mass) after washing were:

1   0.23% for NaOH-activated slag cured at 80°C for six months;
2   0.15% for sodium silicate-activated slag cured at 80°C for six months;
3   0.10% for NaOH-activated slag cured at 20°C for six months;
4   0.10% for sodium silicate-activated slag cured at 20°C for six months;

On the other hand, the slag contains 0.27% (by mass) Na_2O. He also tested the washed materials with ^{23}Na solid-state NMR. The ^{23}Na solid-state NMR spectra, as shown in Figure 4.25 indicated that almost all sodium could be washed away. This implies that Na in alkali-activated slag cement paste is not structurally incorporated into C–S–H, but is adsorbed onto the negatively charged surface of C–S–H.

4.13 Summary

This chapter has discussed the hydration and microstructure of alkali-activated slag cements and how different factors affect the hydration, products and microstructure of hardened pastes. It can be summarized as follows:

1 The characteristics of slag, and the nature and dosage of activator(s) have a significant effect on hydration mechanism, products and microstructure of alkali-activated slag cement pastes and mortars, which are hard to predict.

2 Based on the heat evolution curves, the hydration of alkali-activated slag cements can be classified into three classes. An increase in water-to-slag ratio has a more significant effect on the hydration of alkali-activated slag cement than that of water-to-cement on the hydration of portland cement due to the dilution of activators.

3 The main hydration product of alkali-activated slag cement at room temperatures is C–S–H gel with a low Ca/Si ratio. The minor hydration products are determined by the characteristics of the slag and activator(s) used. Crystallized products, such as tobermotite, xonotlite, etc. can be the main hydration products under autoclave curing conditions.

4 The C–S–H in hydrated alkali-activated slag cement paste can retain more alkali ions than that in portland cement pastes since the former has a lower Ca/Si ratio than the latter.

Chapter 5

Properties of alkali-activated slag cement pastes and mortars

5.1 Introduction

Although concrete, instead of cement pastes or mortars, is used for most applications, cement pastes and mortars determine the properties of concrete. The properties of fresh cement pastes and concretes are important for mixing, transportation, placement and finishing of the concrete mixtures. This chapter discusses the properties of alkali-activated slag cement pastes and mortars.

5.2 Workability

5.2.1 Introduction

Workability is defined as the amount of mechanical work or energy required to compact concrete without segregation. Many terms have been used to describe the properties of fresh cement pastes and concrete for mixing, transportation, placement and finishing: consistency, flowability, mobility, pumpability, finishability and harshness. In many cases, people just use workability to represent all of the properties mentioned above.

Many factors such as the nature of slag and activators, dosage of activator(s), fineness of slag, chemical admixtures, addition of lime, mineral admixtures and timing for the addition of activators have an effect on the rheological properties of alkali-activated slag cement pastes. In general, ultimate resistances to shear or the viscosities of the alkali-activated slag cement pastes are higher compared to those of portland cement pastes. This indicates that the former have higher tendency for the formation of cementitious structure than the latter (Glukhovsky *et al.* 1981). The following sections discuss how these factors affect the rheological properties of alkali-activated slag cement pastes and mortars.

5.2.2 Effect of activators

Activators have a great effect on the rheological properties of alkali-activated slag cement pastes. Jiang (1997) measured the flow curves of slag

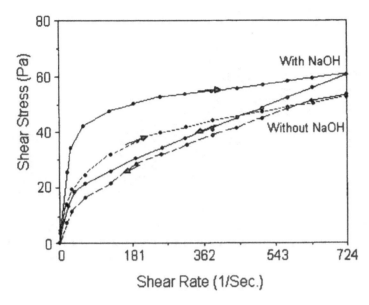

Figure 5.1 Flowability of fresh slag cement paste with/without 5% NaOH at W/S = 0.75 (Jiang 1997).

cement with or without NaOH at water-to-slag ratios of 0.75 and 1.5. The test results are shown in Figure 5.1 and summarized in Table 5.1. At the water-to-slag ratio of 0.75, the addition of 5% NaOH increased the apparent viscosity and yield stress of the cement pastes.

As the water-to-slag ratio is increased to 1.5, it decreased the apparent viscosity and yield stress of the slag cement pastes themselves. However, the addition of NaOH does not show any noticeable effect on the apparent viscosity and yield stress of the cement pastes.

Jolicoeur *et al.* (1992) tested the mini-slump of portland cement and NaOH-activated slag cement pastes with different dosages of NaOH at 10, 30 and 60 minutes, as summarized in Table 5.2. Slag paste without activator showed smaller mini-slump area and quicker slump loss than portland

Table 5.1 Effect of alkaline activator on apparent viscosity and yield stress of cement paste (Jiang 1997)

Material	Water-to-slag ratio	Apparent viscosity (MPa . s)	Yield stress (MPa)
100% Slag	0.75	6.2	32
	1.5	3.2	21
100% Slag + 5% NaOH	0.75	78	690
	1.5	4.2	28

Table 5.2 Mini-slump areas of portland cement and NaOH-activated slag cement pastes
(Jolicoeur et al. 1992)

W/C	W/C = 0.4				W/C = 0.5					
NaOH dosage in Na₂O	PC	0	2	4	PC	0	0.5	I	2	4
10 minutes	40	32	30	35	119	100	92	66	72.5	69
30 minutes	34	20.6	12	20	110	84	23	18	26	43
60 minutes	31	20	2	6	78	70	8	10	7	18

cement paste. At $W/S = 0.4$, the addition of NaOH showed no effect on the mini-slump at 10 minutes, but significantly decreased mini-slump at 60 minutes regardless of NaOH dosage. At $W/S = 0.5$, the mini-slump decreased with the increase of NaOH dosage up to 1%, then showed no obvious changes with NaOH dosage from 1 to 4% at 10 minutes. The mini-slump of NaOH-activated slag cement pastes decreased very quickly as hydration time increased.

The nature of activator also affects the rheological properties of alkali-activated slag cement pastes. It was found that the water requirement for normal consistency for waterglass-activated slag cement paste was 0.28, lower than that for NaOH and Na_2CO_3-activated slag cement paste – 0.30 (Shi 1987, Shi and Li 1989b). For a given water-to-slag ratio, the NaOH dosage, varying from 0.5 to 4% by Na_2O, does not show a significant effect on mini-slump. A similar phenomenon was observed from Na_2CO_3-activated slag cement pastes (Hrazdira and Kalousek 1994). However, the mini-slump decreases with waterglass dosage significantly, especially when the modulus of waterglass is less than 0.5 and higher than 1.5.

Jolicoeur et al. (1992) found that the use of 2% NaCl, Na_2SO_4, $CaSO_4$, Na_2CO_3, $Na_2B_4O_7$, NaAC and Na-NTA as activators did not change the mini-slump areas significantly compared with the control slag paste without any activator. The introduction of additional 0.3% NaOH significantly decreased the mini-slump area of the control paste and the pastes containing NaCl, Na_2SO_4, NaAC or Na-NTA, but the introduction of $Na_5P_3O_{10}$ and Na-Glu significantly increased the mini-slump area of the control paste or the one containing 0.3% NaOH.

For waterglass-activated slag cement, the modulus of the waterglass also has a significant effect on mini-slump (Figure 5.2). When the modulus is lower than 0.5, the workability is low and similar to that of NaOH-activated cement pastes. When the modulus is between 0.5 and 1.0, the workability is very high and the slump loss with time is minimal. As the modulus of sodium silicate is greater than 1, the workability of the paste decreased markedly with the increase of the modulus of the silicate. When the modulus is 2, the workability is lost within few minutes after mixing.

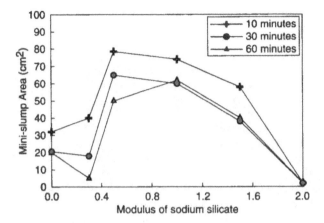

Figure 5.2 Effect of modulus of sodium silicate on mini-slump of alkali-activated slag cement pastes ($W/S = 0.4$) (data from Jolicoeur *et al.* 1992).

5.2.3 Effect of chemical admixtures

Water-reducing admixtures are being widely used in portland cement concrete to reduce the water requirement for a given workability or to increase the workability at a given or lower water-to-cement ratio. Some studies have been conducted on the effect of water-reducing agents on the workability of alkali-activated slag cements. However, contradictory results have been reported because of different activators used.

Two types of superplasticizer – sodium lignosulphonate (LS) and formalin condensates of β-naphthalene sodium sulphonate (NS) – were added into NaOH-activated slag cement paste (Isozaki *et al.* 1986). It was found that LS showed strong plasticizing effects, but NS did not show an obvious effect on the rheological property of the cement paste.

When Douglas and Brandstetr (1990) added 0.2, 0.5 and 1.0% sodium lignosulphonate, or 0.5, 1, 5 and 9% sulphonated naphthalene-based superplasticizer, based on the mass of binder, into sodium silicate-activated slag cement mortars to improve their workability, no change in consistency was observed except for 9% sulphonated naphthalene-based superplasticizer. However, the addition of this superplasticizer decreases the one-day strength very significantly. Gifford and Gillott (1997) observed similar phenomena. They tested four high-range water-reducing admixtures and one normal-range water-reducing admixture for alkali-activated slag cement mortars. When Na_2SO_4 and Na_2CO_3 were used as activators, all these water-reducing admixtures were incompatible. Their use markedly reduce strength development rate. When $Na_2O \cdot SiO_2$ was used as an activator, a very high and mostly impractical dosage had to be used to achieve reasonably placeable mixtures. It seems that sodium lignosulphonate works in NaOH-activated slag cement system, but not in sodium silicate-activated slag cement system.

Jolicoeur *et al.* (1992) studied the effect of activator and water-reducing admixtures on mini-slump of different alkali-activated slag cement pastes. For the pastes without any activator, the addition of sodium poly-naphthalene sulphonate greatly increased the initial mini-slump of the pastes, but the mini-slump decreased very slowly with time. When Na-gluconate was added, the mini-slump of the slag paste was increased initially, and further increased with time.

When sodium poly-naphthalene sulphonate and Na-gluconate were added into NaOH-activated slag cement pastes, both of them showed comparable increase in mini-slump. However, the mini-slump of the pastes containing sodium poly-naphthalene sulphonate was rapidly lost with time and that containing Na-gluconate increased with time. These two plasticizers were also added into slag cement pastes activated with a combination of 2% (by mass of Na_2O) NaOH and 2% Na_2CO_3. Both of them increased the mini-slump of the pastes very significantly. However, the mini-slump was lost in 60 minutes when sodium poly-naphthalene sulphonate was used. The mini-slump of the pastes in the presence of Na-gluconate did not show an obvious change with time. A calcium poly-naphthalene sulphonate was also added into the slag cement pastes activated with a combination of NaOH and Na_2CO_3. However, it was not as effective as sodium poly-naphthalene sulphonate.

For waterglass-activated slag cement, the effect also varies with the modulus of the waterglass (Jolicoeur *et al.* 1992). At low ratios (<0.5), the effect is very similar to that on NaOH-activated slag cement. At the intermediate ratio range (0.5–1.5), both sodium poly-naphthalene sulphonate and Na-gluconate did not affect the mini-slump. In the presence of lime, sodium poly-naphthalene sulphonate or sodium lignosulphonate also did not show an effect on workability (Jolicoeur *et al.* 1992).

Collins and Sanjayan (2001a) investigated the effect of chemical admixtures on the mini-slump of solid sodium silicate-activated slag cement pastes with a water-to-slag ratio of 0.4. The addition of lignosulphonate water-reducing retarder increased the workability alkali-activated slag cement pastes. Increasing additions of naphthalene sulphonate-based superplasticiser moderately improved workability. The combination of lignosulphonate water-reducing retarder and naphthalene sulphonate-based superplasticiser shows a significant improvement in workability. In this study, lignosulphonate water-reducing retarder and naphthalene sulphonate-based superplasticiser worked well probably because solid sodium silicate instead of liquid sodium silicate was used as an activator. A recent study by Puertas *et al.* (2003) indicated that newly developed superplasticizers – vinyl co-polymer- and polyacrylate co-polymer-based superplasticizers – also did not have an

Table 5.3 Slump, setting time and strength of alkali-activated slag cement concrete (Zhu *et al.* 2001)

Batch no	Slump at different time (mm)						Setting time (h:min)		Compressive strength at different ages (MPa)		
	0	1 h	2 h	3 h	4 h	5 h	Initial	Final	3 d	7 d	28 d
C40	230	230	230	230	230	220	6:10	8:10	20.7	30.5	48.9
C50	230	230	230	230	230	220	5:15	7:20	31.6	42.8	68.3
C60	220	220	220	220	220	200	5:05	7:15	41.2	69.4	81.1

obvious effect on the mini-slump of alkali-activated slag cements. However, Zhu *et al.* (2001) reported a very effective retarder, which could retard the setting and keep the workability of the concrete mixtures very well during the first several hours, but the concrete could still give very high strength at three days, as summarized in Table 5.3.

5.2.4 Effect of lime addition

Lime has been added into alkali-activated slag cements for different purposes. Douglas *et al.* (1992) used lime slurry to control the setting of waterglass-activated slag cement concrete. Jolicoeur *et al.* (1992) studied the effect of lime addition on mini-slump of waterglass-activated slag cement pates. When the modulus of the sodium silicate was between 0.5 and 1.0, the addition of 2% lime in either dry or slurry form decreased the mini-slump of the pastes significantly. Lime slurry showed a more obvious effect than dry lime. However, when the modulus of the sodium silicate was increased to 1.5, the addition of lime, whether in the form of slurry or dry powder, slightly increased mini-slump of the pastes. The addition of lime could cause a quicker loss of workability of waterglass-activated slag cement when the modulus is less than 1.5 (Chen and Liao 1992).

Different amounts of lime were added into waterglass (modulus = 1.5)-activated slag cement paste to see its effects on mini-slump areas of the paste (Figure 5.3) (Cheng and Sarkar 1994). The 10-minute measuring point shows that as the lime content increases, so does the pat area, except for the sample with 5% lime. The trend is similar at 30 minutes, although the decrease in pat area begins with lower lime contents. This effect evens out at 90 minutes. Cheng and Sarkar (1994) explained the results based on flocculation of colloidal particles. Flocculation is the principal manifestation of Van der Waals attraction between colloidal particles. According to flocculation kinetics, a potential energy can influence the flocculation rate constant; addition of an electrolyte can vary the barrier height. The effect of ionic concentration is to decrease this rate constant. Therefore, the pat area is larger for higher lime content at the first measuring point. However, since

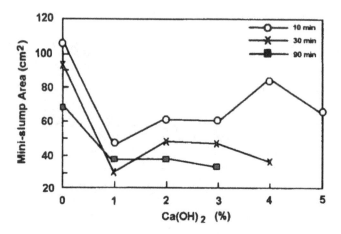

Figure 5.3 Effect of lime addition on mini-slump of waterglass (Ms = 1.5)-activated slag cement pastes (data from Cheng and Sarkar 1994).

the rheology is also partly controlled by the early chemical reactions, it is possible that in the case of 5% lime, chemical reactions supersede Van der Waals attraction force. When molecules are combined in a chemical chain, its structure is stable; it implies a decrease in the pat area. A similar situation arises at 30 minutes in the sample with 4% lime, where one observes a sharp decrease in the pat area from 10 to 30 minutes, and then it becomes measurable.

5.2.5   Effect of mineral admixtures

It is well known that silica fume reduces workability of portland cement concrete largely due to its very high specific surface area, and fly ash increases workability of portland cement concrete mainly due to its spherical shape or its "ball bearing" effect. Several studies have reported that the replacement of slag in alkali-activated slag cement with 10 to 20% silica fume or fly ash can greatly improve the workability of alkali-activated slag cement pastes or concrete (Skvara 1985, Talling and Brandstetr 1989, Gifford and Gillott 1997, Collins and Sanjayan 1999b), as shown in Figure 5.4. Their effects vary with the nature of activators used.

5.2.6   Effect of the timing for the addition of activator

It was found that the workability of alkali-activated slag cement pastes could be increased by two-stage addition of alkaline activators: alkaline solution with a low concentration is mixed with slag at a water-to-slag

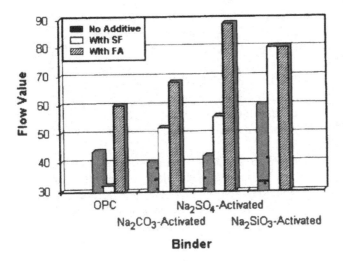

Figure 5.4 Effect of 15% replacement of slag with fly ash or silica fume on Flowability of alkali-activated slag cement mortars (data from Gifford and Gillott 1997).

ratio from 0.1 to 0.15, then mixed with concentrated solution to give the required water and activator (Krivenko 1994b).

5.3   Setting times

As cement hydrates, hydration products form a three-dimensional network. When the network reaches a certain point, fresh pastes lose flowability or plastic consistency and become crumbles under the effect of a sufficiently great external force. This process is called setting. In practice, a certain time is needed for mixing, transporting, placing and finishing cement pastes and concrete. Thus, it is necessary for the mixture to display a certain setting time. Usually, two setting times – initial and final setting times – are specified. There is no well-defined physical meaning to setting or setting times, rather they arbitrarily describe the gradual rheological changes during hydration. Many factors such as the nature of the slag, nature and dosage of activator, and additives will affect the setting of alkali-activated slag cement (Glukhovsky *et al.* 1981, 1988).

5.3.1   Effect of slag on setting times

Purdon (1940) investigated the effect of different parameters on setting times of alkali-activated slag cements. It was found that, for a given activator solution, the setting time's of alkali-activated slag cements varied with different sources of slag even though they have very similar chemical composition. It was noticed that the early strength varied inversely with the setting time, but the 28-day strength is more or less independent of it.

Andersson and Gram (1987) measured the setting time of alkali-activated slag cement mortars consisting of slag with different Blaine finenesses and activators, such as NaOH, Na_2CO_3, $NaO \cdot 0.9SiO_2$ and $Na_2O \cdot 3.35SiO_2$. The setting time of alkali-activated slag cement pastes did not show an obvious change when the fineness of the slag was increased from 350 to $530\,m^2/kg$, but decreased markedly when it was increased from 530 to $670\,m^2/kg$ (Figure 5.5).

Setting time of alkali-activated slag cement pastes increases with the decrease of the basicity of the slag regardless of the nature of activators used. Na_2CO_3-activated slag cement exhibits more increase than NaOH- and waterglass-activated slag cements (Krivenko 1994a). Although blast furnace slag and electrothermal phosphorus slag have the same basicity, the alkali-activated phosphorus slag cements show longer setting times than alkali-activated blast furnace slag cements, especially when NaOH, soda-alkali melt and Na_2CO_3 are used as alkaline activators.

5.3.2 Effect of the nature and dosage of activators

As shown in Figure 5.6, the setting times of alkali-activated slag cement pastes depend on the nature of activators. NaOH- and Na_2CO_3-activated slag cement pastes usually exhibited longer setting time than waterglass-activated slag cements.

The setting times of alkali-activated slag cements usually decrease with the increase of activator dosage. Wu (1999) measured the setting times

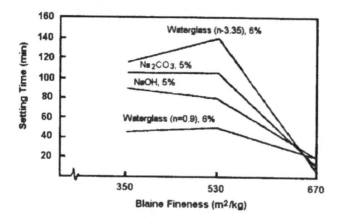

Figure 5.5 Effect of blaine fineness of blast furnace slag on setting time of alkali-activated slag cement pastes (based on Andersson and Gram, 1987).

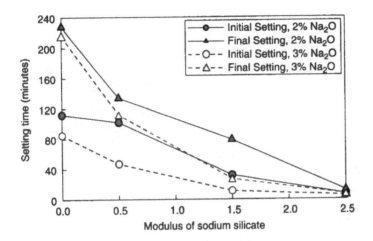

Figure 5.6 Effect of modulus of sodium silicate solution on setting times of alkali-activated phosphorus slag cement pastes (data from Shi and Li 1989b).

of waterglass ($M_s = 1.2$)-activated slag cements with dosage of 2, 4, 6, 8 and 10% by Na_2O. When the dosage was 2 and 4%, it did not show any obvious effect on both the initial and final setting times. When the dosage was increased from 4% to 6%, it slightly decreased the initial setting time from 90 minutes to 76 minutes, but significantly decreased the final setting time from 425 minutes to 245 minutes. The further increase of activator dosage from 6 to 8% markedly decreased the initial setting time from 76 minutes to 34 minutes and final setting time from 245 minutes to 48 minutes. When the dosage is over 8%, the cement set too fast to measure the setting times. Bakharev *et al.* (1999a) suggested that the liquid waterglass dosage should be 4% (based on Na) and a modulus of 0.75 based on both strength development and workability.

The use of liquid sodium silicate usually results in fast setting, which may cause problems for mixing and placement. Several studies have reported the use of solid instead of liquid as an activator for alkali-activated slag cement and concrete (Peng 1982, Collins and Sanjayan 2001a). The use of solid sodium silicate has two advantages: (1) blast furnace slag can be interground with solid sodium silicate and (2) the cements set more slowly than that activated by liquid sodium silicate to give enough time for mixing and placement. However, use of solid sodium silicate usually results in lower early strengths.

The modulus of sodium silicate has a very significant effect on the setting times of alkali-activated slag cements (Zahrada 1986, Shi and Li 1989a, b, Bakharev *et al.* 1999a). Figure 5.6 shows the effect of modulus of liquid sodium silicate on setting times of alkali-activated phosphorus slag cement pastes containing 2% and 3% (by mass of Na_2O) sodium silicate. For a given modulus, the setting time decreased with the increase of sodium silicate dosage. For a given dosage, both the initial and final setting times also decreased linearly with the increase of the modulus of sodium silicate. In another study (Wu 1999), the modulus did not show a noticeable effect on setting times when modulus varied from 0.8 to 1.2, but the setting times decreased rapidly with the increase of the modulus as the modulus was greater than 1.2; and the cements set too fast to measure the setting times when the modulus was greater than 1.6.

When solid sodium silicate is used, a totally different trend is obtained (Figure 5.7) (Peng 1982). For a given activator dosage, both the initial and final setting times increase linearly with the increase of the modulus of solid sodium silicate. This is because both the dissolution rate and solubility of sodium silicate glasses decrease as the modulus of the sodium silicate glass increases, as discussed in Chapter 2. As the dissolution rate decreases, the available sodium silicate in the solution for activation of the slag decreases, which results in longer setting times.

Cheng (2003) examined the setting behavior of waterglass-activated slag cement pastes. For a given waterglass modulus, both the initial and final setting times of the cement pastes decrease as the waterglass dosage increases. However, there is no relationship between setting times and initial pH values of the waterglass solutions when different moduli are used.

5.3.3 Combination of activators

The combination of two or more activators can also change the setting behaviour of alkali-activated slag cement and concrete, depending on the

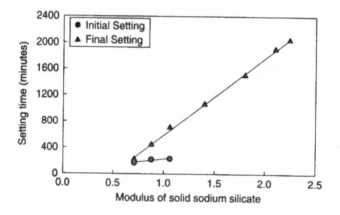

Figure 5.7 Effect of modulus of solid sodium silicate on setting times of alkali-activated slag cements (4% sodium silicate) (data from Peng 1982).

nature and dosage of the activators (Purdon 1940). The alkali-activated slag cement activated by 4% NaOH has an initial setting time of 34 minutes and a final setting time of 55 minutes. The addition of additional 1% Na_2SO_4 does not affect the setting. However, the addition of 1.3% K_2SO_4 or 0.24% $2Na_2O \cdot Al_2O_3$ prolongs the initial setting time to about 100 minutes and final setting to more than 200 minutes (Chen and Liao 1992). In another study, it was noticed that a combination of K_2CO_3 and liquid waterglass with a modulus of 2.0 increased both the initial and final setting time markedly (Gu 1991). The presence of Na_3PO_4 retards the setting of portland cement significantly, but it has no noticeable effect on the setting of NaOH or $Na_2O \cdot SiO_2$-activated slag cements (Shi and Li 1989b). The addition of 0.5% $Na_2B_4O_7 \cdot 10H_2O$ during the production of solid sodium silicate accelerates the setting of solid sodium silicate-activated blast furnace slag because $Na_2B_2O_4$ promotes the dissolution of solid sodium silicate (Peng 1982). Phosphoric acid can exhibit an obvious retarding effect on the setting of sodium silicate-activated slag cement when its concentration is over 0.84 M in the mixing water (Cheng 2003, Cheng et al. 2005). The combination of phosphoric acid and gypsum will shorten the setting times (Cheng et al. 2005).

5.3.4 Effect of additives

As stated above, liquid sodium silicate-activated slag cement can set very fast, especially when high dosage or high modulus is used. Researchers have attempted to use retarders to retard the setting of alkali-activated slag cements.

Wu et al. (1993) investigated the effect of potassium sodium tartrate and molasses on setting times of alkali-activated slag cement. These two substances are typical retarders for portland cement. They noticed these admixtures did not show any effect on the initial setting time, but shortened the final setting time slightly. This difference may be attributed to the difference in the mineral compositions and hydration mechanism between alkali-activated slag and portland cements. Pu et al. (1994) reported that a white inorganic salt, noted as YP-3, could effectively retard the setting of alkali-slag cement. However, the interval between initial and final setting is short.

Brough et al. (2000) investigated the addition of NaCl or malic acid on the setting times of sodium silicate-activated slag cement paste (Table 5.4). When 1 or 4% NaCl is added, it accelerates the setting of the cement. However, the addition of 8% NaCl significantly retards the setting of sodium silicate-activated slag cement paste. The introduction of 0.5% malic acid also retards the setting of sodium silicate-activated slag cement paste.

Table 5.4 Effect of NaCl and malic acid on setting times of alkali-activated slag cements (Brough *et al.* 2000)

Paste	Setting time (h)	
	Initial	Final
Control*	4	5
Control + 1% NaCl	2	2.5
Control + 4% NaCl	3	4
Control + 8% NaCl	10	12
Control + 0.5% malic acid	20	22
Control + 0.5% malic acid + 8% NaCl	>48	>48

* Made with 1.5 M $Na_2O \cdot 2SiO_2$ solution at a ratio of 500 ml solution/1 kg slag.

5.4 Strength

5.4.1 Introduction

The "strength of cement" usually refers to the strength of mortars prepared, cued and tested following some standards. However, the strength of cement pastes is often measured in laboratory research to examine how different factors affect the strength of the cementing systems. The strength of a hardened cement paste is due to the presence of a continuous three-dimensional network of hydration products, which can resist external stress without breaking down (Odler 1998). A hardened cement paste consists of hydration product, non-hydrated material and pores. Hydration product is responsible for attained strength, while pores have negative effects on strength.

The strength of cement and concrete materials is perhaps the most important overall measure of quality, although other properties may also be critical. For alkali-activated slag cement, the following are the main factors determining the strength of cement pastes and mortars:

- the nature of slag and activator(s)
- dosage of activators
- water/slag ratio
- curing temperature
- fineness of slag
- timing of the addition of activator(s)
- other additives
- compaction pressure (for system with very low water/slag ratio)

When the water-to-slag ratio is high, the initial porosity of the system is determined by water-to-slag ratio. However, if the water-to-slag ratio is low, compaction pressure will determine the initial porosity of the system.

5.4.2 The of nature of slag and activator

In addition to the factors which affect the strength development of portland cement, the nature and the dosage of activators play a crucial role in determining the strength of alkali-activated slag cements. Many results confirm that alkaline activators demonstrate selectivity; i.e., the effect that an activator has on strength development may be different for slag of different origins. It is reported that Na_2CO_3 is especially suitable for slags rich in C_2MS, and NaOH is a good activator for slags rich in C_2AS (Malolepszy 1986). Krivenko (1992b) examined the activation of $CaO-SiO_2$, $CaO-Al_2O_3$, $CaO-Al_2O_3-SiO_2$, $CaO-MgO-SiO_2$ and $CaO-Al_2O_3-MgO-SiO_2$ systems with NaOH, Na_2CO_3 and $NaO \cdot SiO_2$. It was found that $Na_2O \cdot SiO_2$ is the most effective for $CaO-SiO_2$, $CaO-Al_2O_3$, $CaO-Al_2O_3-SiO_2$ and $CaO-MgO-SiO_2$ systems, and Na_2CO_3 is the most effective for $CaO-Al_2O_3$ system. Many researches have confirmed that waterglass is a very effective activator (Malolepszy and Petri 1986, Slota 1987, Shi et al. 1989a). Variation of chemical composition among different blast furnace slag results in different hydration products for a given activator, which is the main reason for the selectivity (Shi and Day 1996c).

Figure 5.8 shows strength development of NaOH- and Na_2SiO_3-activated slag cements from two studies (Shi and Day 1996c). Results in Figure 5.8a indicate that Na_2SiO_3-activated slag exhibit a higher strength than NaOH-activated slag cement mortars at both early and later ages. Results in Figure 5.8b are totally opposite. NaOH-activated slag cement mortars

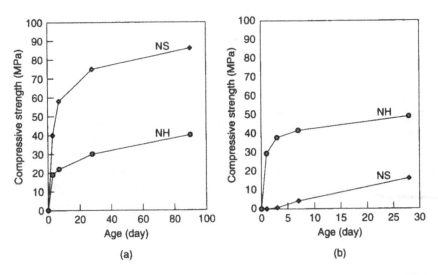

Figure 5.8 Strength development of alkali-slag cement mortars (Shi and Day 1996c) (NS–$Na_2O \cdot SiO_2$ activated slag cement, NH–NaOH activated-slag cement).

exhibit a higher strength than Na_2SiO_3-activated slag cement mortars at both early and later ages.

Table 5.5 shows the effect of combination of different anions on the compressive and flexural strength of alkali-activated slag cement. The results indicated that the combination of any two types of anion or anion group gave higher strength than either one of them, except the combination of $OH^- + SO_4^{2-}$. The effectiveness of the combination of activators also depends on the strength of alkali-slag cement and the nature of activator. It seems that the combination of SiF_6^{2-} with any other anion gave the highest strength increase. In another study (Jolicoeur $et\,al.$ 1992), it was found that the addition of Na_2CO_3 to $Na_2O \cdot 0.5SiO_2$-activated slag increased the one-day strength, but had no observable effect on strength thereafter. However, the addition of Na_2CO_3 decreased the strength of $Na_2O \cdot SiO_2$- and $Na_2O \cdot 1.5SiO_2$-activated slag cements from 1 day to 91 days. It was found that the addition of Na_3PO_4 (from 0.2 to 0.5% P_2O_5 by mass) increased the strength of NaOH- and Na_2SiO_3-activated phosphorus slag cements (Shi and Li 1989b). The authors attributed the strength increase to the composite activation of Na_3PO_4 and NaOH or $Na_2O \cdot SiO_2$. Li and Sun (2000) noticed that the combination of NaOH and Na_2CO_3 gave a higher strength than those cement mortars activated either by NaOH or by Na_2CO_3 alone. Fernandez-Jimenez and Puertas (2003) investigated the combinations of

Table 5.5 Effect of anions or anion groups of activator on strength of alkali-activated slag cement (Krivenko 1994b)

Anion	Compressive strength		Flexural strength	
	Strength (MPa)	Relative strength (%)	Strength (MPa)	Relative strength (%)
OH^-	36.6	100	4.05	100
$OH^- + Cl^-$	38.3	106	4.20	104
$OH^- + SO_4^{2-}$	28.2	77	3.80	94
$OH^- + SiF_6^{2-}$	64.1	175	6.80	168
CO_3^{2-}	45.6	100	5.36	100
$CO_3^{2-} + Cl^-$	48.4	100	5.65	105
$CO_3^{2-} + SO_4^{2-}$	50.2	110	5.80	108
$CO_3^{2-} + SiF_6^{2-}$	55.5	122	5.85	109
SO_3^{2-}	34.8	100	4.20	100
$SO_3^{2-} + Cl^-$	36.7	105	4.38	104
$SO_3^{2-} + SO_4^{2-}$	35.1	101	4.85	115
$SO_3^{2-} + SiF_6^{2-}$	65.9	189	5.20	124
SiO_3^{2-}	81.8	100	8.26	100
$SiO_3^{2-} + Cl^-$	87.8	107	8.40	102
$SiO_3^{2-} + SO_4^{2-}$	95.7	117	8.70	105
$SiO_3^{2-} + SiF_6^{2-}$	100.9	123	9.20	111

waterglass, NaOH and Na_2CO_3 on strength of alkali-activated slag cements and noticed that the presence of silicate anions increased, while carbonate anions decreased the strength of the cements.

Many studies have confirmed that the modulus of the waterglass has a significant effect on the properties of alkali-activated slag cements. Figure 5.9 shows the effect of the modulus of waterglass on strength development of alkali-activated phosphorus slag cement pastes (Shi and Li 1989b). From 3 to 90 days, NaOH-activated phosphorus slag always shows a lower strength than that of $Na_2O \cdot 0.5SiO_2$-activated slag. The modulus does not show an obvious effect on strength as it increases from 0.5 to 1.5. However, the strength decreases almost linearly with the increase of modulus from 1.5 to 2.5 between 3 and 28 days. Based on the strength results, it can be concluded that the optimum modulus for the slag is around 1.0 to 1.5. Another study obtained similar results (Wu 1999).

Bin (1988) found that strength of waterglass-activated slag cements increased with the increase of the ratio up to 1.7, and the ratio did not show an obvious effect thereafter. Wang *et al.* (1994) investigated the activation of acidic, neutral and basic slags with water gass having different moduli. Based on the strength at 28 days, the optimum modulus for the acidic slag was around 1.0 for the neutral slag, and around 1.25 for the basic slag, respectively (Figure 5.10). Those slight differences are not surprising since slag has selectivity for activators as discussed above. Malolepszy and Nocun-Wczelik (1988) found that both compressive and flexural strengths

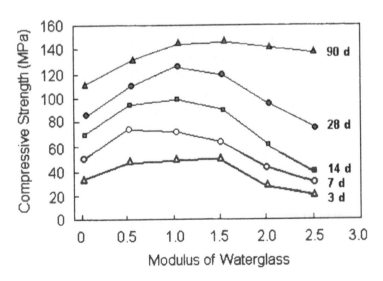

Figure 5.9 Effect of the modulus of waterglass (3% Na_2O based on the mass of the slag) on strength development of alkali-activated phosphorus slag cement pastes (Shi and Li 1989b).

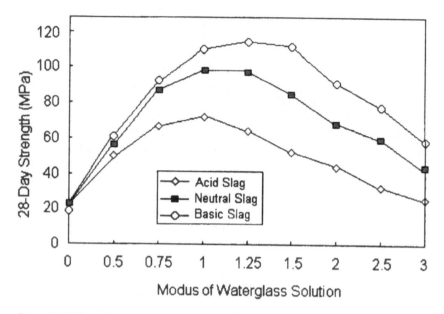

Figure 5.10 Effect of modulus of liquid sodium silicate on 28-day strength of alkali-activated slag cement mortars made with different types of slag (slag fineness is 450 m²/kg, all alkali activator solutions have solid content of 24.4 g/100 ml, alkali solution/slag = 0.41 sand/slag = 2) (Wang et al. 1994).

of sodium silicate-activated slag cement mortars decreased with the modulus as it increased from 1 to 2.5. The strength differences for mortars cured at room temperature were more obvious than those cured at elevated temperature. Bakharev et al. (1999a) found that the modulus of waterglass did not show an obvious effect on strength at early ages at room temperature, but the highest strength was achieved when the modulus of the waterglass was around 1.25, regardless of the activator dosage. Shi and Li (1989b) attributed the optimum modulus phenomena to the formation of "primary C–S–H" and/or polymerisation of silicate anions in the waterglass, which promotes the hydration of slag and the formation of less porous structure.

5.4.3   Fineness of slag

An increase in fineness increases reactivity and strength of cementitious materials (Purdon 1940, Kutti et al. 1982, Isozaki et al. 1986). However, fineness over a certain value may decrease strength of the cementitious material due to increase in water requirement. An early study (Purdon 1940) indicated that the decrease of residue of slag on No. 100 sieve from 53% to 14.7% had a significant effect on both early and later strength of activated slag cement.

The further decrease of the sieve residue to 6.2% had an effect on strength mainly before three days. Three blast furnace slags were ground to the Blaine finenesses of 300, 400, 500 and 600 m²/kg and NaOH was used as an activator (Parameswaran and Chatterjee 1986). The strength results from 1 to 90 days indicated that an evident strength jump could be observed when the Blaine fineness of the slag is increased from 300 to 400 m²/kg. However, the increase of fineness from 400 to 500 m²/kg increased strength slightly, and the further increase of fineness from 500 to 600 m²/kg decreases strength. In another study, it was found that, up to seven days, the strength of NaOH-activated slag cement pastes almost linearly increased with the Blaine fineness of the slag up to 650 m²/kg (Isozaki et al. 1986).

The strength of alkali-activated phosphorus slag cement increases linearly with the Blaine fineness of the slag at early ages (Figure 5.11). However,

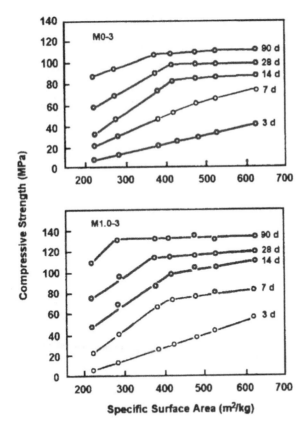

Figure 5.11 Effect of blaine fineness of phosphorus slag on strength development of alkali-activated phosphorus slag cement pastes (activator dosage is 3% Na₂O based on the mass of the slag) (Shi and Li 1989b).

a critical point, corresponding to a Blaine fineness of approximately 400 m²/kg, appears at seven days for Na_2SiO_3 and at 14 days for NaOH-activated granulated phosphorus slag cement when 3% Na_2O is used. The Blaine fineness does not show a noticeable effect on compressive strength when it is higher than the critical value, and the strength still linearly increases with the Blaine fineness when it is smaller than the critical value. In another study, it was found that the critical fineness for strength reduction, varying from 500 to 600 m²/kg, depended on the basicity of the slag (Figure 5.12) (Wang *et al.* 1994). Thus, the Blaine fineness may be controlled to about 400 to 500 m²/kg to obtain proper strength development rate.

5.4.4 Activator dosage

Activator dosage has a significant effect on strength of alkali-activated slag cements, and there is an optimum dosage in most cases. However, the optimum dosage varies with the nature of the slag and alkali activators used, and the curing conditions.

Purdon (1940) investigated the effect of NaOH concentration on strength of alkali-activated slag mortars and concretes, and found that there was an optimum concentration. Prior to the optimum concentration, the strength of both mortars and concretes increases drastically with NaOH concentration.

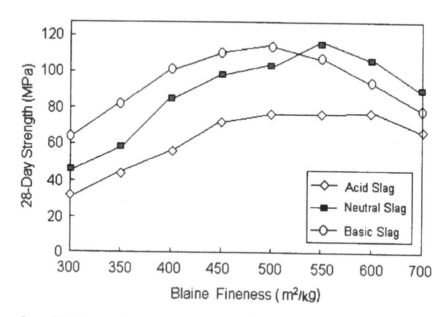

Figure 5.12 Effect of blaine fineness of slag on 28-day strength of alkali-activated slag cement mortars made with different types of slag (5.5% Na_2O as activator in the form of $Na_2O \cdot SiO_2$, sand/slag = 2) (Wang *et al.* 1994).

Then a plateau appears. After the plateau, strength decreases with NaOH concentration. Purdon (1940) suggested that the optimum dosage of NaOH is 5–8% of the mixing water suitable for both mortars and concretes. However, some recent investigations got different conclusions. One found that the addition of less than 5% NaOH (by Na$_2$O) decreases the strength of slag (Jolicoeur *et al.* 1992). Another study indicated that the strength of cement pastes increases with dosage and then a plateau appeared, and no strength drop was measured after reaching the plateau (Chen and Liao 1992).

For waterglass, it was found that both the compressive and flexural strength of activated-slag cements increases almost linearly with activator dosage from 2 to 8% (Na$_2$O) at ages from 3 to 180 days (Bin 1988). The compressive strength of solid sodium silicate-activated slag cement linearly increases with content of solid waterglass regardless of the modulus of the waterglass and ages (Peng 1982).

Under steam curing conditions, the effect of activator dosage on strength seems different from that at room temperature. Tang and Shi (1988) found that, regardless of activators, there is a critical activator dosage below which activators do not show activation effect on granulated phosphorus slag. Above the critical activator dosage, the strength of activated slag increases with activator dosage, reaches a maximum value, and then decreases with activator dosage. The optimum activator dosage was about 3% (by Na$_2$O) regardless of the nature of activators; except for NaOH-activated phosphorus slag cement, its strength did not change with NaOH dosage once above the critical dosage (Tang and Shi 1988).

5.4.5 Timing of the addition of activators

When an activator is liquid, it can only be added during the mixing process. However, if an activator is a solid, it may be added by (1) dissolving it in mixing water and mixing the solution with ground slag; (2) intergrinding the activator with slag and (3) grinding activator and slag separately and blending them before mixing with water. Some studies have reported that method (2) is the most effective because the activator can mix uniformly with slag. Also, some activator(s) may be adsorbed onto the reactive locations on the surface of the slag, which enhances the activation of reactivity of the slag. Note that some activators may hydrolyse during grinding or storage. In such cases, it is better to use the first method to add the activators. Xu and Pu (1999) found that the cement made with Method 3 showed significantly higher strengths than that made with Method 2 from 3 to 28 days. However, the replacement of slag with 10% geothermal silica waste from electricity generation from geothermal resources could significantly increase the strength of NaOH-activated slag cement from 1 to 90 days, increase the strength of waterglass-activated slag cement significantly at 1 and 3 days, but decrease the strength from 7 to 90 days (Escalante-Garcia *et al.* 2003).

5.4.6 Other additives

A variety of additives have been added to alkali-activated slag cements to improve the strength development (Glukhovsky *et al.* 1980, Talling and Brandstetr 1993). For example, the addition of a small amount of portland cement clinker or lime can increase the strength of alkali-activated slag, especially at early ages. It was found that the optimum addition of portland cement clinker is 5–7% (Cheng *et al.* 1991, Chen and Liao 1992, Krivenko 1994a). Results in Figure 5.13 indicate that the addition of 2% lime increases the one-day strength of the alkali-activated slag cement pastes drastically. The increase in lime content from 2 to 6% does not show a marked effect on strength at one and three days, but slightly improves strength at 7 and 28 days.

The addition of silica fume alone decreases strength of alkali-activated slag cements (Chen and Liao 1992). The combination of lime and silica fume results in strength higher than when using lime alone (Douglas *et al.* 1991). When 1 to 3% NaOH was added together with 5% gypsum to a slag, it was found that the strength increases with NaOH dosage at 7 and 29 days (Tango and Vaidergorin 1992). However, the pastes containing 2% NaOH exhibited the highest strength at 63 and 91 days.

The use of chemical admixtures definitely affects the strength development of alkali-activated slag cements. A recent study by Puertas *et al.* (2003) indicated that the addition of vinyl co-polymer superplasticizer drastically decreased the strength of waterglass-activated slag cements at both two and 28 days, while the addition of polyacrylate co-polymer-based superplasticizer had no effect on strength development of the cement.

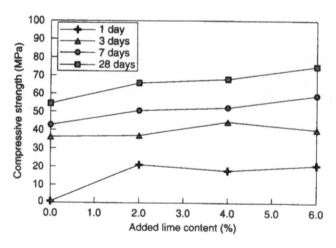

Figure 5.13 Effect of lime addition on strength development of sodium silicate-activated slag cement mortars (5% Na_2O waterglass, M_s = 1.5, W/S = 0.48) (data from Chen and Liao 1992).

5.4.7 Water-to-Slag ratio

Water-to-cement ratio determines the initial porosity of the system. The higher the water-to-cement ratio is, the higher the initial porosity. Thus, for a given hydration degree, the strength of hardened cement pastes decreases as water-to-cement ratio increases. An increase in water-to-cement ratio accelerates the early hydration of portland cement (Bensted 1983). The effect on the hydration of slag due to the change in water-to-slag ratio may vary with the nature of activators used. It was found that the increase in water-to-slag ratio showed a more obvious effect on the strength of Na_2SiO_3-activated slag cement than NaOH-activated slag cement (Shi and Li 1989b). This is in agreement with the heat evolution measurement as discussed in Chapter 4 because of the change in silicate anion species in Na_2SiO_3 solution caused by the increase of water-to-slag ratio, which has a significant effect on the hydration and strength development of alkali-activated slag cements.

5.4.8 Curing temperature

5.4.8.1 Elevated temperature curing under atmospheric pressure

The hydration of cement is accompanied by a series of chemical reactions. Typically, an increase of $10\,°C$ in reaction temperature doubles the rate of a chemical reaction. As discussed in Chapter 4, the apparent reaction activation energy of alkali-activated slag cements is higher than that of portland cement, it can be expected that elevated temperature curing may be very helpful in improving strength development of alkali-activated slag cements.

The strength development of alkali-activated phosphorus slag cement is shown in Figure 5.14. It can be seen that curing temperature has a very

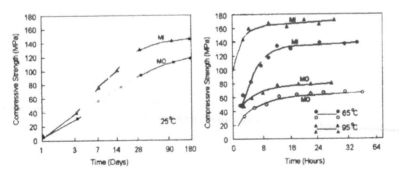

Figure 5.14 Effect of curing temperature on strength development of alkali-activated phosphorus slag cement pastes (MO–NaOH-activated slag, MI-Na_2O · SiO_2-activated slag, activator dosage is 3% Na_2O based on the mass of the slag) (Shi et al. 1991a).

significant effect on strength development. No strength plateau is observed until 180 days at 25 °C. As temperature rises to 65 and 95 °C, a strength plateau appears after 6 to 10 hours of curing. The plateau strengths of NaOH-activated phosphorus slag cement at 65 and 95 °C are significantly lower than that cured for 180 days at 25 °C. However, the plateau strength at 65° is lower than that at 95 °C. For $Na_2O \cdot SiO_2$-activated phosphorus slag cement, the plateau strength at 65 °C is lower but the plateau strength at 95 °C is higher than that cured for 180 days at 25 °C. By comparing these strength development temperatures, it is evident that temperature has a more obvious effect on $Na_2O \cdot SiO_2$-activated than NaOH-activated phosphorus slag cements.

The above results indicate that alkali-activated slag cements can give a very high strength after several hours of steam curing. The other interesting factor is that the strength of alkali-activated slag cement pastes and mortars continuously increases at room temperature after a certain period of steam curing. The increase in strength at room temperature depends on the nature of activators. The high modulus waterglass-activated slag cements show much more significant increase in strength than the low modulus water-glass-activated slag cements (Anderson and Gram 1987).

Deng *et al.* (1989) found that as curing temperature increased from 25 to 40 °C, the strength of low-porosity alkali-activated slag cement compacts increased very significantly from 0.5 to 7 days. A further increase of curing temperature to 80 °C could still increase the strength to some extent from 0.5 to 7 days compared with those cured at 40 °C. However, the strength of compacts decreased as the temperature was further increased to 100 °C.

A strength increase could be observed when curing temperature is increased from 50 to 60 °C, then no obvious temperature effect could be observed up to 90 °C regardless of activator dosage (Talling, 1989). It seems that, under steam curing conditions, the ultimate strength of the alkali-activated slag cement with 3% waterglass (by mass of Na_2O) is always lower than that with 4% (by mass of Na_2O) waterglass-activated slag cement.

5.4.8.2 Autoclave curing

For oil or geothermal well cementing, or radioactive waste treatment applications, the hydrated cement may have to undergo high temperatures. Thus, it is very important that the hydrated cement used exhibits stable hydration products and structure. Figure 5.15 shows the strength development of alkali-activated phosphorus slag, alkali-activated blast furnace slag and portland cement pastes (Shi *et al.* 1991b). The strengths of alkali-activated phosphorus slag and alkali-activated blast furnace slag cement pastes show a plateau after one day, then increase slightly with time. The strength of portland cement pastes reaches the maximum value at one day,

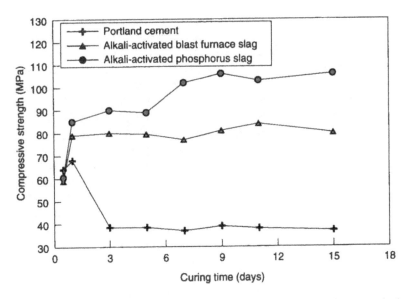

Figure 5.15 Strength development of portland cement and alkali-activated slag cement pastes at 150 °C (data from Shi *et al.* 1991b).

then decreases noticeably. At the age of 15 days, the strength of the portland cement paste is only about one half of its one-day strength. Alkali-activated blast furnace slag and alkali-activated phosphorus slag cement pastes demonstrate much higher strength than portland cement paste after 12 hours, and alkali-activated phosphorus slag cement paste shows higher strength than alkali-activated blast furnace slag paste at all times. Alkali-activated phosphorus slag and alkali-activated blast furnace slag cement pastes appear to be more stable than portland cement pastes at higher curing temperature. The differences in strength are caused by the nature of hydration products and pore structure of hardened pastes, as discussed in Chapter 4.

Sugama and Brothers (2004) investigated how the curing temperature affects the properties of sodium silicate-activated slag cement. It was found that the sodium silicate-activated cements autoclaved at temperatures up to 200 °C displayed an outstanding compressive strength of more than 80 MPa, and a minimum water permeability of less than 3.0×10^{-5} darcy. The combination of C–S–H and tobermorite phases was responsible for strengthening and densifying the autoclaved cement. At 300 °C, an excessive growth of well-formed tobermorite and xonotlite crystals generated an undesirable porous microstructure, causing the retrogression of strength and enhancing water permeability.

5.4.9 Compaction pressure

As mentioned above, the strength of hardened cement pastes depends on the intrinsic property of hydration products and pore structure of the hardened cement pastes. One technique to decrease the initial porosity of a cementing system is to use low water-to-cement ratio. Under high water-to-cement ratio range, as stated above, the strength of cement and concrete depends on the water-to-cement ratio. However, a cementing system cannot consolidate itself if the water-to-cement ratio is lower than certain value. Thus, applied pressure may be needed to consolidate the cement pastes with low water-to-cement ratios. In these cases, the strength of cement pastes will depend on the applied pressure. Due to the low water-to-cement ratio and high applied pressure, the cement pastes will have a very low porosity and exhibit high strength (Roy *et al.* 1972, Roy and Gouda 1973, Roy 1987). A strength of around 500 MPa has been reported using a hot-press process (Roy, *et al.* 1972). The strength of cement pastes with low water-to-cement ratio has been further improved by adjusting the cement particle size distribution, adding polymers and changing cementing components (Lu and Young 1993). These high strength cement products may find non-traditional applications such as solidification of high-level radioactive wastes and as replacements for metals, plastics and inorganic materials (Roy 1987).

The formation and development of structure and properties of low-porosity cementitious materials is determined by the initial porosity or compaction pressure instead of initial water-to-cement ratio (Xu 1988, Xu *et al.* 1992). Figure 5.16 shows the effect of compaction pressure on initial

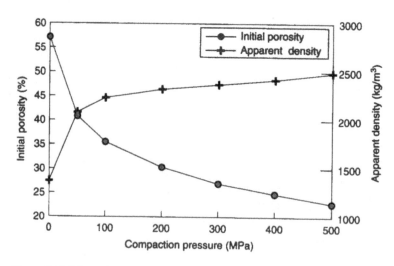

Figure 5.16 Effect of compaction pressure on initial porosity of portland cement pastes (data from Xu *et al.* 1992).

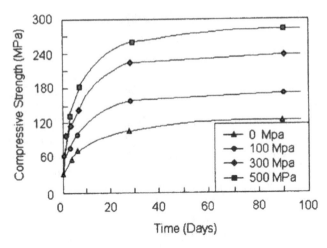

Figure 5.17 Effect of compaction pressure on strength development of alkali-activated slag cement pastes (Xu *et al.* 1993).

porosity and apparent density of the alkali-activated slag cement. The relationship between the initial porosity and compaction pressure can be expressed as follows:

$$P_o = A^* \exp(-Bb_e) + 21 \qquad (5.1)$$

where
P_o = initial porosity;
b_e = compaction pressure;
A, B = constants.

Figure 5.17 shows the effect of compaction pressure on strength development of compacted cylinders. The strength of low-porosity alkali-activated slag cement increases with compaction pressures from 1 to 90 days (Xu *et al.* 1993). The compaction pressure effect is more pronounced at pressures up to 200 MPa and after seven days.

5.5 Shrinkage

Shrinkage is the reduction in volume at constant temperature without external loading. It is an important material property that has significant effects on the long-term performance of designed structures. Shrinkage can be classified into autogenous shrinkage, carbonation shrinkage and drying shrinkage. Autogenous shrinkage refers to volume changes caused by the hydration of cement. Carbonation shrinkage occurs when the hydration products of cement react with CO_2 in the environment. Drying shrinkage results from the drying of cement and concrete materials.

5.5.1 Autogenous shrinkage

Laboratory studies (Deng 1989, Shi *et al.* 1992a, Cincotto *et al.* 2003) have indicated that the autogenous shrinkage of alkali-activated blast furnace slag is higher than portland cement pastes, as shown in Figure 5.18. For a given Na_2O dosage, sodium silicate-activated slag exhibits higher autogenous shrinkage than NaOH-activated slag cement pastes. For a given waterglass modulus, the autogenous shrinkage increases as the waterglass dosage increases. The autogenous shrinkage is well correlated with the progress of hydration of the slag. However, the relationships are dependent on the nature of the slag, and the nature and dosage of the activators used. Given that autogenous shrinkage is the macroscopic effect of both chemical shrinkage and self-desiccation, the volume of mesopores has an important effect on autogenous shrinkage development. It was found that the volume of mesopores in sodium silicate-activated slag cement pastes is much larger than that in sodium hydroxide-activated slag cement pastes (Cincotto *et al.* 2003).

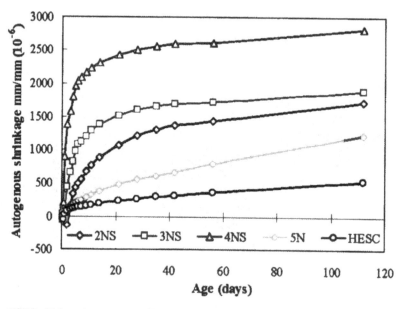

HESC – high early strength portland cement, 2NS – 2.5% sodium silicate (by Na_2O)
3NS – 3.5% sodium silicate (by Na_2O)   4NS – 4.5% sodium silicate (by Na_2O)

Figure 5.18 Autogenous shrinkage of portland cement and sodium silicate ($Na_2O \cdot 1.7SiO_2$)-activated slag cement pastes at 20 °C (from Cincotto *et al.* 2003).

5.5.2 Drying shrinkage

Drying shrinkage is the most important phenomenon since it can cause cracking or wrapping of concrete elements. Moisture loss is the cause for drying shrinkage. The relationship between drying shrinkage and relative humidity can be divided into three stages and is shown in Figure 5.19. It is generally agreed that stage one is attributed to the loss in capillary water, stage two results from the loss of adsorbed water on the surface of C–S–H and stage three represents the loss of water related to the structure of C–S–H.

For alkali-activated slag cements, many internal and external factors can affect the drying shrinkage of hardened cement pastes. The international factors include the nature of the slag, nature of the activator, dosage of activator, water-to-slag ratio and degree of hydration. The external factors include curing temperature, additives, relative humidity, rate of drying and time of drying.

It is well known that alkali-activated slag cement pastes show a significantly higher drying shrinkage than portland cement pastes and the differences increase with the decrease of relative humidity (Bin 1988, Jiang et al. 1997). The effect of activators on drying shrinkage may vary with curing and testing conditions. Bakharev et al. (1999a) noticed that water-glass-activated slag cement pastes exhibited obviously higher drying shrinkage than NaOH-, Na_2CO_3- and Na_3PO_4-activated slag cement pastes.

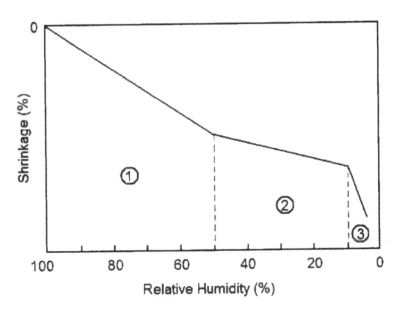

Figure 5.19 Relationship between drying shrinkage and relative humidity (Based on Hansen and Almudaiheen 1987).

Table 5.6 Effect of activators on drying shrinkage of alkali-activated slag cement pastes (Anderson and Gram 1987)

No.	Activator	Water-to-slag or water-to-cement ratio	Shrinkage (%) 231 days at 20 °C and 80% RH	80 °C for 8 hours then 231 days at 20 °C and 80% RH
1	5% NaOH	0.43	0.093	0.067
2	5% Na_2CO_3	0.43	0.111	0.045
3	6% waterglass, $n = 3.35$	0.43	0.229	0.176
4	6% waterglass, $n = 1.80$	0.43	0.133	0.035
5	6% waterglass, $n = 0.90$	0.43	0.146	0.042
6	Portland cement	0.50	0.068	0.048

Anderson and Gram (1987) found that after eight hours of steam curing at 80 °C, NaOH-, Na_2CO_3-, $Na_2O \cdot 0.9SiO_2$- and $Na_2O \cdot 1.8SiO_2$-activated slag cements showed very similar shrinkage from 0.035 to 0.067%, but $Na_2O \cdot 3.35SiO_2$-activated slag cement had a drying shrinkage of 0.176%, which was three to five times higher than cements at 20 °C and 80% relative humidity, as shown in Table 5.6.

The shrinkage difference in alkali-activated slag cement and portland cement can be explained by the differences in pore structure. Alkali-activated blast furnace slag pastes usually show lower porosity than portland cement pastes, but the former mainly contains pores with $r < 100$ Å. Time dependent deformation is mainly attributed to fluid and solid movements within the "gel" components of the material. Since alkali-activated blast furnace slag pastes have a higher proportion of material in this range, one would expect shrinkage strain to be larger than a material, such as portland cement pastes, with a coarser microstructure. Many mechanisms for shrinkage have been proposed for conventional hardened cement and concrete materials (Xi and Jennings 1992), of which, surface tension, capillary tension and disjoining pressure have been mathematically formulated. It is generally agreed that several mechanisms co-exist.

5.6 Pores

5.6.1 Introduction

Hardened cement pastes are porous materials and contain three types of pores: gel pores, capillary pores and air voids. In portland cement pastes, gel pores constitute about 28% of the total C–S–H gel volume and have a size of 1.5 to 2.0 nm, which is the size of the water molecule and will not permit the flow of water. Gel porosity cannot be resolved by SEM and would be included in the volume occupied by C–S–H. Capillary pores make up the portion of the space that was originally filled by water in the fresh cement pastes, and has not been filled by cement hydration products. Air voids are the result of incomplete consolidation or entrapped air, or both. Capillary pores are usually tortuous and tubelike, while air voids are shorter but much larger in size. Table 5.7 describes the classification of gel and capillary pores and their effects on the properties of hardened cement pastes.

The capillary porosity of portland cement pastes, P_c can be calculated as follows (Hansen, 1986):

$$P_c = \frac{W}{C} - 0.36\alpha \qquad (5.2)$$

where W/C is the water-to-cement ratio and α is the degree of hydration of portland cement. As the W/C of the pastes increases, its capillary porosity increases. Figure 5.20 shows the volume relationships among constituents of hydrated portland cement pastes. Table 5.8 shows the approximate age required to discontinue the capillary pores of cement pastes with different water-to-cement ratios. As the water-to-cement ratio is increased to 0.7, those continuous capillary pores within cement pastes will never be segmented.

Table 5.7 Gel pores and capillary pores in hardened cement pastes (Mindess and Young 1981)

Designation	Diameter	Description	Role of water	Paste properties affected
Capillary pores	10 μm–50 nm	Large capillaries	Behaves as bulk water	Strength, permeability
	50–10 nm	Medium capillaries	Moderate surface tension generated	Strength, permeability, shrinkage at high humidity
Gel pores	10–2.5 nm	Small (gel) capillaries	Strong surface tension generated	Shrinkage to RH = 50%
	2.5–0.5 nm	Micropores	Strongly adsorbed water, no menisci form	Shrinkage, creep
	<0.5 nm	Micropores "interlayer"	Structural water involved in bonding	Shrinkage, creep

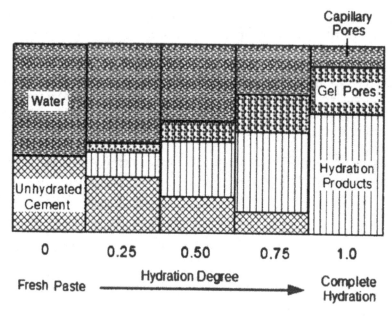

(a) Constituent Water-to-Cement Ratio = 0.5

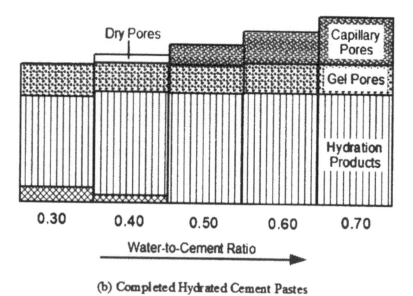

(b) Completed Hydrated Cement Pastes

Figure 5.20 Volume relationships among constituents of hydrated portland cement pastes.

Table 5.8 Approximate time until capillary pore discontinuity (Powers, 1958)

Water/cement ratio (by mass)	Time required (days)
0.40	3
0.45	7
0.50	14
0.60	180
0.70	365
Over 0.70	infinite

5.6.2 Measurement of pores

There are two main methods for measuring the pores in hardened cement pastes: mercury porosimetry and gas adsorption.

Washburn first proposed that the technique of mercury intrusion under pressure could be used to measure pore-size distribution and derived the relationship between the applied pressure and the pore radius for cylindrical pores (Smithwick 1982):

$$P = \frac{-2\gamma\cos\theta}{r} \tag{5.3}$$

where

P = applied pressure;
γ = surface tension of mercury;
r = pore radius at which mercury intrudes at pressure P;
θ = contact angle between solid material and mercury.
Typically, $\gamma = 480\,\text{dyn/cm}$ and $\theta = 140°$, thus:

$$r = \frac{75,000}{P} \tag{5.4}$$

The total pore volume measured by the mercury intrusion is very close to the volume of evaporable water in water-saturated cement pastes.

Adsorption measurement can be used to determine those small pores, especially less than 300 Å in size. Both methods have to assume the geometry of pores and require drying of the samples before measurements. Drying can have very significant effects on C–S–H and affect the pore size distribution in the hardened pastes. At the moment, no method can determine the pores on saturated cement pastes directly.

5.6.3 Pores in alkali-activated slag cement pastes

Many studies have reported that alkali-activated slag cement pastes exhibit lower porosity and smaller pores than portland cement pastes if a proper

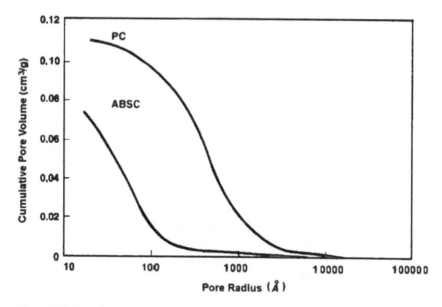

Figure 5.21 Cumulative pore volume of portland and sodium silicate-activated slag cement pastes at 28 days (Shi and Day 1996a).

activator is used. Figure 5.21 represents the mercury intrusion cumulative pore volumes of portland cement and sodium silicate-activated blast furnace slag cement pastes cured 28 days at 25 °C (Shi and Day 1996a). Table 5.9 summarizes their pore structure characteristics. It can be seen that the alkali-activated slag cement paste exhibits not only a lower porosity but also a finer pore structure than portland cement paste. The alkali-activated slag cement pastes contain mainly pores with $r < 100\,\text{Å}$, which can restrict the flow of liquid or diffusion of ions in the pastes. On other hand, a study found that an NaOH-activated blast furnace slag exhibited a higher porosity and a larger portion of capillary pores than a conventional cement paste (Song *et al.* 2000), which agrees with findings from alkali-activated slag cement mortars described in Section 5.6.4.

Table 5.9 Pore structure characteristics of portland cement and alkali-activated blast furnace slag cement pastes cured 28 days at 25 °C (Shi *et al.* 1992b)

Cement paste	Porosity (%)	Pore volume (%)		
		18–100 (Å)	100–1000 (Å)	>1000 (Å)
Portland cement	22.30	18.87	69.68	11.45
Alkali-activated slag cement	12.77	80.98	15.10	3.92

For a given activator, the pores sizes become smaller and porosity is lower as the fineness of the slag increases (Anderson and Gram 1987). It can be explained by the accelerated hydration of slag due to the increase in fineness. Steam cured alkali-activated slag cement pastes also show higher porosity than pastes cured at room temperature (Anderson and Gram 1987).

5.6.4   Pore structure of alkali-activated slag cement mortars

The introduction of aggregate into cement pastes can have a significant effect on the properties of the system. Figure 5.22 gives the pore structure characteristics of PC(III) and alkali-activated slag cement mortars from 3 to 90 days (Shi 1996b). Sodium silicate-activated slag cement mortars (Figure 5.22b) exhibited a significantly lower measured porosity than the PC(III) mortars (Figure 5.22a). The porosity of three-day-old sodium silicate-activated slag cement mortars was even lower than that of 90-day-old PC(III) mortars. The other significant difference between the two mortars was the pore size distribution: PC(III) mortars had a continuous distribution over measured pore size range from 50 to 12,000 Å, while sodium silicate-activated slag cement mortars contained only pores smaller than 100 and greater than 2000 Å.

At a given age, sodium carbonate-activated slag cement mortars (Figure 5.22c) had a slightly lower porosity than PC(III) mortars. At three and seven days, the cumulative pore volume decreased quickly with the increase in pore radius smaller than 1000 Å and the decrease trend slowed down thereafter. At 28 and 90 days, the pore size distribution of sodium carbonate-activated slag cement mortars was very similar to that of sodium silicate-activated slag cement mortars, only pores larger than 300 and smaller than 2000 Å being measured. Sodium hydroxide-activated slag cement mortars (Figure 5.22d) had the highest porosity among the four mortars for a given age, and showed uniformly distributed pores over the measured pore size range.

5.6.5   Pores structure of autoclaved cement pastes

As discussed in Chapter 4, autoclaved alkali-activated cement pastes have very different hydration products and microstructure from those cured under atmospheric pressure. Figure 5.23 shows the pore structure characteristics of alkali-activated blast furnace slag cement (ABSC), alkali-activated phosphorus slag cement (APSC) and portland cement (PC) pastes after one day, five days and 15 days at 150 °C (Shi et al. 1991b). The experimental results indicate that PC paste has the largest porosities among the three

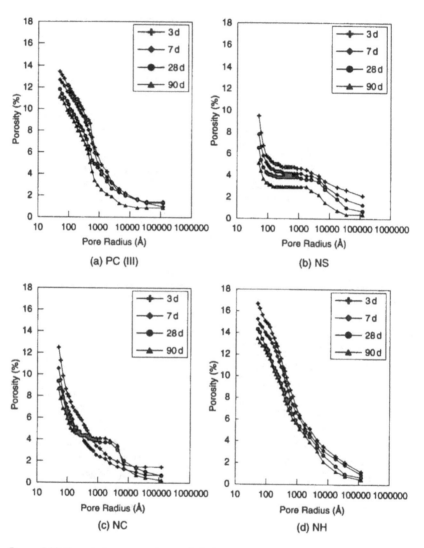

Figure 5.22 Cumulative pore volume of alkali-activated slag and portland cement mortars (Shi and Day 1996b).

cements at any time; this porosity consists mainly of medium and large pores. The content of large pores increases noticeably with curing time.

Alkali-activated blast furnace slag cement paste, hydrated for one day, contains no large pores ($r > 1000\,\text{Å}$), and the content of medium pores and gel pores were about 50% of the total pore volume respectively. When alkali-activated blast furnace slag cement is hydrated for five days, the content

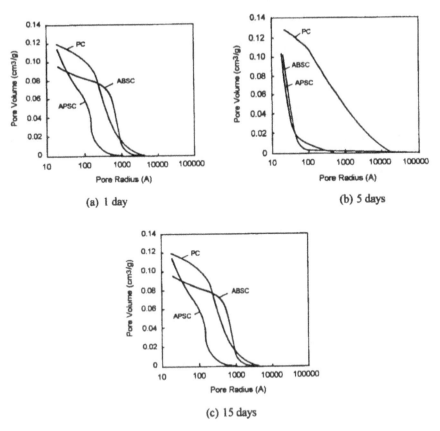

(a) 1 day

(b) 5 days

(c) 15 days

Figure 5.23 Pore size distribution of portland cement and alkali-activated slag cement pastes Cured at 150°C (Shi et al. 1991b).

of small pores ($r < 100$ Å) increases to about 95% of the total pore volume. This is due to the hydration of blast furnace slag particles, which results in an increase of solid volume and a denser paste structure. When alkali-activated blast furnace slag cement paste is hydrated for 15 days, it contains a small number of large pores; this could be due to the increase of the degree of crystallization of hydration products.

As for alkali-activated phosphorus slag cement pastes, they show the lowest porosity among the three pastes at all ages. At the age of one day, the pores in alkali-activated phosphorus slag cement paste consist mainly of medium pores with $100 < r < 1000$ Å. As the curing time increases, total porosity increases, but the proportion of small pores ($18 < r < 100$ Å) increases sharply – changing from 12.55% at one day to 96.85% at five days. The pore structure of a 15-day paste is similar to that of a five-day paste.

5.7 Relationship between porosity and strength

It is generally accepted that the pore structure plays an important role in determining the strength of hardened cement paste and concrete. The relationships between pore structure and strength have been summarized in previous publications (Rossler and Odler 1985, Odler 1991). The pore size distribution, the shape and position of pores are also important, but it is both difficult and impractical to include all these parameters. Many experimental results have confirmed that an acceptable prediction of strength can be obtained by using total porosity. The most common relationships between porosity and compressive strength of portland cement pastes are:

Balshin's equation

$$\sigma = \sigma_o \cdot (1 - P)^A \tag{5.5}$$

Ryshkevitch's equation

$$\sigma = \sigma_o \cdot \exp(-BP) \tag{5.6}$$

Schiller's equation

$$\sigma = D \cdot \ln\left(\frac{P_o}{P}\right) \tag{5.7}$$

and Hasselmann's equation

$$\sigma = \sigma_o \cdot (1 - AP) \tag{5.8}$$

where

σ_o = compressive strength at zero porosity;
P = porosity;
P_o = porosity at zero strength;
σ = compressive strength at porosity P;
A, B, D = experimental constants.

Most other relationships are variations of one of the above four types. Equation 5.6 is especially suitable for low porosity systems and Eqn (5.7) for high porosity systems.

Rossler and Odler (1985) have examined a series of cement pastes with different water-to-cement ratios at different ages, and concluded that the existing relationships between strength and porosity within the common porosity range can be expressed with sufficient accuracy by any of the above equations, but Hasselmann's equation yields slightly more accurate results than others.

All three equations represent a simplification of the system because they consider only total porosity. The pore size distribution, the shape of the pores and the position of pores are also important, but it is both difficult and impractical to include all these parameters. Many experimental results confirm that an acceptable prediction of strength can be obtained by using total porosity. Equation (5.6) is especially suitable for low porosity systems and Eqn (5.7) for high porosity systems. It was found that both alkali-activated slag cement (ASC) and Portland cement (PC) pastes follow the Eqn (5.5) within low porosity range (Xu 1988, Xu et al. 1993):

$$P = 11.49 \exp(-0.031\sigma) \ r = 0.996 (PC) \tag{5.9}$$

$$P = 25.26 \exp(-0.006\sigma) \ r = 0.941 (ASC) \tag{5.10}$$

Below 12% porosity, alkali-activated slag cement paste shows higher strength than portland cement at the same porosity (as shown in Figure 5.24). It seems that hardened alkali-activated slag cement has higher intrinsic strength than portland cement pastes.

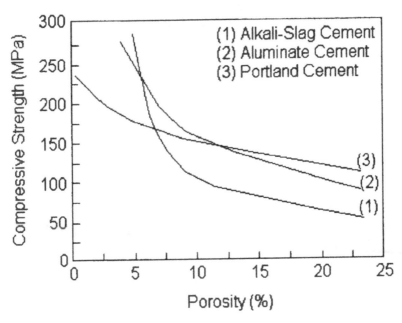

Figure 5.24 Correlation between porosity and compressive strength of alkali-activated slag cement pastes (Xu 1988, Xu et al. 1993).

5.8   Relationship between porosity and strength of alkali-activated slag cement mortars

Shi and Day (1996b) examined the relationships between compressive strength and the mercury intrusion porosity ($r > 50$ Å) of alkali-activated slag and portland cement mortars, as shown in Figure 5.25. It can be seen that all alkali-activated slag mortars, regardless of the nature of activator, follow the same trend, which is significantly different from that which portland cement mortars follow. This may be explained by the similarity and difference in the nature of their main hydration products. Although the variation of activators changes the minor hydration products, the main hydration product of alkali-activated slag is always the same: C–S–H with a low Ca/Si ratio, as discussed in Chapter 4, while the main hydration product of portland cement is C–S–H with a high Ca/Si ratio.

The four strength–porosity relationships, as described above, were used to depict the experimental results. The regression results are summarized in Table 5.10. The four equations yielded less accurate results for portland cement mortars than for alkali-activated slag mortars. The coefficients of variation of regression constants for alkali-activated slag mortars in the four equations ranged from 2.83 to 10.79%. Thus, all the four equations could depict the strength–porosity of alkali-activated slag mortars with sufficient accuracy. Hasselmann's equation showed the most accurate results for both portland cement and alkali-activated slag mortars.

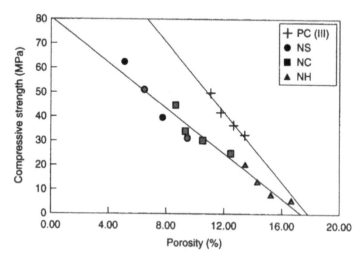

Figure 5.25 Relationship between mercury intrusion porosity ($r > 50$ Å) and strength of alkali-activated slag and portland cement mortars (Shi and Day 1996b).

Table 5.10 Regression results for the strength-porosity relationship of portland cement and alkali-activated slag mortars (Shi 1996b)

Item	Alkali-activated slag			Portland cement		
	Value	Standard error	Coefficient of variation (%)	Value	Standard error	Coefficient of variation (%)
Equation Ryshkevitch's				$\sigma = \sigma_0.(1-P)^A$		
σ_0	131	13.41	10.03	326	65.02	19.97
A	13.83	1.16	8.39	16.15	1.56	9.69
Equation Balshin's				$\sigma = \sigma_0.\exp(-B.P)$		
σ_0	139	15.01	10.79	376	78.01	20.78
B	15.24	1.31	8.67	18.39	1.736	9.436
Equation Schiller's equation				$\sigma = D.\ln\left(\frac{P_a}{P}\right)$		
D	47.03	2.77	5.88	88.35	10.11	11.45
P_a	0.195	0.008	4.32	0.192	0.010	5.26
Equation Hasselmann's				$\sigma = \sigma_0.(1-AP)$		
σ_0	80.72	3.50	4.34	128.2	11.73	9.15
A	5.77	0.164	2.83	5.62	0.236	4.20

5.9 Summary

This chapter has discussed the properties of fresh and hardened alkali-activated slag cement pastes and mortars and can be summarized as follows:

1 The nature and dosage of activator(s) have a significant effect on any property of alkali-activated slag cement pastes and mortars.
2 The addition of mineral admixtures and these commercial chemical admixtures to portland cements have limited or no effect on the workability and times of setting of alkali-activated slag cement pastes and mortars.
3 Hardened alkali-activated slag cement pastes and mortars exhibit larger autogenous and drying shrinkages than portland cement pastes and mortars.
4 Depending on the nature of the activator, alkali-activated slag cement pastes and mortars may exhibit a more or less porous structure than portland cement pastes and mortars. The relationship between porosity and strength for alkali-activated slag cement is different from that for portland cement.

Chapter 6

Properties of alkali-activated slag cement concrete

6.1 Introduction

In the last chapter, the properties of alkali-activated slag cement pastes were discussed. When aggregates are added, cement paste acts as a matrix and holds aggregates together in concrete. Because water forms a layer of film around the surface of aggregates during mixing, the pastes within 20 to 100 μm from aggregate surface have higher water-to-cement ratio, which results in the formation of an interfacial transitional zone between aggregate and cement paste. In portland cement concrete, this interfacial transitional zone can make up one-third or more of the matrix and is the most porous and weakest portion of the concrete (Popovics, 1998). The properties and quantity of aggregate will affect the properties of the concrete. This chapter will discuss the properties of fresh and hardened alkali-activated slag cement concrete.

6.2 Workability of fresh concrete

6.2.1 Introduction

Properties of fresh concrete are very important since they affect the choice of equipment needed for handling and consolidation. Many terms such as 'consistency', 'flowability', 'mobility', 'pumpability', 'compactibility', 'finishability' and 'harshness' have been used to describe the properties of fresh concrete. 'Workability' is often used to represent all those properties of fresh concrete. It is defined as the amount of mechanical work or energy required to produce fully compacted concrete without segregation. A large number of tests have been proposed for the measurement of workability. The common ones include: (1) slump test, (2) compaction test, (3) flow test and (4) Vebe test.

The rheological properties of alkali-activated slag cement pastes were discussed in Chapter 5. Of course, any factors that affect the rheological properties of cement pastes will affect the workability of the concrete. However, the amount and properties of aggregates and the relative proportions of fine and coarse aggregate can have a great effect on the workability of fresh concrete. For a given condition, the increase in the aggregate/cement ratio will decrease the workability. Also the shape and texture of aggregate affect the workability of concrete. The more nearly spherical the particles are, the more workable the concrete. It is important to have a proper relative proportion of fine and coarse aggregate in order to obtain a concrete with good workability. A deficiency in fine aggregate will result in a harsh concrete mixture. On the other hand, an excess of fine aggregate will lead to a rather more permeable and less economical concrete, although the mixture will be easily workable (Mindess and Young 1981).

6.2.2 Effect of activator on slump and slump loss

The nature of activators plays an important role in determining the slump and slump loss of alkali-activated slag concrete. Figure 6.1 indicates the workability of AASC activated by dry powdered sodium silicate is significantly greater than companion AASC activated by liquid activators. AAS1 is an alkali-activated slag cement concrete activated by solid sodium silicate and displays much higher initial slump and reduced slump-loss than OPC, for the same mixing proportions. The slump of AAS1 increases slightly with time initially and decreases thereafter. The slump increase can be attributed to the dissolution of solid sodium silicate, which is slow and endothermic. The slump loss of AAS1 within the overall testing period of 120 minutes was smaller than OPC. AAS2 uses liquid sodium silicate and H/C uses

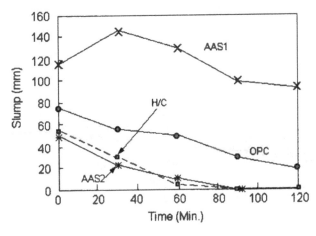

Figure 6.1 Effect of activator on slump and slump loss of alkali-activated slag cement concrete (Collins and Sanjayan, 1999a).

a combination of sodium hydroxide and sodium carbonate as activators. AAS2 and H/C show similar initial slump and slump loss during the testing period. However, they have a lower initial slump and slightly higher slump loss than OPC, and significantly less workability than AASC1.

6.2.3 Effect of ultra-fine additives

The replacement of portland cement with a small portion of ultrafine blast furnace slag or fly ash can adjust the particle size distribution of the cementing materials and reduce the initial porosity of the system, which may result in an improvement of workability of fresh concrete. Figure 6.2 shows the effect of the replacement of conventional ground blast furnace slag with 10% of different ultrafine materials such as ultrafine ground blast furnace slag (AAS/UFS), ultrafine fly ash (AAS/UFA) and condensed silica fume (AAS/CSF) on initial slump and slump loss within the first two hours from the time of mixing of alkali-activated slag cement concrete (Collins and Sanjayan, 1999b). AAS concrete displayed considerably better workability than OPC concrete (115 mm initial slump compared with 75 mm initial slump for OPC concrete). At 30 minutes, AAS concrete demonstrates higher slump than the initial value and at 120 minutes the slump loss of AAS concrete is minimal compared with OPC concrete, which loses 73% of the original slump.

AAS/CSF concrete demonstrated significantly lower slump than AAS concrete, with a reduction in initial slump from 115 to 37 mm. This type of

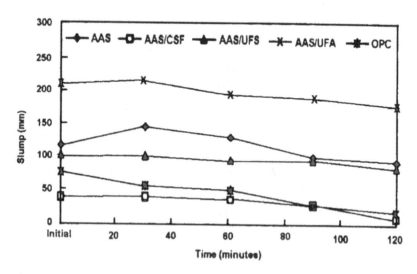

Figure 6.2 Effect of ultrafine materials on slump of alkali-activated slag cement concrete (data from Collins and Sanjayan 1999b).

concrete requires the use of a superplasticizer to overcome the low workability. There was a minor improvement in slump at 30 minutes; however, considerable slump loss was measured between 60 to 120 minutes. Nevertheless, the rate of slump loss was not as high during this period as for the OPC concrete. AAS/UFS concrete has a slightly lower slump than AAS (13% reduction in initial slump), but higher than OPC concrete and showed minimal slump loss over two hours. AAS/UFA concrete showed the highest slump, with 83% increase in the initial slump than AASC, followed by 13% slump loss over two hours. Following the elapse of two hours, the slump is superior to the initial slump of all of the mixes.

6.2.4 Other factors

Other factors such as the fineness of slag and temperature can also affect the workability of the concrete. The flow of alkali-activated slag cement concrete increased with the fineness of the slag within certain range (Hakkinen *et al.* 1987). Temperature also affects the slump of alkali-activated slag cement concrete. It was found that at $W/S = 0.45$, the slump of alkali-slag concrete increased with temperature then decreased with temperature; at $W/S = 0.33$, the slump kept constant at 180 mm from 5 to 20 °C, but dropped to 80 mm at 60 °C (Gjorv 1989).

6.3 Air entrainment

6.3.1 Introduction

The conversion of water into ice results in volume increase. If excess water can readily escape into adjacent air-filled voids, damage of concrete will not happen. This is the underlying principle of air entrainment. The volume of capillary pores should be minimized so the volume of freezable water would not exceed that which could be accumulated by the entrained air voids.

Entrained air produces discrete, nearly spherical bubbles having a typical diameter of 50 um in the cement paste. Those bubbles never become filled with the products of hydration of cement as gel can form only in water. Air entrainment does not increase the permeability of concrete (Neville 1996).

The air bubble should be very close to the escaping water so it can move into the air bubbles easily. It is usually recommended that the thickness of the hardened cement paste between adjacent air voids, or the spacing of bubbles, should be less than 200 um.

The air bubbles are drawn in by simple mechanical stirring action. The air-entrainer allows the formation of stable bubbles. Many factors such as cement content, water-to-cement ratio, mixing time and speed, type of mixture and agent affect the air-entraining process.

6.3.2 Compatibility between air-entraining agents and alkali-activated slag cement

There are three major types of air-entrainers: animal and vegetable fats and oils, natural wood resins or "Vinsol resin" and sulphonated organic compounds. It is well known that there is a compatibility issue between cement and air-entrainers. Thus, it is important to understand the compatibility between an air-entraining agent and alkali-activated slag cement in order to introduce a proper amount of air into alkali-activated slag cement concrete.

Several studies have tried to use air-entrainment admixtures to achieve a certain amount of air content. Douglas et al. (1992) reported that conventional air-entrainment admixtures for portland cement concrete did not work in alkali-activated slag cement concrete mixtures (Douglas et al. 1992). Gifford and Gillott (1996b) found that a commercially available air-entrainment admixture consisting of an aqueous solution of modified sulphonated hydrocarbon worked well in the portland cement mixture, but was ineffective in alkali-activated slag cement concrete mixtures. However, an air-entrainment admixture designed for use in low-slump concrete or for high alkali cements was found to be effective in alkali-activated slag cement concrete.

6.4 Strength

6.4.1 Introduction

The strength of cement and concrete materials is perhaps the most important overall measure of quality, although other properties may also be critical. In Chapter 5, the strength of alkali-activated slag cement pastes and a variety of factors affecting the strength of cement pastes were discussed. However, the nature and amount of aggregate used can have a very significant effect on the strength of concrete.

6.4.2 Effect of aggregate

Figure 6.3 shows the compressive strength of sodium silicate-activated slag cement paste, mortar and concrete for a given water-to-slag ratio and sodium silicate dosage. It can be seen that at early ages, mortars and concrete demonstrate much higher strength than the cement pastes. The strength development rate of concrete and mortars slows down significantly while the strength development rate of the paste still remains very high after

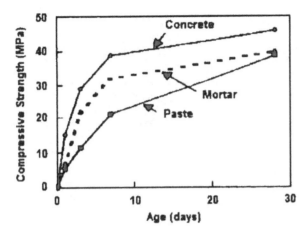

Figure 6.3 Comparison of compressive strength of sodium silicate-activated slag cement paste, mortar and concrete w/s = 0.5, sodium silicate ($Na_2O \cdot SiO_2$) dosage = 4% by mass of Na (Collins and Sanjayan 2001a).

seven days. The difference in strength between the paste and mortar or concrete narrowed significantly at 28 days.

In Chapter 4, it was pointed out that for a given activator dosage, an increase in water-to-slag ratio also results in a dilution effect on activator, which decreases the hydration rate of the slag and strength of the system. When aggregates are introduced into a mortar or concrete, the aggregates will adsorb some water and consume some water for wetting their surfaces, which leads to a lower effective water-to-slag ratio. This is why alkali-activated slag mortar and concrete can display much higher early strength than alkali-activated slag cement paste for a given water-to-slag ratio and sodium silicate dosage. Thus, it is not difficult to understand that alkali-activated slag concrete can reach compressive strengths of 60 to 150 MPa very easily. Table 6.1 lists the compressive strengths of some alkali-activated slag cement concrete batches reported by Pu *et al.* (Wang *et al.* 1995). It can be seen that some batches could reach a strength of over 60 MPa at one day, and many batches reached over 100 MPa at one year. Such a rapid strength gain and high strength are difficult to obtain from conventional portland cement concrete.

6.4.3 Effect of curing conditions

The properties of concrete are dependent on the hydration of cement and time. Powers reported that water in capillary pores will be lost

Table 6.1 Compressive strength of alkali-activated slag cement concrete (Wang et al. 1995)

Mix. no	Slump (cm)	Aggregate (coarse/fine)	Density (kg/m³)	Compressive strength (MPa)				
				1 d	3 d	7 d	28 d	365 d
JK4	1	Limestone/ Fine sand	2490	10.7	46.1		82.4	91.4
JK1	3	Limestone/ Fine sand	2500	36.0	–	63.8	77.0	109.9
JK15	1	Limestone/ Fine sand	2510	4.6	36.2	54.3	80.2	92.7
Jk16	8	Limestone/ Fine sand	2470	3.3	8.8		62.5	93.1
Jk26	1	Limestone/ Fine sand	2520	24.8	41.8		76.7	103.7
JK30	3	Limestone/ Sand	2550	56.1	71.3		102.3	122.4
Jk31	9	Limestone/ Sand	2520	60.9	80.8		99.0	116.0
JK32	2	Granite/ Fine sand	2550	61.5	79.0		99.0	114.4
Jk33	1	Granite/ sand	2470	68.1	96.2		117.0	132.2
JK103	2	Granite/ Fine sand	2520	12.1	46.0		66.3	86.5

and the hydration of portland cement is significantly decreased when the relative humidity within the capillary pores drops below 80%. The water loss will not only affect the strength development but also increase the plastic shrinkage and permeability and reduce durability of the concrete. Thus, water loss by evaporation from the capillary pores must be prevented. Furthermore, water lost internally due to self-desiccation (due to hydration of cement) has to be replaced by ingress of external water into the concrete in order to obtain good quality concrete (Neville 1996). This means proper curing is critical in order to obtain quality concrete.

Figure 6.4 shows the strength development of alkali-activated slag cement concrete cured in water bath, sealed and exposed to air conditions (Collins and Sanjayan 2001b). When the concrete is cured in water, compressive strength of the concrete keeps increasing until the end of the testing period of 365 days. However, if the specimens are cured in sealed condition, the strength stopped increasing at about 90 days. This may be attributed to the lack of moisture available for the hydration of slag inside the specimen. Also, sealed specimens demonstrated lower strength than the specimens cured in water bath even at very early ages.

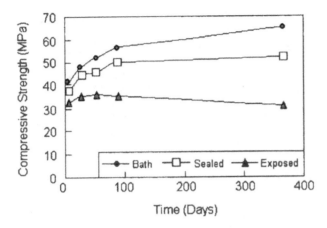

Figure 6.4 Effect of curing conditions on strength development of alkali-activated slag cement concrete (Collins and Sanjayan 2001b).

The specimens exposed to air exhibit the lowest strength all the time and strength retrogression at ages greater than 28 days. The strength reaches a maximum after 14 to 28 days of hydration, then starts to decrease. For portland cement concrete, the effect of inadequate water on strength is greater at higher water to cement ratio, lower strength development and in the presence of fly ash or slag (Neville 1996).

6.4.4 Effect of ultrafine materials

A partial replacement of portland cement with ultrafine material like silica fume can improve the strength and other properties of concrete significantly. The replacement of portland cement with a small portion of ultrafine blast furnace slag or fly ash has been widely reported. The use of the ultrafine material can adjust the particle size distribution of the cementing materials and reduce the initial porosity of the system, which will result in an improvement of workability of fresh concrete and strength of hardened concrete.

Figure 6.5 shows the effect of the replacement of conventional ground blast furnace slag with 10% of different ultrafine materials such as ultrafine ground blast furnace slag (AAS/UFS), ultrafine fly ash (AAS/UFA) and condensed silica fume (AAS/CSF) on strength development of alkali-activated slag cement concrete with time (Collins and Sanjayan 1999b). AAS refers to using 100% conventional ground blast furnace slag. AAS concrete showed higher strength than OPC concrete at all ages beyond one day. Between 56 and 91 days, the strength gain of the OPC control concrete levels off, whereas AAS concrete continues to gain strength. AAS/UFS shows marginally higher strength at all ages than both OPC and AAS concretes.

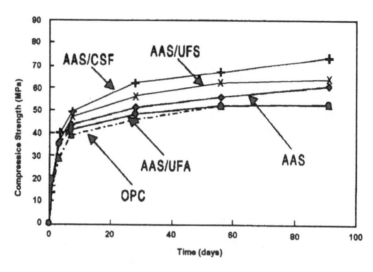

Figure 6.5 Effect of ultrafine materials on strength development of alkali-activated slag cement concrete (Collins and Sanjayan 1999b).

Although the strength improvement of AAS/UFS concrete is better than AAS concrete between 7 and 56 days (10% higher than AAS concrete at 28 and 56 days), the improvement declines to 5% at 91 days. AAS/CSF concrete demonstrated the best strength development at all ages beyond three days. The 91-day strength of AAS/CSF concrete is 74.2 MPa, which is the highest strength achieved on all the mixes and is 12% higher than the corresponding strength of the AAS concrete. Between 28 and 91 days, the slope of the strength-growth curve for AAS/CSF concrete is almost identical to AAS concrete, indicating the improvement in strength is mostly due to the improved hydration of the binder. Bakharev *et al.* (2000) also noticed that the replacement of 5% ordinary slag with ultrafine slag improved both the workability and strength of the concrete.

Although one-day strength of AAS/UFA concrete is higher than OPC and AAS concretes, the later age strengths are not as high as those for AAS/CSF and AAS/UFS concretes. At 56 days and beyond, the strength of the AAS/UFA concrete was identical to OPC concrete. This may be due to the lack of $Ca(OH)_2$ in the binder to promote the pozzolanicity of the UPA (Collins and Sanjayan 1999b).

6.5 Stress–strain relationship and modulus of elasticity

When a load is applied to a concrete material, it always deforms. The deformation is called strain. Stress–strain relationship is very important in structural design. Concrete, like many other construction materials, shows elastic behaviour only to a certain degree. Strictly speaking, the Young's modulus of elasticity is only applied to the straight part of stress–strain curve. The tangent to the curve at the origin is called initial tangent modulus.

The deformation occurring during loading is elastic and the sequent increase in strain is regarded as creep. The secant modulus is a static modulus. For comparative purpose, the maximum stress applied for determination of the secant modulus is specified as 40% of ultimate strength in ASTM C 469 (2003). To eliminate creep, at least two cycles of preloading are required in ASTM C 469. Hakkinen (1986) measured the stress–strain relationship of portland cement concrete and F-concrete and found that there was no obvious difference in the stress–strain curves from the two concretes (Figure 6.6). Thus, it is suggested that the secant modulus of elasticity of F-concrete E in GPa can be estimated from the compressive strength of the concrete f_c in MPa as recommended in ACI 318 for conventional cement concrete.

$$E = 4.73\sqrt{f_c} \qquad\qquad (6.1)$$

Table 6.2 lists the compressive strength, and measured and calculated modulus of elasticity. The calculation indicates that the measured results are very much in agreement with calculated results.

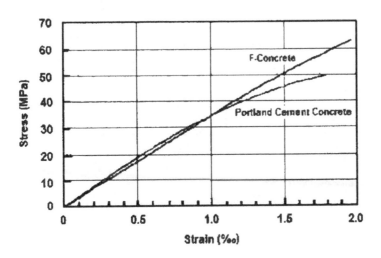

Figure 6.6 Stress–strain curves of F-concrete and portland cement concrete (data from Hakkinen, 1986).

Table 6.2 Compressive strength and modulus of elasticity of alkali-activated slag cement concrete (Douglas *et al.* 1992)

Batch no.	W/B	Sodium silicate/Slag	Compressive strength (MPa)		Measured modulus of elasticity (GPa)		Calculated modulus of elasticity based on ACI 318 (GPa)	
			28 d	91 d	28 d	91 d	28 d	91 d
1	0.48	0.48	42.4	46.3	30.2	31.1	30.8	32.2
2	0.48	0.48	43.4	51.6	29.7	31.2	31.2	34.0
3	0.48	0.39	36.2	42.3	30.7	32.3	28.5	30.8
4	0.48	0.37	39.0	43.9	33.7	34.8	29.5	31.3
5	0.48	0.33	39.0	43.9	33.7	34.8	29.5	31.3
6	0.41	0.32	41.2	48.4	29.4	32.1	30.4	32.9
7	0.49	0.48	36.3	43.4	28.3	30.9	28.1	31.2

6.6   Fracture of alkali-activated slag cement concrete

Ji (1991) measured the fracture energy of two alkali-activated slag cement concretes using the RILEM TC50-FMC method, which is schematically illustrated in Figure 6.7. The mixing proportion for the concrete is slag:sand:coarse aggregate:water = 1:0.52:2.22:0.34. The compressive strength of the two concrete were 69.7 and 73.7 MPa respectively. He also tested two other notches of 17 mm and 34 mm in addition to the notch of 50 mm as recommended by RILEM TC50-FMC (1985). The fracture

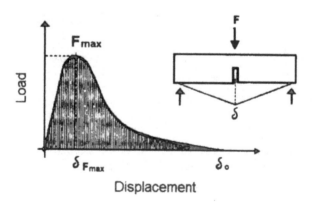

Figure 6.7 Scheme of the RILEM TC50-FMC test for measurement of fracture energy G_F.

energies for the two concrete were 167.4 and $186.8\,J/m^2$ and the size of notch does not show any obvious effect on the fracture energy.

6.7 Interface

6.7.1 Introduction

The interfacial zone between portland cement paste and aggregate or reinforcement, which is characterized by the prevalence of calcium hydroxide and the higher porosity, is the weakest region which controls many important properties of concrete such as strength, permeability and durability (Breton *et al.* 1993). The thickness of the transition zone, as determined by measurement of the orientation of $Ca(OH)_2$ crystal or microhardness, ranges from 50 to 100 um (Maso 1980, Sun *et al.* 1986).

A typical microstructure of the interfacial zone between the cement paste and quartz sand in portland cement mortars is shown in Figure 6.8. A thin layer of products, which is regarded as the duplex film, adhered on the surface of the sand grain. Beyond it, there is a porous transition zone with a thickness of 40 to 60 um. A high density of $Ca(OH)_2$ cluster is observed in the porous transition zone. No crystalline morphology, as would be expected of AFt, was observed although it was detected in powder X-ray diffraction test. As age proceed, the porous transition zone became thinner and denser.

It is believed that the dense transition zone between cement paste and aggregate is one of the main contributions to the better properties of alkali-activated slag cement and concrete. This section discusses the feature of interfacial zone between alkali-activated slag cement paste and aggregate.

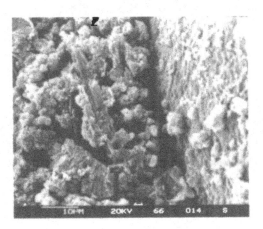

Figure 6.8 SEM observation of the transition zone between portland cement paste and quartz (Shi and Xie 1998).

6.7.2 Evolution of interfacial transitional zone in alkali-activated slag cement mortars and concrete

Brough and Atkinson (2000) used backscattered electron image to quantitatively compute the composition in the interfacial transitional zone in alkali-activated slag cement mortar. The evolution of the interfacial properties in the mortars is shown in Figure 6.9. Initially, the anhydrous material is packed against the sand grains, leaving a narrow interfacial zone of about 20 um width containing less anhydrous slag than that in the regions of bulk paste. At both one and seven days, significant amount of hydrates is found in the system, but this does not effectively fill the voids in the interface, where a higher level of porosity than in the bulk paste is still found. To some extent, the extra space at the interface is the result of shrinkage cracking on drying for sample preparation; but nevertheless, this indicates a paste that is more porous and fragile in the region of the interface. Further hydration to 14 days leads to a system with improved microstructure, with the hydrates much more effectively filling in the space in the system. Considerable amounts of hydrates are found in a region where there was originally very little anhydrous slag, indicating considerable mass transport over a significant distance.

Microscopic observations have confirmed that the interface between alkali-activated slag cement paste and sand in cement mortars appears to be very dense and uniform even after 28 days of hydration (Shi and Xie 1998). The examination of a 27-year-old alkali-activated slag concrete indicated not only a dense and uniform interface, but also trace of chemical reactions between cement paste and aggregate, as shown in Figure 6.10b (Deja 2002a).

The analysis of the distribution of main chemical elements in the interfacial zone between alkali-activated slag cement paste and perlite showed significantly increased concentrations of Na and K in a range up to $70\,\mu m$ from the interface (Figure 6.11). However, the Ca concentration was lower

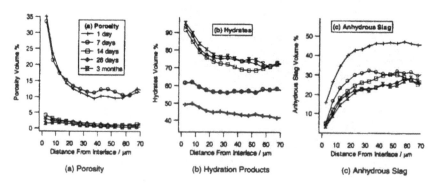

Figure 6.9 Evolution of interfacial transition zone in alkali-activated slag cement mortar (Brough and Atkinson 2002).

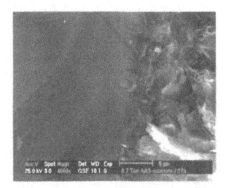

Figure 6.10 Interface between alkali-activated slag paste and aggregate, in a concrete after 27 years (Deja 2002a).

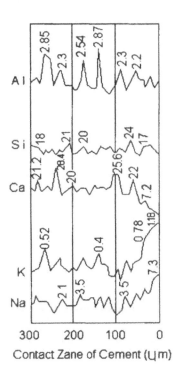

Figure 6.11 The distribution of elements in the interfacial zone between alkali-activated slag cement paste and expanded perlite (Krivenko 1994b).

in the transitional zone than that in bulk paste, which favours the formation of low-basic calcium hydrosilicates. Si and Al are uniformly distributed in the interfacial zone. Several studies have confirmed that alkalis concentrate in the alkali–aggregate reaction rims around the reactive aggregate (Brouxel 1993, Dove and Rimstidt 1995). Another study has found that alkalis can also concentrate in the transition zone around non-reactive aggregate (Breton *et al.* 1993).

6.7.3 Microhardness

Measurement of microhardness is an useful tool to characterize the inter-facial zone. Figure 6.12 shows the variation of microhardness of interfa-cial zones between alkali-activated slag cement pastes and several different aggregates after one year of hydration. The largest width of the contact zone (100–150 μm) is observed in case of using the quartz sand, medium (60–80 μm) for granulated slag and agloporit, and the least (20–30 μm) for sandstone aggregates.

The uniform and dense interfacial transitional zone, which is approx-imately 160 μm, contains products resulting from reactions between the

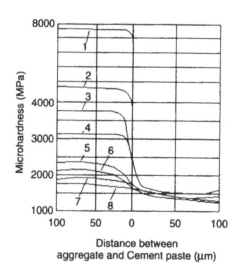

1 – quartz sand;   2 – sandstone;   3 – ferrous alevrites and argillites;
4 – non-granulated slag;   5 – agloporit;   6 – granulated slag;
7 – clay shales and argillites; 8 – expanded clay.

Figure 6.12 Microhardness of layers of the contact zone "slag alkaline cement–aggregate" (Krivenko 1994b).

substances in binder and on the surface of aggregates. This is confirmed by the analysis of the concentration of main chemical elements in the interfacial zone between cement pastes and aggregate (Figure 6.11) (Rumyna 1974, Glukhovsky 1981). The interfacial transition zone forms through dissolution-precipitation and solid-state reactions like the hydration of alkali-activated slag cements, and the main reaction products are low-basic calcium silicate hydrates and alkaline-alkali-earth aluminosilicate hydrates (Table 6.3) (Glukhovsky 1979). The ability of the alkali-activated slag cements to interact with clay minerals provides a possibility to use a wide range of low-quality aggregates such as sand, sand loam, and loam soils, shell limestone, fuel, non-granulated blast furnace slags, intrusive and effusive natural rocks, pyroactivated and partially glassy aluminosilicate substances. Crystalline aggregates may act as nuclear seeds for the faster formation of reaction products in the interfacial zone than amorphous aggregates.

6.7.4 Bonding between aggregate and alkali-activated slag cement paste

The bonding between the alkali-activated slag cement paste and aggregate depends upon properties and nature of the hydration products in the interfacial zone, and the physical and chemical interaction between cement paste and aggregate (Gordon 1969). In order to quantitatively measure the bonding between cement paste and aggregate, a coefficient of efficiency of aggregate in concrete α is proposed, which is defined as a ratio of direct tensile bonding strength of a cement paste or mortar with an aggregate to the direct tensile strength of the cement paste or mortar (Volyansky 1958):

$$\alpha = R_{bt}/R_{cem}t_{cem} \tag{6.2}$$

where
  R_{bt} = direct tensile strength of the concrete in MPa
  R_{cem} = direct tensile strength of cement paste or mortar in MPa;
  t_{cem} = share of fracture area of the cement stone projected by a normal to a breaking load.

Two types of fracture modes have been observed during the direct tensile testing: (a) fracture through the bulk cement pastes and aggregates and (b) fracture along the interface between cement paste and aggregate.

The test results from concretes made with natural aggregates are given in Table 6.4. Crushed granite give the highest coefficient of efficiency of

Table 6.3 Hydration products in the interfacial transitional zone (Glukhovsky 1979)

Aggregate	Dominant phase/mineral in aggregate		Crystalline phase of contact zone after	
	Phase	Mineral	1 cycle of steam curing	long-term hardening
Sandstone quartz sand	Crystalline	Quartz, tridymite	Low basic calcium hydro-silicates, gyrolite	Low basic calcium hydro-silicates
Ferrous alevrites and argillites	Dehydrated amorphisized substance	Montmorillonite, quartz, Orthoclase	The same	Low basic calcium hydro-silicates, alkaline and alkaline-alkali-earth hydro-aluminosilicates
Agloporit	Glass, amorphisized substance	Quartz, mullite	The same	The same
Expanded clay	The same	Quartz, spinel	The same	The same
Non-granulated slag	Crystalline glassy	Gehlenite, melilite	Low basic calcium hydro-silicates, gyrolite	The same
Granulated slag	Glass	Melilite	Low basic calcium hydrosilicates	The same
Clay shales and argillites	Dehydrated amorphisized substance	Montmorillonite, kaolinite, hydromica	The same	The same

Note: In case of using of calcined soda or soda-alkali melt as an alkaline component, a calcite is formed additionally in the contact zone (along with the above listed phases).

Table 6.4 Effect of coarse aggregate on bonding between coarse aggregate and alkali-activated slag cement mortars (slag:sand = 1:3) (Volyansky 1958)

Aggregate type	Tensile strength (MPa)		Mean value of t_{agg}/t_{cem} ratio	Coefficient of efficiency of aggregate in concrete (α)
	R_{aggr}	R_{cem}		
Crushed granite	4.660	4.110	0.750	1.180
Crushed marble	6.860	2.500	0.300	0.754
Quartz gravel	4.250	3.440	0.678	1.120

aggregate in concrete (α) of 1.18 and quartz gravel gave a value of 1.12. Crushed marble showed the lowest coefficient of efficiency of aggregate in concrete −0.754. Thus, when the crushed marble was used as aggregate, Mode A dominates the failure.

The increase in alkali activator dosage (expressed as the density of the alkali component solution) results in an increase of fracture Mode A and an increase in the mean value of the ratio t_{aggr}/t_{cem} as well (Table 6.5).

When fine aggregates with high modulus of elasticity like basalt sand are used, the coefficient of efficiency of the aggregate α increased significantly. The results in Tables 6.5 and 6.6 confirm that the bonding between alkali-activated slag cement paste and aggregate is significantly higher than that between portland cement paste and aggregate, which explains the fact that fracture of alkali-activated slag cement concrete takes place along the aggregate.

Some very interesting results have been obtained from the specimens made with synthetic granular aggregates (Table 6.6) (Rumyna 1974). The very high coefficient of efficiency ($\alpha = 1.63$) from No. 11 in Table 6.6 indicated that these alkali-activated slag cement concretes behaved similarly as the cement paste, which implied a dense interfacial zone. The specimens made with expanded clay gravel as aggregate (No. 13) had the lowest value of coefficient of 0.7.

Taking into account that bonding is determined by chemical interactions between cement paste and aggregate, the α value will vary with the isolation of the aggregate. Thus, a coefficient φ, which would reflect the influence of chemical interactions on the bonding, is introduced express see how

Table 6.5 Bonding strength between the slag-sand mortar with coarse aggregate (crushed granite) vs alkaline component density ($Na_2O \cdot SiO_2$) (Volyansky 1958)

Fine aggregate	Density (kg/m^3)	Bonding strength (MPa)	Percentage of specimens destroyed % of fracture mode			Mean value of $t_{agg}t_{cem}$ ratio	Coefficient of efficiency of aggregate in concrete (α)
			A	A+B	B		
Quartz sand	1180	4.80	50	–	–	0.55	1.23
(grading	1180	4.80	–	–	50	0.28	0.90
modulus = 2.20)	1220	6.70	89	–	–	0.74	1.35
	1220	6.70	–	11	–	0.52	0.85
	1260	6.50	89	–	–	0.76	1.42
	1260	6.50	–	–	11	0.48	0.85
Basalt sand	1180	3.60	–	–	50	0.42	1.19
(grading	1180	3.60	50	–	–	0.68	1.70
modulus = 2.50)	1220	4.00	56	–	–	0.75	1.65
	1220	4.00	–	22	22	0.50	1.20
	1260	4.50	89	–	–	0.80	1.75
	1260	4.50	–	11	–	0.50	1.17

Note: Aggregate strength – 3.10 MPa.

Table 6.6 Coefficient of efficiency of various aggregates in alkali-activated slag cement concrete

Nos	Aggregate	Tensile strength (MPa)			Portion of fracture area		α	φ
		R_{bt}	R_{aggr}	R_{cem}	t_{aggr}	t_{cem}		
1	Synthetic granular material of sand	4.71	4.80	4.80	0.40	0.60	1.63	4.1
2	The same, covered with paraffin	1.01	4.80	4.80	0.38	0.60	0.40	4.1
3	Synthetic granular material of shell limestone	3.39	1.42	4.80	0.45	0.55	1.29	3.0
4	The same, covered with paraffin	0.93	1.42	4.80	0.54	0.46	0.43	3.0
5	Synthetic granular material of perlite	1.85	1.20	4.80	0.68	0.30	1.21	3.1
6	The same, covered with paraffin	0.76	1.20	4.80	0.55	0.45	0.35	3.1
7	Synthetic granular material of expanded clay	2.78	3.84	4.80	0.44	0.56	1.05	3.3
8	The same, covered with paraffin	1.20	3.84	4.80	0.35	0.65	0.30	3.3
9	Crushed shell limestone	2.41	0.70	4.80	0.40	0.60	0.87	2.7
10	The same, covered with paraffin	0.98	0.70	4.80	0.38	0.60	0.30	2.7
11	Expanded perlite	3.14	0.40	4.80	0.35	0.65	1.00	2.3
12	The same, covered with paraffin	1.15	0.40	4.80	0.40	0.60	0.43	2.3
13	Expanded clay	1.80	1.70	4.80	0.47	0.53	0.71	1.8
14	The same, covered with paraffin	0.89	1.70	4.80	0.48	0.52	0.40	1.8

the coefficient of efficiency of aggregate in concrete varies without consideration of chemical interactions between cement paste and aggregate:

$$\varphi = \alpha_e / \alpha_{ei} \qquad (6.3)$$

where

α_e = coefficient of efficiency of the non-isolated aggregates in the concrete;

α_{ei} = coefficient of efficiency of the isolated aggregates in the concrete.

As predicted, the most intensive interaction is observed at the contact with the aggregates that are similar in composition to the mortar constituent of concrete, for which $\varphi = 4.1$ (Nos 1 and 2). Shell limestone, perlite and expanded clay gave very high φ values between 3.0 and 3.3, which indicates intensive chemical interactions taking place between the aggregate and alkali-activated slag cement mortar.

6.8 Shrinkage

6.8.1 Introduction

In Chapter 5, the shrinkage of cement pastes was discussed. The introduction of aggregate decreases the shrinkage of the system. The reduction of shrinkage caused by aggregate depends on the amount of aggregate in the concrete (Figure 6.13), the stiffness of aggregate and the maximum size of aggregate. Lightweight aggregate itself is dimensionally stable. However, since lightweight aggregate has a low modulus of elasticity, it can be expected that lightweight aggregate concrete would exhibit higher shrinkages than normal weight concrete. The shrinkage relationship between paste ε_p and concrete ε_{con} can be expressed as follows:

$$\varepsilon_{con} = \varepsilon_p (1 - A)^n \qquad\qquad (6.4)$$

where A is the aggregate content and n is a constant between 1.2 and 1.7.

6.8.2 Effect of activator

The nature of activator is one of the most important factors determining the properties of alkali-activated slag cements and concrete. Of course, it

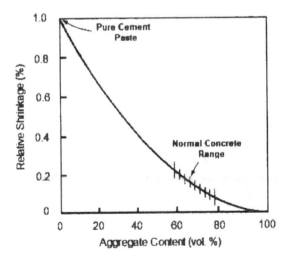

Figure 6.13 Effect of aggregate on shrinkage of concrete.

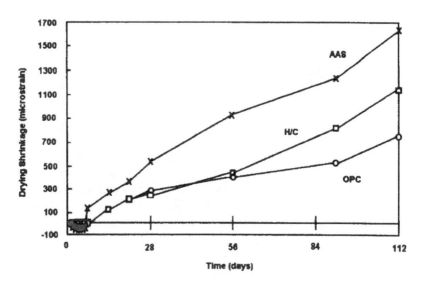

Figure 6.14 Effect of activator on drying shrinkage of alkali-activated slag cement concrete (Collins and Sanjayan 1999c).

will have a significant effect on the drying shrinkage of alkali-activated slag cement and concrete. Figure 6.14 shows the drying shrinkage of portland cement concrete, sodium silicate-activated slag cement concrete and sodium hydroxide-sodium carbonate-activated slag cement concrete (Colling and Sanjayan 1999c). Those specimens were cured in water during the first seven days and showed some expansion. Once they were exposed to drying, sodium silicate-activated slag cement concrete displayed considerably higher dry shrinkage than the ordinary portland cement concrete. The drying shrinkage of sodium hydroxide-sodium carbonate-activated slag cement concrete was similar to that of portland cement concrete up to 56 days, but became significantly higher than that of portland cement concrete between 56 and 112 days. Thus, it is important to measure the long-term drying shrinkage of the concrete. It was found (Bakharev *et al.* 2000) that lignosulphonate-based admixture caused a slight reduction in shrinkage, while naphthalene-based superplasticiser significantly increased the shrinkage and reduced the strength of alkali-activated slag cement concrete. Air-entraining and shrinkage reducing admixtures, as well as gypsum (6%) were effective in reducing shrinkage of alkali-activated slag cement concrete.

Douglas *et al.* (1992) investigated the drying shrinkage of alkali-activated slag cement concrete with different amounts of sodium silicate with a ratio of 1.47. They found that, after seven days of water curing, there were no obvious differences for the drying shrinkage of those alkali-activated slag cement concretes containing different amounts of sodium silicate (Table 6.7). However, those drying shrinkages were obviously higher than

Table 6.7 Drying shrinkage of alkali-activated slag cement concrete (Douglas *et al.* 1992)

Batch no.	W/B	Sodium silicate/slag	Shrinkage strain ($\times 10^{-6}$)					
			7 d	14 d	28 d	56 d	112 d	224 d
1	0.48	0.48	284	380	511	674	746	986
2	0.48	0.48	305	404	525	645	738	940
3	0.48	0.39	298	408	557	646	791	929
4	0.48	0.37	203	305	397	500	589	745
5	0.48	0.33	245	319	440	525	663	844

portland cement concrete or portland slag cement concrete with equivalent water-to-cementing materials ratio and workability (Douglas *et al.* 1992).

6.8.3 Effect of aggregate

Collins and Sanjayan (1999c) measured the autogenous shrinkage of alkali-activated slag and portland cement concretes made with basalt aggregate (Figure 6.15). Alkali-activated slag cement concrete displayed much higher autogenous shrinkage than portland cement concrete. This is in agreement with the results from pastes as discussed in Chapter 5. Air-cooled blast furnace slag aggregate was also used for production of another batch of alkali-activated slag cement concrete; the two batches of alkali-activated slag cement concretes did not show any differences in autogenous shrinkage.

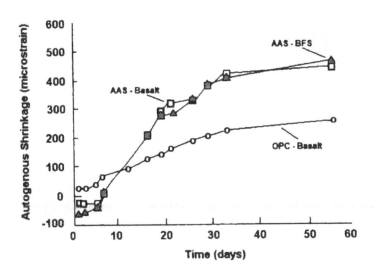

Figure 6.15 Autogenous shrinkage of portland cement and alkali-activated slag cement concretes (Collins and Sanjayan 1999c).

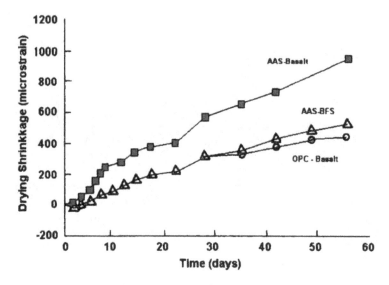

Figure 6.16 Drying shrinkage of alkali-activated slag cement concretes (Collins and Sanjayan 1999c).

Figure 6.16 is the drying shrinkage of alkali-activated slag cement concretes made with basalt and blast furnace slag aggregate. AAS made with basalt aggregate showed much higher drying shrinkage than the one made with blast furnace slag aggregate, while the later was only slightly higher than the portland cement concrete made with basalt aggregate. They attributed the difference in drying shrinkage between the concretes made with basalt aggregate and slag aggregate to the moisture released from the slag aggregate during drying. However, the water absorption of the basalt during 24 hours was only 1.2%, and that of the slag was only 4.4%.

6.8.4 Effect of curing conditions

However, it was found that continuous curing in lime-saturated water resulted in expansion rather than shrinkage (Douglas *et al.* 1992). This may be contributed to the ingress of Ca^{2+} from the curing water into the concrete specimens and the formation of additional C–S–H. The specimens after seven days of bath curing followed by exposed curing at 23°C and 50% RH exhibited only slightly lower drying shrinkage than those exposed from day one (Collins and Sanjayan 1999d). Elevated temperature curing at early ages decreases the drying shrinkage of alkali-activated slag cement concrete, while waterglass-activated slag cement concrete still exhibited higher drying shrinkage than NaOH-activated slag cement concrete (Bakharev *et al.* 1999b).

6.8.5 Mechanism of shrinkage

Drying shrinkage is caused by the loss of evaporable water. It is generally agreed that there are at least four phenomena involved in the drying shrinkage of the cement paste: capillary stress, disjoining pressure, changes in surface free energy and movement on interlayer water. Collins and Sanjayan (2000b) examined the relationship between pore size distribution and drying shrinkage. They noticed that although AASC showed a less porous structure and lost less mass than portland cement concrete during drying, the former had greater shrinkage than the latter.

The drying shrinkage behaviour of concrete materials can be described by the pore size distribution and thermodynamic behaviour of water in the pores. Pore size distribution can be characterized by r_s, which is defined as the radius of the pores where the meniscus forms; i.e., the pores whose radii are smaller than r_s are assumed to be filled with liquid water while pores larger than this are dry. As the drying progresses, the parameter r_s would decrease. The smaller the pore size is, the larger the capillary tensile forces set up at the meniscus, hence higher the resulting shrinkage.

6.9   Cracking tendency

In concrete, the potential shrinkage of cement paste is restrained by the aggregate, which results in some internal stress. The non-uniform shrinkage within the concrete also results in some stress. When the total stress is greater than the strength of cement paste, cracking will happen. The formation of cracks decreases the flexural strength of the materials and makes it easier for the corrosive substances to ingress into the concrete (Byfors *et al.* 1989). Thus, any factors that increase the shrinkage will increase the cracking tendency of the concrete.

As pointed out in the last section, alkali-activated slag cement concrete can display much higher shrinkage than ordinary portland cement concrete. This means that the former will be much easier to crack than the latter when exposed to an atmosphere at low relative humidity. Collins and Sanjayan (2001b) used a crack-detection microscope, which has a 4 mm range, 0.01 mm division and 40 × magnification, to examine the cracking of alkali-activated slag cement concrete under different curing conditions. They noticed that the concrete specimens cured in water or sealed conditions did not show visible surface cracks. However, the specimens exposed to an environment at 50% relative humidity and 20 °C had a lot of surface cracks. Figure 6.17 shows the crack frequency (numbers of cracks per sample) at 3, 7, 28 and 56 days after one day of moist curing. The majority of cracks have a size of 0.01 mm. The frequency of the 0.01 mm crack reached the maximum at seven days. The width of cracks can be as high as 0.3 mm.

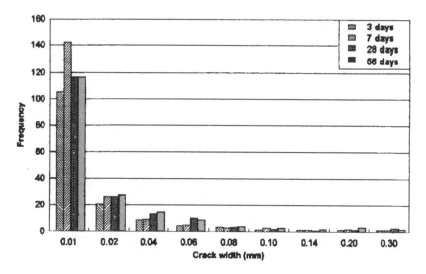

Figure 6.17 Crack frequency of alkali-activated slag cement concrete exposed to 50% RH at 20°C after one day of moist curing (based on Collins and Sanjayan 2001b).

Collins and Sanjayan (2000a) also investigated the cracking tendency of alkali-activated slag concrete using restrained beams (Figure 6.18). If alkali-activated slag concrete beams are exposed to 50% RH at 23°C after one day in mould following casting, they cracked within one day and the cracks grew to 0.97 mm at 175 days, while portland cement concrete beams cracked within nine days and their cracks grew to 0.33 mm at 175 days.

Alkali-activated slag concrete beams that were bath-cured in lime-saturated water for 3 and 14 days prior to exposure to 50% relative humidity and 23°C temperature cracked respectively at 2 and 44 days after the exposure; however, the magnitude of the crack width was considerably less than alkali-activated slag concrete beams which had no curing. The crack width was comparable to the restrained ordinary portland cement concrete beams. Attention to good curing is therefore essential when utilizing alkali-activated slag concrete. The alkali-activated slag concrete made with slag aggregate showed a much better cracking tendency performance than the one made with basalt aggregate, even better than portland cement concrete. They attributed the superior performance of the AASC/BFS beams to lower magnitude of drying shrinkage, superior tensile strength and lower elastic modulus than the alkali-activated slag concrete with normal weight coarse aggregate. Lower stiffness of AASC/BFS could explain why this concrete type accommodates more strain before cracking. AASC/BFS shows no surface crazing in contrast with the alkali-activated slag concrete when exposed

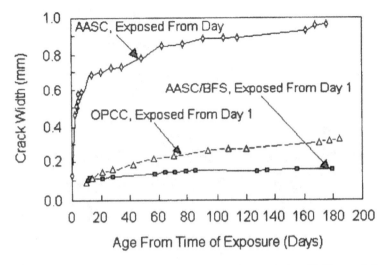

Figure 6.18 Cracking of different concretes during drying (Collins and Sanjayan 2000a).

from day one onwards. The absence of crazing eliminates the possibility of these types of cracks serving as initiators of drying shrinkage cracks.

The use of shrinkage reducing agent decreases the drying shrinkage of alkali-activated slag concrete when it is not cured properly. It also decreases the early strength of the concrete (Collins 1999). When, adequate bath curing is provided, the introduction of chemical shrinkage reducing agent shows little effect on the cracking tendency of alkali-activated slag concrete beams.

6.10   Creep

6.10.1   Introduction

Creep refers to the deformation of concrete under loads at a constant temperature. There are two types of creep: basic creep and drying creep. Basic creep refers to the creep generated under constant humidity conditions, and drying creep is the creep during drying. The creep of cement paste increases at a gradually decreasing rate, approaching a value several times larger than the elastic deformation (Figure 6.19).

On unloading, deformation decreases immediately due to elastic recovery. This instantaneous recovery is followed by a more gradual decrease in deformation due to creep recovery. The remaining residual deformation, under equilibrium conditions, is called "irreversible creep". Creep of concrete is normally evaluated using unsealed loaded and unloaded companion

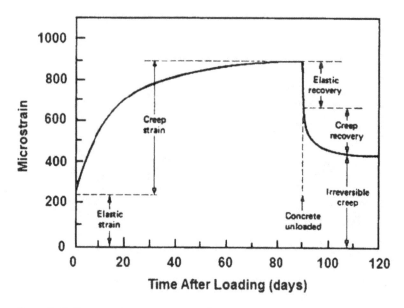

Figure 6.19 Creep and creep recovery of cement paste.

specimens exposed at a constant drying environment. Thus, the total defor-mation may be separated into elastic compression, basic creep and drying creep (moisture loss, autogeneous and carbonation shrinkage).

Creep coefficient, specific creep or creep compliance is generally used to describe creep strain by different mathematical prediction models. The creep coefficient is defined as the ratio of creep strain (basic plus drying creep) at a given time to the initial elastic strain. The specific creep is defined as the creep strain per unit stress. The creep compliance is defined as the creep strain plus elastic strain per unit stress, whereas the elastic strain is defined as the instantaneous recoverable deformation per unit length of a concrete specimen during the initial stage of loading.

6.10.2 Creep of alkali-activated slag cement concrete

Hakkinen (1986) measured the creep of F-concrete and portland cement concrete (Figure 6.20). It can be seen that for a given amount of binder, F-concrete with a water-to-binder ratio of 0.35 showed even higher creep than portland cement concrete with a water-to-cement ratio of 0.50. On the other hand, it can be seen that a decrease in binder content from $400 \, kg/m^3$ to $320 \, kg/m^3$ decreased the creep very significantly. This means that alkali-activated slag cements have higher creep than portland cement for given conditions.

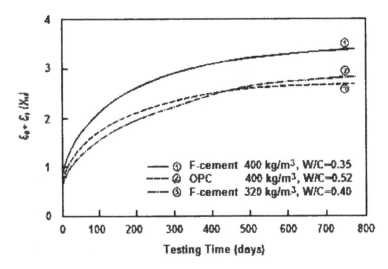

Figure 6.20 Creep and momentary elastic deformation of portland cement and F-concrete (Hakkinen 1986).

Collins and Sanajayan (1999a) also compared the creep of alkali-activated slag concrete with portland cement concrete. They noticed that portland cement concrete showed larger creep than alkali-activated slag cement concrete during the first three days. However, portland cement concrete showed smaller creep than alkali-activated slag cement concrete after three days. After 112 days of loading, alkali-activated slag cement concrete had a creep of 42 microstrain/MPa compared with 37.5 microstrain/MPa for portland cement concrete.

On other hand, it is noticed that the variation of water content, binder and binder content do not show a noticeable effect on the creeping coefficient of the concrete (Figure 6.21).

6.10.3 Factors controlling creep of concrete

The deformation due to creep is attributed to the movement of water between the different phases of the concrete. When an external load is applied, it changes the attraction forces between the cement gel particles. This change in the forces causes an imbalance in the attractive and disjoining forces. However, the imbalance is gradually eliminated by the transfer of moisture into the pores in cases of compression, and away from the pores in cases of tension.

Factors which contribute to the dimensional changes may be categorized as mixture composition, curing conditions, ambient exposure conditions

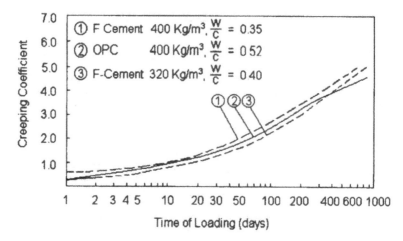

Figure 6.21 Ratio between creep and momentary elastic deformation of portland cement concrete and F-concrete (Hakkinen 1986).

and element geometry. Generally, concretes with aggregates that are hard, dense, and have low absorption and high modulus of elasticity are desirable when concrete with low creep is needed. Aggregates with lower absorption will therefore produce concrete with lower creep and shrinkage characteristics. Concrete with higher elastic modulus will produce lower creep values. Thus, aggregates affect concrete deformation through water demand, aggregate stiffness and volumetric concentration and paste/aggregate interaction. Concrete with a large maximum aggregate size and lower paste content will have lower creep and shrinkage characteristics.

High early strength cement typically shrinks and creeps more than normal cement. Low heat and portland-pozzolan cements produce larger percentages of gel compared to normal portland cement, thus causing an increase in shrinkage and creep. Thus, it is expected that alkali-activated slag cement concrete will demonstrate more creep than portland cement concrete under the same mixing proportions and testing conditions.

Generally, finer cement particles exhibit less shrinkage under moist conditions. The lower the fineness of a cement, the higher the creep in the concrete. Cement fineness has little influence on the amount of creep of concretes containing ordinary cement. When a constant W/C ratio is maintained, creep increases as the slump and cement content increases or as the amount of cement paste is increased. The specific creep, the creep strain per unit of applied stress, decreases with decreasing water content for the conditions of a constant aggregate-to-cement ratio.

Designs typically use one of the two code models to estimate creep and shrinkage strain in concrete, ACI 209 model recommended by the American Concrete Institute Committee 209 (ACI 209R-92 1993) or the CEB 90 Eurocode 2 model recommended by the Euro-International Committee. The ASSHTO LRFD is based on the ACI 209 model. Three other models are the B3 model, developed by Bazant; the GZ model, developed by Gardner; and the SAK model developed by Sakata (Meyerson 2001).

6.11 Summary

This chapter has discussed the properties of fresh and hardened alkali-activated slag concrete and can be summarized as follows:

1   As discussed seen in Chapter 4, the nature and dosage of activators have a significant effect on the properties of fresh and hardened alkali-activated slag cement concrete.
2   The addition of mineral admixtures and commercial chemical admixtures to portland cements has limited or no effect on the workability and time of setting of alkali-activated slag cement concretes.
3   The relationship between compressive strength and the modulus of elasticity of alkali-activated slag cement concrete follows that for portland cement concrete.
4   There is no porous transitional zone between cement paste and aggregate or reinforcement in alkali-activated slag cement concrete as that in portland cement concrete. The bonding between cement paste and aggregate or reinforcement in alkali-activated slag cement concrete is much better than that in portland cement concrete.
5   Hardened alkali-activated slag cement concrete exhibits larger autogenous and drying shrinkages than portland cement concrete.

Chapter 7

Durability of alkali-activated cements and concretes

7.1 Introduction

Durability is one of the most important desired properties of concrete. Concrete is inherently a durable material. However, concrete is susceptible to attack in a variety of different exposures unless some precautions are taken. It is generally agreed that alkali-activated cement and concrete exhibit better corrosion resistance, but a larger shrinkage than conventional portland cement and concrete. This chapter discusses various durability properties of alkali-activated slag cement and concrete.

7.2 Water permeability and chloride diffusion

7.2.1 Water permeability

Permeability of concrete plays an important role in durability of portland cement concrete because it controls the rate of entry of moisture that may contain aggressive chemicals and the movement of water during heating or freezing. For a given cement, water-to-cement ratio (W/C) has the largest effect on permeability of concrete. A decreased W/C increases the strength of concrete and hence improves its resistance to cracking from the internal stresses that may be generated by adverse reactions.

The permeability of mature cement paste is very low even though it has a high total porosity since water cannot move through very small discontinuous gel pores. Figure 7.1 shows how the water-to-cement ratio affects the water permeability coefficient of hardened cement pastes and concrete. As W/C ratio increases, the total porosity and portion of continuous capillary pores increase, which results in higher permeability. For a given W/C ratio, the water permeability coefficient of concrete is about two orders higher than that of the cement paste due to the effects from the internal cracking and continuous pores in the cement paste–aggregate interfacial zone. Also, the water permeability of concrete increases with the increase of the maximum aggregate size.

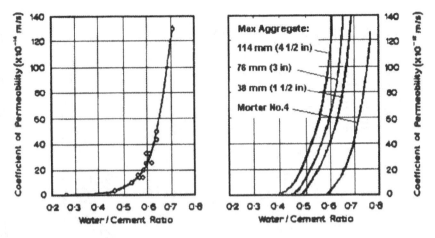

Figure 7.1 Effect of water-to-cement ratio on permeability coefficient of portland cement paste (Powers *et al.* 1954) and concrete (based on US Bureau of Reclamation, 1975).

Since the microstructure and properties of alkali-activated cement and concrete depend on the characteristics of alkaline activator(s) and cementing component(s) used, it is not surprising that the permeability of hardened cement and concrete varies from one research to another. Many researchers have confirmed that, for a given water-to-cement or water-to-slag ratio, a properly designed alkali-activated slag cement paste can exhibit a significantly lower total porosity and a lower portion of capillary pores than a portland cement paste does. It leads to the expectation that the former will have a lower permeability than the latter. However, for a given activator dosage, an increase in water-to-slag ratio will have a more obvious effect on the pore structure and permeability of alkali-activated slag cement than that of water-to-cement ratio on portland cement since it also dilutes the activation effect on the slag.

Davidovits (1994) reported a permeability of 10^{-11} m/s for alkali-activated cement paste compared with 10^{-12} m/s for portland cement. Wu *et al.* (1993) reported that the permeability of alkali-activated slag cement concrete is several times lower than that of portland cement concrete. Shi (1996b) measured the water permeability of three alkali-activated slag cement mortars using NaOH (NH), Na_2CO_3 (NC) and Na_2SiO_3 (NS) as activators, and one ASTM Type III cement (PC(III)) as a reference. A limited amount of water permeability results were obtained because some specimens broke during cutting. The water permeability test results are plotted in Figure 7.2. NS mortars showed the lowest while PC (III) showed the highest water permeability at seven days.

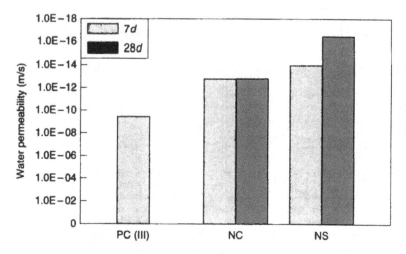

Figure 7.2 Water permeability of portland and alkali-activated slag cement mortars (Shi 1996b).

As mentioned above, the introduction of aggregate into portland cement paste increases its permeability due to the interface effect. The interface between alkali-activated slag cement paste and aggregate is totally different from that between portland cement paste and aggregate. Very porous zone and coarse crystals can be easily observed between portland cement paste and sand under scanning electronic microscope, but no difference could be noticed between alkali-activated slag cement paste and sand (Shi and Xie 1998). Microhardness measurement indicates that the interfacial transitional zone in alkali-activated slag cement concrete is stronger and less porous than that in portland cement concrete (Ilyin 1994). Thus, it can be expected that the inclusion of aggregate may not have any negative effect on permeability of alkali-activated slag cement concrete.

7.2.2 Transport and binding of chlorides

Chloride ions have a special ability to destroy the passive oxide film of steel even at high alkalinities. The penetration of chlorides in sufficient amounts to the depth of reinforcing steel will cause corrosion of the reinforcing steel and damage of reinforced concrete structures. Chloride-induced corrosion is a very common cause of concrete deterioration along sea coasts and in cold areas where deicing salts are used. The repair of concrete structures damaged by chloride-induced corrosion is very expensive. Thus, selection of a quality concrete material with low chloride permeability is very important in construction of a durable steel reinforced concrete structure where chloride-induced corrosion is a concern.

7.2.2.1   Chloride diffusion

The diffusion cell test developed by Page *et al.* (1981) is often used to measure the apparent diffusivity of Cl⁻ through hardened cement pastes and concrete. Chloride diffusion test on both F-cement and portland cement pastes indicated that the diffusion rate of Cl⁻ through F-cement pastes is about 30 to 40 times slower than that through portland cement pastes for a given water-to-cement or -slag ratio (Table 7.1). It was also noticed that cracks of 10–50 microns wide, induced by drying below 50% relative humidity did not show any influence on the Cl⁻ diffusion rate of the specimens. However, specimens with cracks wider than 50 microns running through the specimens had very higher Cl⁻ diffusion rate.

The replacement of portland cement with blast furnace slag decreases the chloride diffusion in the hardened cement pastes, mortars and concrete (Roy 1989). A recent study by Roy *et al.* (2000) has indicated that the addition of alkalis to blended portland slag cement not only increases the strength of the cement, but also decreases the chloride diffusion in the cement paste significantly as shown in Figure 7.3. The other clear trend is that the diffusion rate of chloride decreases with increased slag replacement.

7.2.2.2   Rapid chloride permeability test results

The rapid chloride ion permeability test method – ASTM C 1202 (2003) or AASHTOT 227 (1990) – is being widely used to determine the permeability of a variety of concretes to chloride ions. The test method has been severely criticized by many researchers and scientists in the world because of its severe testing conditions, which may cause both physical and chemical changes (Pfeifer *et al.* 1994, Shi *et al.* 1998). Actually, the rapid chloride ion permeability test is essentially a measurement of electrical conductivity, which depends on both the pore structure and the chemistry of pore

Table 7.1 Comparison of chloride diffusion in portland cement and alkali-activated slag cement pastes (Hakkinen *et al.* 1987)

No.	Paste	Water-to-cement (slag) ratio	Diffusion coefficient (cm²/s)
1	Portland cement	0.23	321×10^{-12}
2	Alkali-activated slag	0.23	75×10^{-12}
3	Portland cement	0.35	6390×10^{-12}
4	Alkali-activated slag	0.35	240×10^{-12}

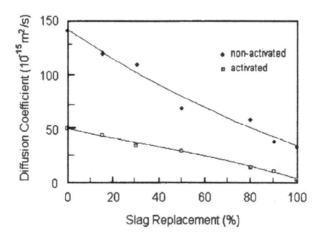

Figure 7.3 Effect of alkali activation on chloride diffusion in portland slag cement (Roy *et al.* 2000).

solution. For a given specimen size and applied voltage, the recorded initial current can be regarded as a representative of electrical conductivity of the specimen. For different concrete mixtures, their electrical conductivity can be significantly affected by the change in pore solution composition. Recently, analyses and calculations based on electrochemical theories indicate that the replacement of portland cement with supplementary cementing materials may decrease the passed charges by more than 90% due to the change in pore solution composition, which has little effect on the transport of chloride ions (Shi *et al.* 1998, Shi 2004a).

Douglas *et al.* (1992) used the rapid chloride permeability test to measure the passed coulombs of six batches of sodium silicate-activated slag cement concrete. They found that the passed charges varied from 1311 to 2547 coulombs at 28 days and 676 to 1831 coulombs at 91 days, which are low and correspond to results obtained from a low water-to-cement ratio normal concrete or concrete incorporating 50% slag as cement replacement with a water-to-cementitious materials ratio of about 0.55. The other significant phenomena is that the passed charges increase with the increase of sodium silicate-to-slag ratio, which indicates that passed charges may be related to the alkali concentration in the concrete pore solution.

Mercury intrusion measurement indicated that Na_2SiO_3-activated slag mortars exhibited much lower porosity and finer pore structure than ASTM type III portland cement mortars, Na_2CO_3-activated and NaOH-activated slag mortars. The results from the rapid chloride permeability test have indicated that PC(III) mortars exhibited a passed charge of about 24,000 coulombs and NS exhibited a passed charge of 21,000 coulombs at three days (Shi 1996b). The charge passed through PC(III) mortars decreased significantly from three to seven days, varied from 12,000 to 15,000 coulombs from Seven to 90 days. The charge passed through NS mortars decreased significantly from 3 to 28 days, and remained almost constant at 5,000 coulombs thereafter. NC and NH mortars exhibited a very narrow range of the passed charge from 3 to 90 days: 2,000 to 3,000 coulombs. Although the pore solutions were not analyzed, it is speculated that the chemistry of the pore solution appears to contribute more to the electrical conductivity or the passed charge than the pore structure does for alkali-activated slag cement mortars and concrete.

7.2.2.3 Chloride binding

It is well known that chlorides, no matter whether added during mixing or later transported into concrete, can be bound to the hydration products of cement in the concrete by physisorption and chemisorption (Tang and Nilsson 1993). Several studies have reported that the chloride-binding capacity of concrete is independent of water-to-cement ratio (Tutti 1982, Tritthart 1989, Tang and Nilsson 1993), but is strongly dependent on the content of C–S–H gel in the concrete (Tang and Nilsson 1993). It is also found that the chloride binding isotherm obeys Freundlich equation at high free chloride concentration ($>0.01\,mol/l$) and Langmuir equation at low concentration ($<0.05\,mol/l$).

The chloride binding is also important when an industrial by-product containing chlorides is used as an activator. Stark *et al.* (2001) and Charchenko *et al.* (2001) tried to use cement kiln dust, which contains high concentration of alkali chlorides and sulphates, as an activator. However, portland cement needs to be added to achieve satisfactory results. The chloride and sulphate ions may cause corrosion of reinforcement in concrete. Free chloride ions have to be converted to calcium chloride and could then be incorporated into the Friedel's salt ($3CaO \cdot Al_2O_3 \cdot CaCl_2 \cdot 10H_2O$) by adding calcium aluminate into the system. The optimum Cl-to-Al molar ratio for the Friedel's salt to form is 1. In order to fully use the alkalis as alkaline activators and to fix all chlorides in the cement kiln dust, the ratio of alkali metal oxides ($K_2O + Na_2O$) content in the cement kiln dust to aluminium oxide (Al_2O_3) content in the slag $\left(\frac{K_2O + Na_2O}{Al_2O_3} \right)$ should be equal to 1.

7.3 Frost resistance

The frost resistance of concrete is of considerable importance in areas where freezing happens. The moisture in concrete will freeze when concrete is cooled to certain temperatures. Due to the capillary effect, water in capillary pores freezes at a lower temperature than bulk water. Kubelka (Powers and Brownyard 1947) derived a relationship between the pore radius and freezing temperature from thermodynamic considerations:

$$\frac{T_o - T}{T_o} = \frac{2v_f \sigma_f}{rq} \qquad (7.1)$$

where
  T_o = normal freezing temperature, 273.15 K;
  T = freezing temperature in a capillary pores with a radius of r;
  v_f = molar volume of water, 18.018×10^{-6} m^3/mol;
  σ_f = surface tension of water, 75.62×10^{-3} J/m;
  q = latent heat of fusion of water, 6017 J/mol.

Figure 7.4 shows the theoretical relationship of Eqn (7.1). Due to the capillary effect, the water in capillary pores freezes at a lower temperature than normal water. A radius of 300 Å corresponds to -4 °C and a radius of 150 Å corresponds to -8 °C.

Unprotected cement pastes dilate as they are frozen, which results in internal tensile stresses and cracking. Several mechanisms (Ramachandran *et al.* 1981, Stark and Wicht 2001) have been proposed to explain the paste behaviour during freezing: generation of hydraulic pressure by ice formation, volume increase (9%) due to the conversion of water to ice, desorption of water from C–S–H, and segregation of ice. Frost destruction of water-saturated portland cement pastes happens mainly during a spasmodic freezing of micro-capillary pore water from the freezing temperature to around -20 °C (Ramachadran *et al.* 1981, Paschenko *et al.* 1991).

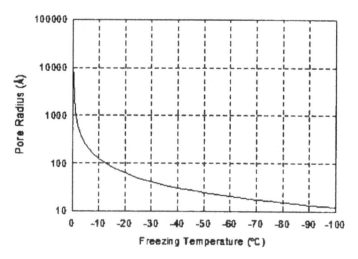

Figure 7.4 Relationship between pore radius and freezing temperature.

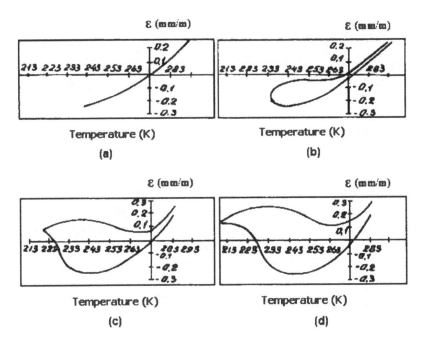

Figure 7.5 Deformations of the alkali-activated slag cement pastes (water-to-slag ratio = 0.3) with moisture content 19.3% during freezing–thawing (a) −243 K or −30 °C; (b) −234 K or −39 °C; (c) −219 K or −54 °C and (d) −212 K or −61 °C).

The deformation of alkali-activated slag cement pastes during freezing, as shown in Figure 7.5, is different from that of portland cement, which may be attributed to different electrolytes in their pore solutions. The solution in capillary pores of portland cement pastes freezes spasmodically and the meniscus disappears, whereas the solution in capillary pores of alkali-activated slag cement paste freezes gradually and the meniscus remains during freezing.

The freezing point of the pore solution is affected considerably by the nature of alkaline activator used. For example, when potassium carbonate is used, the freezing point of the eutectic mixture is 237 K (or −36 °C), while that of sodium carbonate is only 270.9 K (or −2.1 °C). The other reason for the lower freezing point of alkali-activated slag cement pastes, compared with portland cement pastes, could be attributed to the differences in their pore structures. As discussed in Chapter 4, alkali-activated cement pastes contain more gel pores while portland cement pastes contain more capillary pores for a given water-to-cement or -slag ratio. Like portland cement concrete, the frost resistance of alkali-activated slag cement concretes, to a great extent, depends upon the pore solution chemistry, freezing rate and curing conditions.

Activators have a great effect on the frost resistance of alkali-activated slag cement concrete. The sodium silicate-activated slag cement concrete usually has the least porous structure, highest strength and best frost resistance (Glukhovsky 1979, Glukhovsky et al. 1988). Experiment results indicated that the sodium silicate-activated slag cement concrete could resist 300 to 1300 freezing–thawing cycles, depending upon manufacturing process and some other factors. A concrete with proper slag and using soda-alkali melt as an activator was not as good as sodium silicate-activated slag cement concrete, but still could resist 200 to 700 freezing–thawing cycles, which is even superior to portland cement concrete for a given strength.

The nature of the slag does not have any effect on frost resistance of steam-cured alkali-activated slag cement concrete, but the basicity of the slag does have a significant effect on the frost resistance of the concretes cured at room temperatures. The concrete made with acidic slag exhibited the lowest frost resistance (Glukhovsky et al. 1988).

The frost resistance of lightweight alkali-activated slag cement concrete made with porous aggregates such as expanded perlite, expanded clay and shell limestone is better than lightweight portland cement concrete for a given strength grade. This may be attributed to the differences in their structures and interfaces between cement paste and aggregate. Experimental results from Gerasimchuk (1982) indicated that lightweight alkali-activated slag cement concrete had lower porosity than corresponding lightweight portland cement concrete. On the other hand, Gerasimchuk (1982) also noticed that the interfacial zone in alkali-activated slag cement concrete contained different hydration products like calcium silicate hydrate and alkali-earth-alkali aluminosilicate hydrates, which contributed to the less porous structure and better frost resistance of the concrete.

$Na_2O \cdot SiO_2$-activated slag cement concrete, regardless of the basicity of the slag used, also showed much better frost resistance in corrosive salt solutions than Na_2CO_3-activated slag cement concrete and portland cement concrete (Timkovich 1986). It was also noticed that sodium silicate-activated slag cement concrete destructed differently from sodium carbonate-activated slag cement concrete. The former deteriorated very slowly over the surface, while the latter destructed over the whole specimen. Timkovich (1986) contributed the destruction of sodium carbonate-activated slag cement concrete specimen to its more porous structure so corrosive substances could penetrate into the specimen more easily than sodium silicate-activated slag cement concrete. In another study, it was found that no visible deterioration could be observed on $Na_2O \cdot SiO_2$-activated slag cement concrete after 45, 60 and 100 freezing–thawing cycles in 5% NaCl, $MgSO_4$ and $CaCl_2$, respectively (Skurchinskaya and Belitsky 1989). The frost resistance of alkali-activated slag cement concrete tends to decrease with the increase of salt concentrations. However, low concentration salt solution seems to have more obvious destructive influences during the initial period probably due to a quicker penetration of the salt solution into concrete.

Freezing–thawing testing (Hakkinen 1986) indicated that alkali-activated slag cement concrete without air-entrainment could display better resistance to freezing–thawing cycles than portland cement concrete; however, air-entrained alkali-activated slag cement concrete showed more damage than portland cement concrete during the freezing–thawing testing. A salt scaling study (Byfors *et al.* 1989) on F-concrete indicated that the lost mass of concrete during the test was dependent on the water-to-slag ratio rather than the air content in the concrete.

Pu *et al.* (1991) conducted more than 200 freezing–thawing cycles on alkali-activated slag cement concrete and could not measure noticeable mass loss from the specimens. Douglas *et al.* (1992) measured changes in relative dynamic modulus of elasticity, resonant frequency, length, pulse velocity and mass change of alkali-activated slag cement concrete during the ASTM C 666 (2003) freezing–thawing test. According to the ASTM C 666, the test should be discontinued when the relative dynamic modulus of elasticity reaches 60% of the initial value. For the optional length change test, a 0.10% expansion may be used as the end of the test. Experimental results indicated that four batches displayed changes well below the specified values even after 500 cycles, but one batch failed before 300 cycles. Measurement of void spacing factor in alkali-activated slag cement concrete indicated that it ranged between 0.27 and 0.37 mm, higher than the limit of 0.22 mm for concrete resistant to freezing–thawing cycles (Douglas *et al.* 1992). Nevertheless, four batches of alkali-activated slag cement concrete still showed good resistance to freezing–thawing cycles in spite of high void spacing factor. The failed batch has the highest void spacing factor of 0.37 mm, which probably explained the lower resistance to freezing–thawing cycles. Actually, they observed that air could be entrained in those alkali-activated slag cement concretes easily. However, a comparison between fresh and hardened concrete indicated that the air voids in fresh alkali-activated slag cement concretes are not stable and some were lost with time.

Gifford and Gillott (1996a) found that commercially available air entrainment agents can be used for alkali-activated slag cement concrete, but a higher dosage was required in order to get the same air content as in ordinary portland cement concrete. Freezing–thawing durability of alkali-activated slag cement concrete should be at least as good as ordinary portland cement concrete for given adequate air void parameters.

7.4   Acid attack

Hydrated cement paste is an alkaline material and can be attacked by acidic solutions very easily. The acid corrosion of hardened cementing materials has drawn more and more attention recently due to the corrosion of

concrete sewer pipes and concrete structures in municipal wastewater treatment plants (Flemming 1995, Synder *et al.* 1996), the impact of aggressive
substances from animal feed and manure (Belie *et al.* 1996) and concerns
regarding the acid corrosion resistance of cement-solidified wastes (Stegemann *et al.* 1997).

Figures 7.6 and 7.7 show the corroded depth of cement pastes with time in
pH 3 nitric and acetic acid solutions (Shi and Stegemann 2000, Shi 2003a).

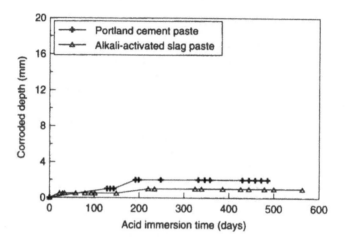

Figure 7.6 Corrosion of cement pastes in pH 3 nitric acid solution (Shi 2003a).

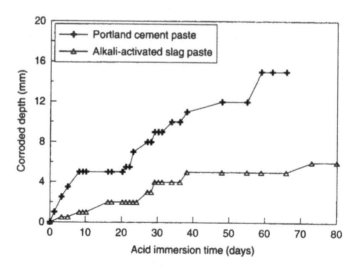

Figure 7.7 Corrosion of cement pastes in pH 3 acetic acid solution (Shi 2003a).

Alkali-activated slag cement showed much less corroded depths than the portland cement pastes.

After 580 days of immersion in pH 3 nitric acid solutions, the portland cement pastes were corroded approximately 2.5 mm, while the alkali-activated slag cement pastes were corroded only about 1.3 mm. High aluminum cement is well known for its good resistance to acid corrosion. Alkali-activated cement can even show much better corrosion resistance in HCl and H_2SO_4 solutions than high aluminum cement (Davidovits 1994).

It appears that as acetic acid is much more aggressive than nitric acid the difference in acid corrosion resistance became more obvious. After 60 days of immersion, 15 mm of portland cement but only 5 mm of alkali-activated slag cement pastes were corroded. This is further confirmed by Bakharev et al. (2003). However, it has been reported that, at the same concentration, mineral acids are more corrosive to hardened cement pastes than weak acids (Pavlik 1994). The contradiction can be attributed to the different testing conditions–constant pH was used in this study and a constant concentration was used by Pavlik (1994).

At room temperature, a fully hydrated PC paste consists of 50–60% C–S–H with a high Ca/Si ratio of 1.5 to 1.8, 20–25% $Ca(OH)_2$ and 15–20% calcium sulphoaluminates (AFt and AFm) by volume, while ASC pastes consist mainly of C–S–H with a Ca/Si ratio of around 1 (Wang and Scrivener 1993). $Ca(OH)_2$ decomposes at a pH below 12 and calcium sulphoaluminates decompose at a pH below 11. C–S–H releases Ca^{2+} as the pH drops. When the pH is below 9, C–S–H has released most lime, and has left a layer of silica and aluminosilicate gels that prevents uncorroded cement pastes from further corrosion. Further leaching of calcium and inward movement of acid to the corrosion front then becomes controlled by diffusion through this layer (Pavlik 1994). The dissolution of $Ca(OH)_2$ and calcium sulphoaluminates, and the decalcification of C–S–H with a high Ca/Si ratio in hardened portland cement pastes leave a very porous corroded layer, while the low lime content in activated-blast furnace slag results in a dense silica gel protective layer. It was noticed that for a given activator, the lower the basicity of the slag, the better the acid corrosion resistance of the alkali-activated slag cement (Pu et al. 1999).

The other factor that affects the acid resistance of cement is the acid consumption for the complete matrix destruction of the cementing material. Table 7.2 lists the required amount of different acid solutions to destruct portland cement and alkali-activated slag cement. Alkali-activated slag cement has a lower amount of consumed acid in pH 5 CH_3COOH solutions, but shows a much higher amount of consumed acid in pH 3 HNO_3 and H_3COOH solutions. This means that alkali-activated slag cement has a much higher acid neutralization capacity at lower pH values.

Table 7.2 Amount of acid required for complete matrix destruction (Shi and Stegemann 2000)

Corrosion media	Consumed acid (mmol of acid/g of dry hardened paste)	
	Portland cement	Alkali-activated slag cement
pH 3 HNO₃	19	25
pH 3 CH₃COOH	160	360
pH 5 CH₃COOH	21	12

Results from several studies (Iler 1979) have indicated that the solubility of amorphous silica at room temperature is almost constant, varying from 100 to 150 ppm, when pH is below 9, then it increases rapidly when pH is above 9, as shown in Figure 7.8. Thus, the silica gel left after acid corrosion should be as much as possible in order to increase the corrosion resistance of the cement.

A partial replacement of blast furnace slag with steel slag does not show an obvious effect on the acid corrosion resistance of the material (Shi 1999). However, a partial replacement of blast furnace slag with low calcium fly ash can significantly increase its acid corrosion resistance (Blaakmeer 1994). The strength of the cement decreases with the increase of fly ash replacement, but the corrosion resistance increases with the increase in the fly ash replacement (Blaakmeer 1994). This further confirms that acid resistance of

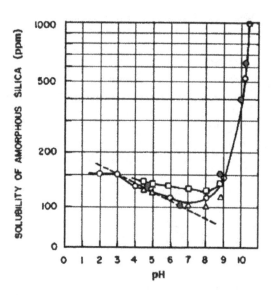

Figure 7.8 Effect of pH on solubility of amorphous silica (Iler 1979).

cement concrete materials is determined mainly by the nature of hydration products.

The introduction of dehydrated magnesium silicate into alkali-activated slag cement could enhance the corrosion resistance of the hardened cement pastes by 1.5–2 times (Krivenko *et al.* 1991b, Brodko 1992). The corrosion resistance was evaluated based on the ratio of flexural strength of specimens in a boiling solution of H_2SO_4 for 50 hours to the flexural strength of the specimens immersed in water. The enhanced acid resistance by adding additives may be attributed to the acceleration of the binding alkali into insoluble compounds, synthesis in the hydration products of the secondary products of serpentinite formation: antigorite, serpentine, amphibole asbestos minerals of the anthophyllite and actinolite types that are known to possess high resistance in sulfuric acid solution. The composition with the highest corrosion resistance (corresponding to a coefficient of corrosion resistance = 1.2) is composed of 14–27.5% magnesium silicate additive, 48.8–80% dehydrated serpentinite and 20–52.2% dehydrated talc. With increasing quantities of serpentinite, the range of compositions with coefficient of corrosion resistance = 1.2 becomes larger; however, its strength falls. Long-term tests by the Japanese company "Nippon Steel Corporation" confirmed high corrosion resistance of alkali-activated slag cement concrete (Table 7.3).

The addition of NaA, NaX and NaY zeolites also increases the acid corrosion resistance of alkali-activated slag cements using waterglass with silicate modulus (Ms) = 2.8 and density = 1300 kg/m³. Cement with the best acid resistance contains 30% NaA or 20% NaX zeolites, which give coefficients of corrosion resistance = 1.22 and 0.98 respectively.

Zeolites are classified into 7 groups and each group contains similar structures connecting the tetrahedra (Si, Al)O_4 (Breck 1974). Table 7.4 lists several natural zeolites with SiO_2/Al_2O_3 ratios of 3:4, 4:5, and 4:10 used for investigation. The acid resistance of the cement with dehydrated high-silica zeolite (Group III) consisting predominantly of mordenite showed the highest coefficient of acid resistance of 1.3–1.56, as shown in Figure 7.9,

Table 7.3 Coefficient of corrosion resistance of alkali-activated slag cement concrete in sulphuric acid solution with a concentration of 10 g/l

Cement	Coefficient of corrosion resistance after immersion in H_2SO_4 different period of time	
	6 months	12 months
Alkali-activated slag cement	0.88	1.18
Sulphate resistant portland cement	0.65	0.69
Quartz silicofluoride acid resistant cement	0.80	1.00

Table 7.4 Mineralogical composition of zeolite rocks

Type	Mineral	Major minerals	Minor minerals	Classification group number
I	Chabazite $Ca_2(Al_4Si_8O_{24})12H_2O$	+		4
	Natrolite $Na_{16}(Al_{16}Si_{24}O_{80})16H_2O$	+		5
	Stilbite $NaCa_2(Al_5Si_{13}O_{36})14H_2O$		+	7
	Mordenite $Na_8(Al_8Si_{40}O_{96})24H_2O$		+	6
II	Chabazite $Ca_2(Al_4Si_8O_{24})12H_2O$	+		4
	Stilbite $NaCa_2(Al_5Si_{13}O_{36})14H_2O$	+		7
	Clinoptilolite $Na_6(Al_6Si_{30}O_{72})24H_2O$		+	4
	Analcime $Na(AlSi_2O_6)H_2O$		+	I
	Natrolite $Na_{16}(Al_{16}Si_{24}O_{80})16H_2O$		+	5
	Mordenite $Na_8(Al_8Si_{40}O_{96})24H_2O$		+	6
III	Chabazite $Ca_2(Al_4Si_8O_{24})12H_2O$	+		4
	Mordenite $Na_8(Al_8Si_{40}O_{96})24H_2O$	+		6
	Analcime $Na(AlSi_2O_6)H_2O$		+	I

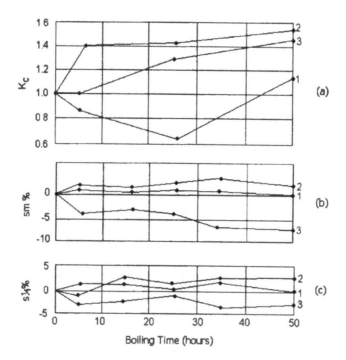

Figure 7.9 Change in (a) coefficient of corrosion resistance, (b) mass and (c) volume of the specimens made from a alkali-activated slag cement containing different amounts of dehydrated group III natural zeolite (1–10%; 2–20%; 3–30% by mass).

which can be attributed to structural characteristics of the mordenite that can incorporate sulphate ions.

7.5 Chloride salts

It is well known that a high concentration of $CaCl_2$ solution deteriorates portland cement concrete. The chloride deterioration mechanism of cement concrete is attributed to the formation of $CaO \cdot 3CaCl_2 \cdot 12H_2O$, which substantially increases the volume and causes expansion of the concrete (Monosi and Collepardi 1990).

Deja and Malolepszy (1989, 1994) immersed NaOH-activated copper slag and portland cement mortars in water and in a corrosive solution containing 230 g/l NaCl, 64 g/l $MgCl_2$, 15 g/l KCl and 14 g/l $MgSO_4$, The changes in compressive strength of these cement mortars with time are shown in Figure 7.10. The strength of the NaOH-activated copper slag mortars slightly increases with time regardless of whether it is immersed in water or in the corrosive solution. However, the strength of portland cement mortars immersed in water slightly increases with time and decreases with time when immersed in the corrosive solution. This means that the NaOH-activated copper slag mortars have better corrosion resistance than portland cement mortars. Kurdowski *et al.* (1994) found that Na_2CO_3-activated blast furnace slag cement also showed excellent resistance to the similar corrosive

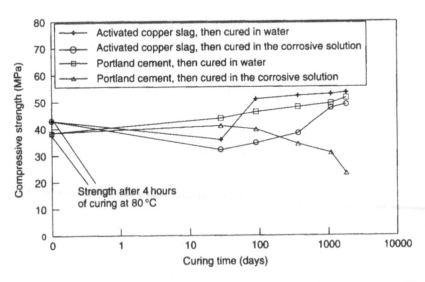

Figure 7.10 Strength of alkali-activated copper slag and portland cement mortars (Data from Deja and Malolepszy 1989, 1994).

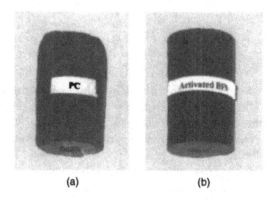

(a)   (b)

Figure 7.11 Appearance of portland cement mortar (PC) and alkali-activated slag mortar after 75 days in 30% CaCl₂ solution (a) portland cement paste; (b) alkali-activated slag paste) (Shi 2003a).

solution and even no visual trace of corrosion could be observed after 7 years of immersion in the solution.

In another study (Shi 2003a), alkali-activated slag and portland cement mortars, after 28 days of curing in a moisture chamber, were suspended in a 30% CaCl₂ solution. Both the corroded depth and mass change during the testing were measured. For a given immersion period of time, the corroded depth of portland cement mortars is much greater than that of alkali-activated slag cement mortars. Figure 7.11 shows the appearance of portland cement mortar and alkali-activated slag mortars after 75 days in 30% CaCl₂ solution. Visual observation during the testing indicated that small cracks occurred first along the corners of the portland cement mortar sample, followed by swelling and spalling of the surface, while alkali-activated slag mortar sample demonstrate excellent resistance to CaCl₂ attack.

7.6 Alkali corrosion resistance

Although hydrated cement has a pH usually higher than 13.5, hydration products can still be disintegrated in strong alkali solutions. Figure 7.12 shows the strength of portland cement and alkali-activated slag cement pastes in 5% NaOH and other corrosive solutions. The reference samples were immersed in water. It can be seen that strength of portland cement pastes decreases significantly after immersion in the NaOH and the other corrosive solutions, but alkali-activated slag cement pastes, after immersion in the NaOH and the other corrosive solutions, shows basically the same strength when immersed in water.

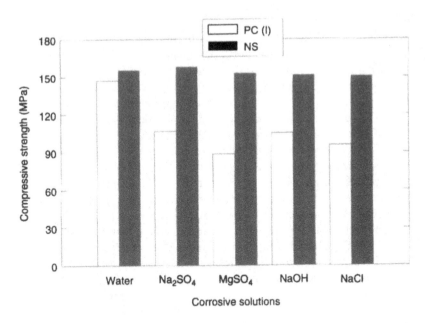

Figure 7.12 Strength of portland cement and alkali-activated slag cement pastes in various 5% corrosive solutions (data from Deng *et al.* 1989).

7.7 Sulphate attack

Sulphate attack refers to the deterioration of concrete resulting from chemical reactions occurring when concrete is exposed to a solution containing a sufficiently high concentration of dissolved sulphates. This is particularly prevalent in arid regions where naturally occurring sulphate minerals are present in the water and groundwater. Deterioration of concrete due to sulphate attack is generally attributed to reactions of portland cement hydration products with sulphates to form expansive reaction products after hardening, which produces internal stress and a subsequent disruption of the concrete, or to form products having little cementing value, and thereby turn the concrete into mush. The detailed discussions on sulphate attack of portland cement concrete and deterioration mechanisms of the concrete can be found in reference books (Mindess and Young 1981).

Douglas *et al.* (1992) reported that the changes in dynamic modulus of elasticity, pulse velocity, weight and length of sodium silicate-activated slag cement concrete after 120 days of immersion in 5% sodium sulphate solutions. They noticed that the changes are even smaller than those observed from the controlled specimens immersed in lime-saturated water or tap water for the time period.

Hakkinen (1986, 1987) evaluated the sulphate resistance of both alkali-activated slag cement and portland cement in (1) 1% Na_2SO_4 solution, (2) 10% Na_2SO_4 solution, (3) 1% $MgSO_4$ solution and (4) 10% $MgSO_4$ solution by measuring their compressive and flexural strengths. All specimens did not show any corrosion after two years of immersion in 1% Na_2SO_4 solution. The portland cement samples were destroyed in 10% Na_2SO_4 solution, while the alkali-activated slag cement samples survived well. After about two years of immersion in 1% $MgSO_4$ solution, those portland cement specimens lost about 25% strength compared with the strength of those cured in water, but the alkali-activated slag samples did not show any strength decrease. All specimens were destroyed after about one year of immersion in 10% $MgSO_4$ solution. Actually, Na_2SO_4 is also a good activator for slag and other cementing components. Thus, it is not surprising that alkali-activated slag cement can behave well in concentrated Na_2SO_4 solutions. In another study (Bakharev et al. 2002), it was noticed that alkali-activated slag concrete lost less strength than portland cement concrete after immersion in 5% Na_2SO_4 and $MgSO_4$ solutions. The difference was more obvious when they were immersed in the $MgSO_4$ solutions.

Krivenko (1986) investigated the corrosion resistance of alkali-activated slag cement concretes, made with different types of slag and activator, in different sulphate solutions by measuring the flexural strength of concrete specimens (Table 7.5). Regular portland cement and sulphate-resistant cement were used as references. The results indicated that corrosion resistance of the concrete depends on both the nature and concentration of the sulphate salt and the composition of the alkali-activated slag cement. In the sodium sulphate solutions, alkali-activated slag cement concrete behaved very well regardless of cement composition. This may be attributed to the lack of condition of formation of other products deteriorating the structure of the concrete. Ammonium and those bivalent or trivalent metal sulphates such as magnesium, manganese, aluminium, nickel, copper, zinc, etc. showed severe corrosion on the concrete. The immersion of alkali-activated slag cement concrete in those sulphate solutions would result in the formation of hydrosilicate, hydroxides and $Ca_2SO_4 \cdot 2H_2O$, which do not have any cementing property. The corrosion of concrete in sulphate solutions increased with the nature of cation in the following order: $Na^+ > Zn^{2+} > Cu^{2+} > Ni^{2+} > Al^{3+} > NH^{4+} > Mg^{2+} > Mn^{2+}$.

Regardless of its composition, alkali-activated slag cement concrete is superior to regular portland cement concrete in all these sulphate solutions, and even superior to sulphate-resistant portland cement concrete in sodium sulphate solutions. In other sulphate solutions, alkali-activated slag cement concrete made with acidic slag and high modulus waterglass

Table 7.5 Variations of flexural strength of the alkali-activated slag cement concrete specimens (10 mm × 10 mm × 60 mm) in different 6% (by mass) sulphate solutions (Krivenko 1986)

Sulphate	Alkaline activator	Flexural strength, MPa				
		Slag alkali-activated cement concrete			Reference concrete	
		Basic slag $Mb = 1.18$	Neutral slag $Mb = 1.05$	Acidic slag $Mb = 0.75$	Sulphate resistant portland cement	Portland cement
		Duration of testing, months				
		3/6	3/6	3/6	3/6	3/6
Na_2SO_4	Na_2CO_3	11.23/11.44	–	12.01/13.14	8.76/7.67	6.34/3.71
	Na_2SiO_3	16.23/17.58	10.11/14.66	18.25/19.87		
$ZnSO_4$	Na_2CO_3	11.92/9.40	–	9.06/2.18	9.43/13.20	6.88/1.34
	Na_2SiO_3	17.02/17.55	8.07/13.33	19.84/20.39		
$CuSO_4$	Na_2CO_3	2.04/2.81	–	10.76/5.22	10.58/11.92	7.23/2.16
	Na_2SiO_3	17.14/2.01	8.63/4.67	16.55/16.74		
$NiSO_4$	Na_2CO_3	6.80/0.00	–	8.26/6.46	9.01/11.11	6.50/1.16
	Na_2SiO_3	14.85/15.41	9.66/9.17	17.38/12.40		
$Al_2(SO4)_3$	Na_2CO_3	0	–	6.19/9.38	8.21/7.39	0
	Na_2SiO_3	7.82/6.65	2.73/2.50	7.82/6.65		
$(NH_4)_2SO_4$	Na_2CO_3	0	–	0	8.98/6.92	0
	Na_2SiO_3	2.10/0.00	4.76/2.27	0		
$MgSO_4$	Na_2CO_3	0	–	0	9.63/8.83	0
	Na_2SiO_3	0	9.32/13.54	0		
$MnSO_4$	Na_2CO_3	0	–	0	9.06/6.07	0
	Na_2SiO_3	0	13.17/9.91	0		

solutions is similar to or even better than the sulphate-resistant portland cement concrete Krivenko (1986). The deterioration of alkali-activated slag cement in $MgSO_4$ solution is similar to that of regular portland cement pastes: strengthening, crack appearance, predominantly perpendicular to longitudinal axis of the specimens, bending of the specimens, increase in linear dimensions and volume, failure.

The nature of activator also has a great effect on the corrosion resistance of the alkali-activated slag cements. For a given density, the corrosion resistance of the cement increased with activator in the following order: $NaOH < Na_2CO_3 \approx Na_2SiO_3 < Na_2Si_2O_5 < Na_2O \cdot 2 \cdot 65SiO_2$. On the other hand, the quantity of alkali based on unit mass of slag decreased in that order. The reduction of alkali content seems to increase corrosion resistance of alkali-activated slag cement in $MgSO_4$ solution. The alkali-activated slag cements made with acidic slags and sodium silicate showed similar resistance to bivalent or trivalent metal sulphates as sulphate-resistant portland cement (Table 7.6).

The (alkaline activator) solution-to-slag ratio has a significant effect on the corrosion resistance in $MgSO_4$ solutions. An increase in the solution-to-slag ratio decreased the corrosion resistance of the specimens although it also increased the dosage of the alkali activator in the cementing system. This means the negative effects due to the increase of water-to-slag ratio surpassed the positive effects due to increase in the dosage of the alkali activator. The steam curing of the specimens made with waterglass with a modulus from 1 to 3 decreased the corrosion resistance compared to the specimens cured under normal conditions. The use of non-silicate alkaline

Table 7.6 Resistance of the alkali-activated slag cement concrete in magnesium sulphate solutions (Krivenko 1986)

Slag (basicity)	Alkaline activator	Resistance in $MgSO_4$ solutions of							
		5% Concentration				6% Concentration			
						Age (in months)			
		1	*3*	*6*	*12*	*1*	*3*	*6*	*12*
Blastfurnace	NaOH	1.48	0.87	0.00	–	1.76	0.00	–	–
(Mb = 1.18)	Na_2CO_3	1.01	0.97	0.00	–	1.20	0.07	0.00	–
	$Na_2O \cdot SiO_2$	1.16	0.77	0.45	0.00	1.53	0.00	–	–
	$Na_2O \cdot 2SiO_2$	1.24	1.57	1.34	0.00	1.24	2.05	0.00	–
	$Na_2O \cdot 3SiO_2$	1.70	2.17	1.88	0.00	2.97	3.00	0.00	–
Electrothermo-	NaOH	1.19	1.09	0.88	0.00	1.29	0.77	0.27	0.00
phosphorous	Na_2CO_3	2.42	1.00	1.00	1.00	3.60	1.00	1.00	1.00
(Mb = 1.00)	$Na_2O \cdot SiO_2$	1.05	0.88	1.00	0.75	1.21	1.11	0.75	0.00
	$Na_2O \cdot 2SiO_2$	1.58	3.02	3.26	3.51	2.58	4.41	3.59	3.33
	$Na_2O \cdot 3SiO_2$	4.65	7.17	14.4	6.63	1.88	25.9	35.90	9.26
Blastfurnace	NaOH	1.36	0.91	0.98	0.00	1.47	0.00	–	–
(Mb = 1.15)	Na_2CO_3	1.08	0.96	0.85	0.00	1.10	0.38	0.00	–
	$Na_2O \cdot SiO_2$	1.04	0.76	0.81	0.00	1.24	0.41	0.00	–
	$Na_2O \cdot 2SiO_2$	1.15	1.41	1.22	0.91	1.75	2.19	1.95	0.47
	$Na_2O \cdot 3SiO_2$	1.28	2.52	3.48	1.53	1.57	3.32	2.96	1.07
Blastfurnace	NaOH	1.02	1.09	0.63	0.00	1.15	0.00	–	–
(Mb = 0.78)	Na_2CO_3	1.10	0.74	0.65	0.00	1.18	0.42	0.29	0.00
	$Na_2O \cdot SiO_2$	1.01	1.04	1.24	0.00	1.49	0.43	0.27	0.00
	$Na_2O \cdot 2SiO_2$	1.37	1.28	3.60	1.35	1.51	1.40	2.87	1.07
	$Na_2O \cdot 3SiO_2$	2.97	4.49	9.90	3.09	3.62	9.05	13.96	2.50

activators increased the corrosion resistance of the alkali-activated slag cement in the $MgSO_4$ solution.

When the specimens are half-immersed in solutions, the resistance of alkali-activated slag cement depends upon the nature of alkaline activators and decreases in the following order: $(85\%NaOH + 15\%KOH) < Na_2O \cdot SiO_2 < Na_2O \cdot 2SiO_2$, which is attributed to a different rate of flow of the condensation-crystallization processes taking place during formation of the alkali-activated slag cement stone structure. The resistance of the alkali-activated slag cement increases with the increase of the modulus of waterglass. This is attributed to the introduction of colloidal silica from waterglass. Some mix designs and curing conditions for alkali-activated slag cement concrete were recommended for use in different corrosive environments, as shown in Table 7.7.

Table 7.7 The recommended mix design of the alkali-activated slag cement concretes for resistance to different sulphates (Krivenko 1986)

Aggressive chemical		Cement compositions			Recommended curing conditions
Salt	Content, kg/m³	Slag	Alkaline activator		
			Name	Density (kg/m³)	
Sodium sulphate, magnesium chloride and nitrate	Up to 80	All types of blastfurnace and electrothermo-phosphorous	any	Not limited	Steam curing and normal conditions
Manganese, magnesium, aluminium, and ammonium sulphates	Up to 7	Blastfurnace acid and electrothermo-phosphorous	Sodium silicates (Ms >= 2)	No limit	Steam curing and normal conditions
	Up to 9	"	"	Up to 1200	
	Up to 13	"	"	Up to 1200	Normal conditions
Nickel, copper, and zinc sulphates	Up to 15	"		Not limited	Steam curing and normal conditions
	Up to 20	"		Not limited	Steam curing and normal conditions

7.8   Corrosion resistance in liquid organics

Durability of alkali-activated slag cement concrete in liquid organics is determined by the composition of cement, characteristics of pore structures of the concrete and the nature of the organic fluids (Gontcharov 1984). Results in Table 7.8 indicate that alkali-activated slag cement concrete shows much better corrosion resistance than portland cement concrete after two years of immersion in some liquid organics such as kerosene, diesel fuel, mineral oil, petroleum. Alkali-activated slag and portland cement concretes showed similar corrosion resistance in low-molecular organics such as glycerine, acetic and milk acids.

Large molecule liquid organics are generally much less aggressive to alkali-activated slag cement concrete than to portland cement concrete. After two years of immersion in mineral oil, animal fat and 30% sugar solution, it was found that the corrosion resistance of alkali-activated slag cement concrete was more than three times higher than that of portland cement concrete except the alkali-activated slag cement concrete made from acid blast furnace slag and sodium waterglass with a modulus of 2.5 (Table 7.9). Alkaline activators play an important role in determining the corrosion resistance of alkali-activated slag cement concrete in aggressive gasoline media and increased the corrosion resistance in the following order: $NaOH > 90\%Na_2CO_3 + 10\%NaOH > Na_2CO_3 > Na_2O \cdot SiO_2 > Na_2O \cdot 2SiO_2$.

Table 7.8 Corrosion resistance of alkali-activated slag cement concrete in various liquid organics (Gontcharov 1984)

Liquid organics	pH-value	Titratable acidity, mol/l	Coefficient of resistance after 720 days of immersion	
			Alkali-activated slag cement	Portland cement
Gasoline	–	0.0005	0.98–0.99	0.98–0.99
Benzol	–	0.0007	0.96–1.00	0.70–0.90
Kerosene	–	0.0008	0.92–0.99	0.60–0.78
Mineral oil	–	0.0011	0.64–0.96	0.50–0.70
Diesel fuel	–	0.0009	0.72–0.94	0.50–0.67
Sulphur petroleum	–	0.0013	0.5–0.97	0.56–0.40
Animal (Pig) fat	–	0.0500	0.56–0.97	Destroyed
Sugar solution (30% concent.)	–	0.0100	0.68–1.18	0.30–0.64
Salt brine of meat-packing plant	–	–	0.68–1.18	0.30–0.64
Acetic acid (10% concent.)	2.8	–	0.25–0.45	0.15–0.24
Milk acid (10% concent.)	3.45	–	0.3–0.79	0.20–0.35

Table 7.9 Effect of cement composition, testing time and corrosive media on corrosion resistance of alkali-activated slag cement concrete

Cement composition		Coefficient of resistance						
Blast furnace slag	Activator	Mineral oil		Animal (pig) fat			30% sugar solution	
		1 year	2 years	1 year	2 years	3 years	1 year	2 years
		Alkali-activated slag cement concrete						
Basic	Na_2CO_3	0.89	0.85	0.72	0.70	0.75	0.72	0.72
Mb = 1.18	$Na_2O \cdot SiO_2$	0.92	0.90	0.76	0.72	0.74	0.87	0.85
	$Na_2O \cdot 2 \cdot 5SiO_2$	1.02	1.00	0.77	0.74	0.76	0.90	0.87
Neutral	Na_2CO_3	0.95	0.90	0.85	0.80	0.77	0.75	0.71
Mb = 1.05	$Na_2O \cdot SiO_2$	0.97	0.95	0.88	0.84	0.75	0.96	0.92
	$Na_2O \cdot 2 \cdot 5SiO_2$	0.95	0.93	0.90	0.88	0.80	0.95	0.89
Acidic	Na_2CO_3	1.03	0.99	0.99	0.85	0.80	1.02	0.99
Mb = 0.78	$Na_2O \cdot SiO_2$	1.02	1.02	0.98	0.96	0.78	0.98	0.98
	$Na_2O \cdot 2.5SiO_2$	0.60	0.62	0.96	0.95	0.79	1.06	1.03
Sulphate resistant portland cement concrete (M400)(Reference)								
		0.93	0.72	0.36	Destroyed		0.71	0.48

7.9 Alkali–aggregate reaction

Stanton first observed the expansion and cracking of concrete caused by alkali–aggregate reaction in the late 1930s. There are two types of alkali–aggregate reaction which can lead to cracking of concrete: (a) alkali–silica reaction and (b) alkali–carbonate reaction. Alkali–silica reaction is a chemical reaction within concrete between specific siliceous constituents, which sometimes occur in the aggregate and the alkalis in the concrete mixture. The reaction product is alkali silicate gel that swells after absorbing moisture and causes expansion and cracking of concrete. Alkali–carbonate reaction is an expansive de-dolomitization process resulting from the reaction between dolomite in carbonate aggregate and alkalis in the concrete mixture.

A high concentration of alkalis in the concrete mixture is one of the critical conditions for alkali–aggregate reaction to happen. Usually, 3–5% Na_2O based on the mass of slag is used in alkali-activated slag cement and concrete. There is therefore a concern about alkali–aggregate reaction in alkali-activated slag cement concrete when alkali reactive aggregate is used.

It is generally agreed that the replacement of portland cement with a proper amount of blast furnace slag can reduce or eliminate alkali–aggregate reaction in the portland blast furnace slag cement concrete. (Hogan 1985,

Duchesene and Berube 1994, Thomas and Innis, 1998). The incorporation of ground granulated blast furnace slag into portland cement results in the formation of C–S–H with a low Ca/Si ratio that can bind more alkali ions than the C–S–H with a high Ca/Si ratio. Also, use of slag decreases the permeability of the concrete and reduces the mobility of alkalis in the concrete.

German Cement Association investigated 40 portland and blast furnace slag cements and proposed a relationship between the effective alkali content A_{tot}, minimum slag content H and the total alkali content A_{tot}, over which the cement will cause expansion due to alkali–aggregate reaction (Smolczyk 1980):

$$A_{eff} = A_{tot} \left[1 - 1.8 \left(\frac{H}{100} \right)^2 \right]$$

(7.2)

Figure 7.13 is the plot of the Eqn (7.2) to produce an effective alkali content of 0.6. It can be seen that the total alkali content increases drastically as the slag content reaches over 60%. It allows a total alkali content of around 6% when the slag content is 70%.

Recently, Thomas and Innis (1998) tested four alkali–silica reactive aggregates, with slag replacement of portland cement level from 25 to 65%. They found that their test results did not fit the Eqn (7.3) well, but could be better described by the following Equation:

$$A_{eff} = A_{tot} \left(1 - \frac{H}{100} \right)$$

(7.3)

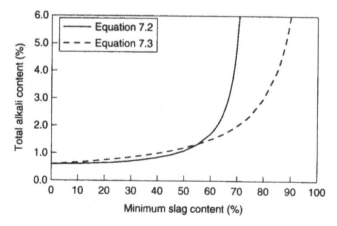

Figure 7.13 Effect of slag content on total alkali content for alkali–aggregate reaction.

Equation 7.3 is also plotted in Figure 7.13. It can be seen that there is no significant difference between the two equations when the slag level is below 55%, but the difference increases with slag content when the slag content is over 55%. Based on Eqn 7.3, it allows a total alkali content of 6% when the slag content is 90%.

The alkalis in alkali-activated slag pastes can exist in several forms (Shi 2003b): (1) incorporated into C–S–H; (2) physically adsorbed on the surface of hydration products and (3) free in pore solution. C–S–H with low Ca/Si ratio has a higher uptake of R^+ than C–S–H with high Ca/Si. Thus C–S–H formed in alkali-activated slag pastes can bind more R^+ than that formed in PC pastes. It can therefore be expected that the free R^+ available for alkali–aggregate reaction is less than that originally assumed. Actually, the alkali–aggregate reaction expansion of alkali-activated cement will be dependent on the nature of the cementing components and activator(s) used, the dosage of the activator(s) and the alkali-reactive aggregate.

Mesto (1982) measured the alkali–aggregate reaction expansion of alkali-activated slag cement using mortar bars containing opal as reactive aggregate. He noticed that the expansion of alkali-activated slag cement depended on the nature of the slag used and the opal content, and observed a maximum expansion when the opal content was about 5%. In another study, Hakkinen (1986) reported the alkali–aggregate expansion of alkali-activated cement increased as its alkali content increased from 2.2% to 5.3% for an opal content of 8%. When the alkali content of F-concrete is about 2% based on the mass of the slag (a typical alkali content for F-concrete), its expansion is not higher than a conventional portland cement. However, the expansion also increased with opal content from 0, 3, 8 to 15% for a given alkali content of 2.4%. It is obvious that the expansion is associated with the pessimum SiO_2/Na_2O ratio of alkali–silica reaction gel. Zhang and Groves (1990) found that the pessimum SiO_2/Na_2O ratio was around 12–13, while Hobbs (1988) noticed that the pessimum ratio varied with the aggregate-to-cement ratio.

Shi (1988) used the rapid autoclave method (Tang *et al.* 1989) to measure the expansion of alkali-activated portland phosphorus slag cement. Although the expansion of the two alkali-activated slag cements do not exceed the limit value 0.1%. SEM observation noticed that most opal is dissolved and many holes are observed. Pu and Yang (1994) used the same method to examine the expansion and microstructure of alkali–silica reaction in alkali-activated slag cement. They also found that alkali–silica reaction took place in alkali-slag cement concrete when reactive aggregate is contained, but they noticed that the expansion differed from activator to activator. They suggested that no destructive expansion would happen when NaOH is used as an activator and the reactive aggregate content is less than 15%, while the maximum reactive aggregate content can be 50% when Na_2CO_3 or $Na_2O \cdot nSiO_2$ is used as an activator.

The autoclave method does not seem suitable for evaluating AAR in alkali-activated slag cement because the AAR reaction occurs before the activation of slags. Also, because alkali-activated cement contains a high content of alkali, even much higher than high alkali portland cement, the common mix proportions used to evaluate potential expansion are not suitable for alkali-activated cements.

Gifford and Gillott (1996b) examined the dimensional change of sodium silicate- or sodium carbonate-activated slag cement concrete incorporating either an innocuous control aggregate, an aggregate that is known to be reactive by way of alkali–silica reaction or alkali–carbonate reaction in portland cement concrete, or an aggregate that contains reactive silica. The equivalent alkali content in the reference portland cement was 1.2% and was 6% for the alkali-activated slag cement. It was found that after one year of storage, portland cement concrete containing alkali-reactive carbonate aggregate lead to excessive expansion well beyond the recommended maximum value of 0.04%. Alkali-activated slag cement concrete containing this type of aggregate not only displayed nominally twice the expansion than that of the portland cement concrete, but exhibited excessive cracking within a few weeks of storage. In all cases of alkali–silica reaction, alkali-activated slag cement concrete exhibited much less expansion than those displayed by the portland cement concrete. They concluded that alkali-activated slag cement concrete is more susceptible to deleterious expansion due to alkali–carbonate reaction and less susceptible to deleterious expansion due to alkali–silica reaction.

Bakharev *et al.* (2001a) measured the expansion of ordinary portland cement concrete and alkali-activated slag cement concrete prisms containing alkali-reactive aggregate following the ASTM C 1293 testing procedure. The activator used was sodium silicate with a mass ratio $SiO_2/Na_2O = 0.75$ and its dosage was 4% of Na, based on the mass of slag. The alkali-activated slag cement concrete prisms showed an expansion of 0.045% after 12 months of curing, which is higher than the limit of 0.04%, and they continued to expand to 0.1% at 22 months; while, the portland cement concrete prisms showed an expansion of about 0.02% at 12 months and 0.03% at 22 months. In another study (Fernandez-Jimenez and Puertas 2002), the accelerated mortar bar test following ASTM C 1260 indicated that NaOH-activated mortars with 4% Na_2O expanded much less than portland cement mortars even up to 140 days.

Several researches (Pu and Yang 1994, Yang 1997, Yang *et al.* 1999, Krivenko *et al.* 1998) conducted extensive studies on the alkali–aggregate reaction and expansion of mortar bars made with different alkali-activated cementing systems, including alkali-activated slag cements, alkali-activated portland slag cement, using NaOH, Na_2CO_3 and $Na_2O \cdot nSiO_2$ as activators. Yang (1997) also investigated several other factors such as, the basicity of the slag, partial replacement of slag with fly ash or silica fume, clinker content, the dosage of activators, amount of silica glass, water-to-cementing material ratio and the combination with other salts. The main features of alkali–aggregate reaction resulting from alkali-activated slag cement can be summarized as follows (Yang 1997; Yang *et al.* 1999):

1   Regardless of alkaline activators, expansion mainly develops during first 30 to 60 days, and reaches a plateau thereafter. The expansion increases with increase of alkali dosage, amount of reactive aggregate up to 30% and the basicity of slag, but with the decrease of water-to-cementing materials ratio of the system.

2   Among the $Na_2O \cdot nSiO_2$-, Na_2CO_3- and NaOH-activated slag cements, $Na_2O \cdot nSiO_2$-activated slag cement exhibits the highest and NaOH-activated slag cement exhibits the lowest expansion for given Na_2O dosages and testing conditions. Figure 7.14 shows the expansion of mortars made with alkali-activated slag cements containing 6% Na_2O, of $Na_2O \cdot nSiO_2$, Na_2CO_3 and NaOH and different amounts of silica glass.

3   The modulus of waterglass also has a great effect on the expansion of alkali–aggregate reaction of waterglass-activated slag cement. The pessimum modulus of waterglass for expansion is around 1.8 (Figure 7.15).

4   When reactive aggregate content is less than 5%, the expansion of alkali-activated cement systems is within the expansion limit regardless of alkali dosage and the nature of activators.

5   The immersion of alkali-activated cement mortar bars containing silica glass in saturated $CaCl_2$, NaCl and Na_2SO_4 solutions increases the expansion, while Na_2SO_4 has the most effect and $CaCl_2$ has the least effect on the expansion of alkali-activated cementing systems. Those salt solutions also show more obvious effects on alkali-activated portland slag cement systems than on alkali-activated slag cementing systems.

6   As expected, the expansion of alkali-activated portland slag cement systems decreases with the increase of slag content.

Thus, it can be concluded that alkali–aggregate reaction definitely happens in alkali-activated slag cement concrete when the aggregate is alkali-reactive. Of course, the expansion will be dependent on the aggregate and the composition of the cement.

However, the replacement of slag with pozzolanic materials such as fly ash, silica fume and metakaolin can reduce or even eliminate the alkali–aggregate reaction expansion. Pu and Chen (1991) noticed that the use of silica fume could even eliminate the expansion from alkali–aggregate

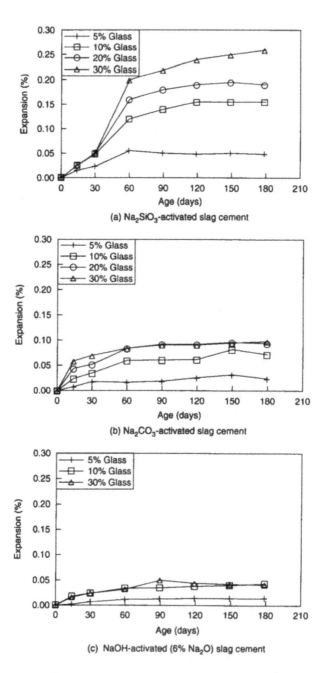

Figure 7.14 Effect of the nature of activator and silica glass content on mortar bar expansion (data from Yang 1997).

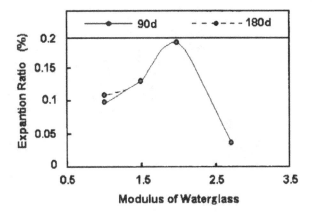

Figure 7.15 Effect of modulus of waterglass (5% Na_2O) on alkali–silica reaction expansion of waterglass-activated slag cement mortars containing 10% silica glass (data from Yang et al. 1999).

reaction. The extensive study by Yang (1997) indicated that the replacement of slag with pozzolanic materials such as coal fly ash and silica fume decreases the expansion of alkali-activated slag and portland slag cementing systems. Use of 30 to 50% low calcium fly ash can reduce the expansion of alkali-activated slag cement below the accepted limit. When 10% silica fume is used, no obvious expansion can be measured from both $Na_2O \cdot nSiO_2$- and Na_2CO_3-activated slag cements. This can explain why Malek and Roy (1997b) found that the 28-day mortar expansion of several alkali-activated cementitious materials is obviously lower that that of high alkali cements and ASTM Type I cement added with potassium hydroxide, and Davidovits (1994) also reported a very low alkali–aggregate reaction expansion since a high volume of metakaolin was used in their cementing system.

7.10 Carbonation

Carbonation refers to the reaction of hydrated cement with carbon dioxide in the environment. All cement hydration products can react with carbon dioxide to form calcium carbonate and other products. Moisture is an important factor in carbonation. Carbonation does not happen in dry cement paste or cement paste at 100% relative humidity. The optimum condition for carbonation appears to be around 50% relative humidity (Neville 1994).

Carbonation of concrete reduces the pH of the material from original >13 down to about 8, which destroys the passivation and induces the corrosion of steel reinforcement in the concrete. Also, carbonation is usually accompanied with a shrinkage, which may produce cracking or surface crazing of concrete.

Byfors *et al.* (1989) studied the carbonation of alkali-activated slag concrete and compared with portland cement. The carbonation rate of an alkali-activated slag concrete with a water-to-slag ratio of 0.35 was similar to that of portland cement concrete with a water-to-cement ratio of 0.60. Thus, alkali-activated slag concrete can be carbonated much faster than portland cement concrete. They attribute the faster carbonation rate of alkali-activated concrete to both its low alkalinity and cracking during the carbonation testing. Optical microscopic examination of thin sections of the specimen after carbonation indicated the presence of cracks and strong carbonation along the walls of cracks. Carbonated alkali-activated cement pastes contained a higher portion of capillary pores than that of the uncarbonated. As expected, the carbonation rate of alkali-activated slag concrete at 50% relative humidity is faster than that at 80% relative humidity.

Two studies (Malolepszy and Deja 1999, Deja 2002a) reported the carbonation of Na_2CO_3- and $Na_2O \cdot 1 \cdot 5SiO_2$-activated slag cement mortars in 0.1 MPa pressure of CO_2 at 20°C and 90% relative humidity after 28 days of curing. One batch of Na_2CO_3-activated slag cement mortars were steam-cured first then cured at room temperature until 28 days. After two years of carbonation testing, the carbonation depth of Na_2CO_3-activated slag cement mortars was about 8 to 9 mm, and 6.5 mm for $Na_2O \cdot 1 \cdot 5SiO_2$-activated slag cement mortars. Both the flexural and compressive strengths of the carbonated alkali-activated slag cement mortars increased compared to those cured in natural conditions. These smaller carbonation depths could be attributed to the high relative humidity used in the experiment.

Bakharev *et al.* (2001b) investigated the carbonation of alkali-activated slag cement concrete in 0.352 M sodium bicarbonate ($NaHCO_3$) solution and in an atmosphere containing 20% CO_2 at 70% relative humidity. $NaHCO_3$ solution was used to simulate carbonated ground water. A conventional concrete with the same water-to-binder ratio and similar strength at 28 days was used for comparison purpose. Both strength and carbonated depth were measured and plotted in Figure 7.16. It was found that the strength of both alkali-activated slag cement concrete and portland cement concrete increased with time, but lower than that of those immersed in water. The difference between the two alkali-activated slag cement concrete was about 10%, but was only about 4% for the portland cement concrete. After 12 months of immersion in the sodium bicarbonate solution, the alkali-activated slag concrete showed a carbonated depth of 20 mm, and portland cement concrete a depth of 12 mm. A soft surface could be observed on those carbonated specimens in the solution. The differences between the two concretes can be attributed to the difference in products in the two concretes before and after carbonation.

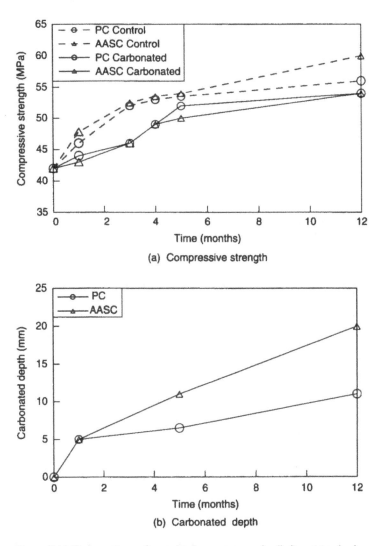

Figure 7.16 Carbonation of portland cement and alkali-activated slag cement concrete in 0.352 M NaHCO₃ solution (Bakharev *et al.* 2001b).

It is well known that the main hydration product of alkali-activated slag is C–S–H with a low Ca/Si ratio. The carbonation of C–S–H can be expressed as follows:

$$C\text{–}S\text{–}H + CO_3^{2-} \rightarrow CaCO_3 + XSiO_2 \cdot nH_2O + 2OH^- \qquad (7.4)$$

Fully hydrated portland cement pastes, consist roughly of 70% C–S–H with a high Ca/Si ratio, 20% Ca(OH)$_2$, 7% AFt and 3% AFm. The carbonation of Ca(OH)$_2$ is a solid volume increase process:

$$Ca(OH)_2 + CO_3^{2-} \rightarrow CaCO_3 + 2OH^- \qquad (7.5)$$

In an atmosphere containing 20% CO$_2$ at 70% relative humidity, alkali-activated slag cement concrete was carbonated much faster than portland cement concrete, as shown in Figure 7.17. Carbonation did not show an

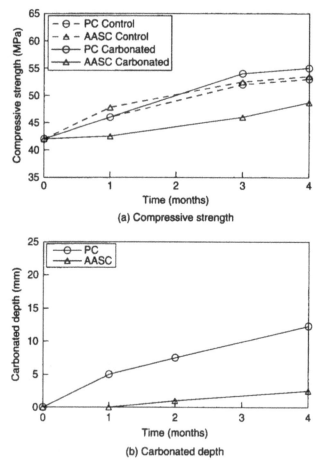

(a) Compressive strength

(b) Carbonated depth

Figure 7.17 Carbonation of portland cement and alkali-activated slag cement concrete in an atmosphere containing 20% CO$_2$ at 70% relative humidity (Bakharev *et al.* 2001b).

effect on the strength of portland cement concrete, but carbonated alkali-activated slag cement concrete showed a lower strength as compared with the reference concrete. The fast carbonation of alkali-activated slag cement concrete may be attributed to the formation of cracks due to both drying and carbonation during the carbonation test.

Recently, Shi (2003a) noticed that, after 75 days of being exposed to an atmosphere containing 15% CO_2 at 53% relative humidity, visual observation indicated that the alkali-activated slag paste specimen started to crack a few days after the start of carbonation testing. Figure 7.18 shows the nest cracks on the surface of the specimens after carbonation test. The cracking could be attributed to both drying shrinkage and carbonation shrinkage. Hakkinen (1993) also observed serious cracking during the carbonation of alkali-activated slag cement concrete and could not find any relationship between the carbonated depth and time, as for portland cement and portland slag cement concretes. However, Ionescu and Ispas (1986) reported that the carbonation of sodium silicate-activated slag/fly ash cement was similar to that of portland cement concrete, and Xu and Pu (1999) even reported that alkali-activated slag cement concrete had a lower carbonation rate than portland cement concrete.

As discussed above, hydrated alkali-activated slag pastes consist mainly of C–S–H with a low Ca/Si ratio and contain almost no crystallized compounds. Under such drying condition, it can be expected that the drying shrinkage of alkali-activated slag pastes is very large as reported previously. Actually, cracking is often observed under scanning electronic microscope in hardened alkali-activated cement and concrete due to drying, which results in fast carbonation, especially at lower relative humidity.

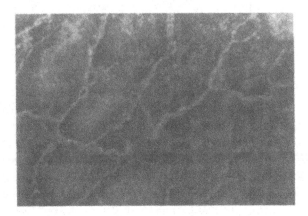

Figure 7.18 Appearance of hardened alkali-activated slag cement pastes after carbonation testing at 23 °C and 53% relative humidity (×25) (Shi 2003a).

7.11   Corrosion of steel reinforcement in alkali-activated cement concrete

The pore solution of a fully hydrated portland cement usually has a pH over 13 and consists mainly of alkali hydroxides. The high pH results in the formation of a passive layer on the surface of steel reinforcement in the concrete, which is a very dense, impenetrable film preventing the further corrosion of the steel. However, the passive layer can be destroyed due to the carbonation of the concrete, which will drop the pH down to around 8, or due to the transport of chlorides to the surface of the steel. Once the passive layer is destroyed, the steel can be corroded very quickly. Section 7.2 discussed the transportation of chloride ions in alkali-activated cement and concrete and the carbonation of alkali-activated cement and concrete. This section will mainly focus on how the cement and concrete materials affect the behaviour of steel in the concrete.

Deja *et al.* (1991) and Malolepszy *et al.* (1994) investigated the corrosion of steel in alkali-activated slag mortars cured in water and in 5% $MgSO_4$ solution by measuring polarization curves, corrosion current and mass loss of the reinforcement. They found that the shapes of anodic polarization curves at 28 days and 336 days have only slight difference when cured in water. This means that there is a passivation layer on the surface of the steel reinforcement. However, an obvious difference can be observed for those samples cured in the $MgSO_4$ solution, which means that the passivation has been decreased.

The measurement of corrosion current indicated that the steel in the alkali-activated slag mortar had much higher corrosion current than that in Portland cement mortar, but it decreases with time in both mortars.

After 336 days of immersion, the steel in both mortars cured in water did not show any mass loss at all, while the one in the alkali-activated slag cement mortar cured in 5% $MgSO_4$ solution showed 0.19% mass loss and the one in the Portland cement mortar showed 0.42% mass loss.

Alkaline by-products can also be used as activators. However, some by-products contain chlorides and sulphates, which may initiate corrosion of steel in concrete. Krivenko and Pushkaryeva (1993) investigated the corrosion of steel in alkali-activated slag cement concrete using two mixed activators, one consisted of 90% $Na_2CO_3 + 10\%$ NaOH and the other one consisted of 45% $Na_2CO_3 + 40\%$ $Na_2SO_4 + 15\%$ NaCl. As shown in Table 7.10 the corrosion of steel depends on the nature of slag, nature and dosage of alkaline activator, and carbonation of the concrete. When a mixture of 90% $Na_2CO_3 + 10\%$ NaOH is used as an activator, the mass loss increased with the concentration of activator. The use of a mixture of

Table 7.10 Mass loss of rebar in alkali-activated slag cement concrete made from different slags and alkaline activators under wetting and drying cycles (Krivenko and Pushkaryeva 1993)

Slag	Alkaline activator solution density, kg/m³	Mass loss of reinforcement (g/m²) at different ages (months)							
		A mixture of 90% Na₂CO₃ + 10% NaOH				A mixture of 45% Na₂CO₃ + 40% Na₂SO₄ + 15% NaCl			
		6	9	12	18	6	9	12	18
Basic $Mb > 1$	1100	0.52	0.53	0.52	0.52	0.00	0.00	0.00	0.00
	1150	0.70	0.73	0.71	0.72	0.89	0.91	0.90	0.91
	1200	0.98	0.96	0.98	0.97	1.07	1.09	1.07	1.06
Acid $Mb = 1$	1100	0.00	0.00	0.00	0.00	0.00	0.00	0.00	0.00
	1150	0.41	0.43	0.40	0.42	0.63	0.61	0.62	0.62
	1200	0.71	0.70	0.72	0.71	0.58	0.60	0.59	0.59
Neutral (electrothermo-phosphorous) $Mb = 1$	1100	0.00	0.00	0.00	0.00	0.00	0.00	0.36	0.36
	1150	0.71	0.76	0.73	0.72	46.91	74.73	78.54	84.10
	1200	1.12	1.14	1.11	1.13	59.12	85.73	86.04	87.40

Table 7.11 Effect of additives on the mass loss of the reinforcement in alkali-activated phosphorous slag cement concrete with an activator containing 45% $Na_2CO_3 + 40\%$ $Na_2SO_4 + 15\%$ NaCl ($\rho = 1150\,kg/m^3$) under wetting–drying cycles (Krivenko and Pushkaryeva 1993)

Additive		Mass loss of reinforcement (g/m^2)		
Name	mass %	6 months	12 months	18 months
Reference cement (without additive)		46.91	78.54	84.10
Sodium hydroxide	5	18.01	31.12	42.10
OPC clinker	5	0.41	0.38	0.39
Wastes from enrichment of polymetallic ores	7	0.72	0.73	0.72
Slag from production of ferroniobium + calcium fluorite (CaF_2)	10+5	8.81	0.52	0.51

45% $Na_2CO_3 + 40\%$ $Na_2SO_4 + 15\%$ NaCl as an activator did not show obvious difference from the mixture of 90% $Na_2CO_3 + 10\%$ NaOH, except the two batches made with phosphorus slag and activator solutions with density of 1150 and 1120 kg/m^3, due to the carbonation of the specimens.

The addition of additives can densify the structure of hardened concrete and change the pore solution chemistry. Table 7.11 and Figure 7.19 show the effects of additives on the mass loss of the reinforcement in the alkali-activated slag cement concrete made with granulated phosphorous slag and an activator containing 45% $Na_2CO_3 + 40\%$ $Na_2SO_4 + 15\%$ NaCl ($\rho = 1150\,kg/cub.m$) under wetting–drying cycles or in different corrosive environments.

7.12 Fire resistance

Concrete is a non-combustible material. It has good fire resistant property and is often used to protect steel from the effects of fire. However, concrete can be damaged by exposure to high temperatures and will suffer loss of strength, cracking and spalling. Hydrated portland cement contains a considerable portion of free calcium hydroxide, which decomposes above 400–500 °C, leaving calcium oxide (quick lime). This will result in significant strength loss. If the calcium oxide is cooled down and becomes wetted after cooling, it rehydrates to calcium hydroxide accompanied by an expansion in volume and may disrupt concrete, which is not disintegrated during the fire. In hydrated alkali-activated cements, free $Ca(OH)_2$ usually does not exist. Thus, it can be expected that disruption of concrete due to the rehydration of CaO will not happen. However, the behaviour of alkali-activated cement and concrete does depend on the composition of the cement. Of course,

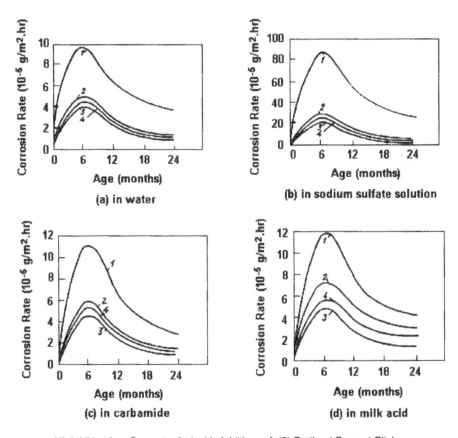

(a) in water

(b) in sodium sulfate solution

(c) in carbamide

(d) in milk acid

(1) Additive-free Concrete; 2–4 with Additives of: (2) Portland Cement Clinker,
(3) Slag from Production of Ferroniobium and Calcium Fluorite (CaF2),
(4) Wastes from Enrichment of Polymetallic Ores.

Figure 7.19 Effect of additives on the mass loss of the reinforcement in the alkali-activated slag cement concrete made with electrothermal phosphorous slag and an activator containing 45% Na_2CO_3 + 40% Na_2SO_4 + 15% NaCl ($\rho = 1150\,kg/m^3$) in different corrosive environments.

aggregate expansion or decomposition on heating will affect the mechanical or integrated properties of the concrete.

An investigation on dehydration kinetics of alkali-activated cementing components from systems such as $CaO–SiO_2$, $CaO–Al_2O_3$, $CaO–Al_2O_3–SiO_2$, $CaO–MgO–SiO_2$ and $CaO–Al_2O_3–MgO–SiO_2$ establishes a principle for designing fire-resistant concrete (Krivenko 1994b). It reveals that many cementing materials from different systems can show very good fire resistance, such as alkali-activated γ- or β-C_2S in the $CaO–SiO_2$

system, alkali-activated CA and CA$_2$ in the CaO–Al$_2$O$_3$ system, alkali-activated CAS$_2$ in the CaO–Al$_2$O$_3$–SiO$_2$ system and alkali-activated CMS$_2$ in the CaO–MgO–SiO$_2$ system. It has been found that the transformation of C–S–H to anhydrous minerals during heating will not cause any serious structural deterioration when the Ca/Si ratio of C–S–H rages from 0.5 to 0.9 (Krivenko 1994b).

Jumppanen *et al.* (1986) conducted an extensive research on the properties of alkali-activated slag cement and concrete at high temperatures using two ordinary portland cements as references. The three cement pastes displayed different thermal expansion behaviour. It was difficult to tell the differences between the alkali-activated slag cement paste and the two portland cement pastes up to 800 °C. As expected, the thermal expansion of concrete was mainly determined by aggregates.

Pore structure measurements indicated that the gel porosity of ordinary portland cement paste decreases significantly as temperature increases, while the gel porosity of alkali-activated slag cement pastes increases with temperature up to 250 to 350 °C, and decreases with temperature thereafter.

The stress–strain relationships at different temperatures for ordinary portland cement concrete and alkali-activated slag cement concrete are shown in Figures 7.20 and 7.21. When temperature is lower than 350 °C (for ordinary portland cement concrete) or 250 °C (for alkali-activated slag cement), the strength of the concrete is increased due to accelerated hydration and strengthening effect caused by drying. As the temperature

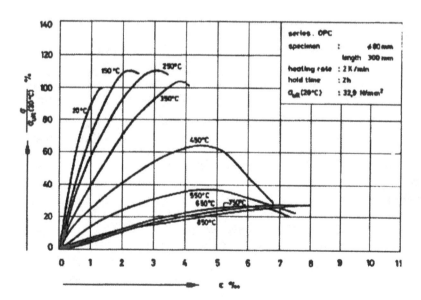

Figure 7.20 Stress–strain relationship of ordinary portland cement concrete at high temperatures (Jumpanen *et al.* 1986).

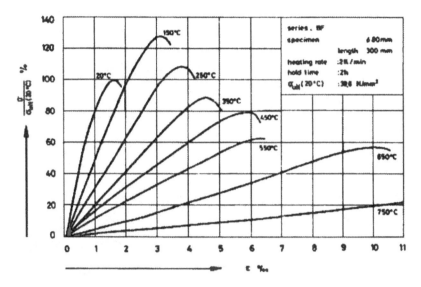

Figure 7.21 Stress–strain relationship of alkali-activated slag cement concrete at high temperatures (Jumpanen *et al.* 1986).

increases further, aggregate and cement pastes show different expansions, which results in strength loss. When the temperature reaches 450°C, $Ca(OH)_2$ in portland cement concrete decomposes and causes a significant strength drop, which does not happen to alkali-activated slag cement concrete, while, at 750°C, alkali-activated slag cement concrete shows larger shrinkage than the portland cement concrete.

By comparing the stress–strain relationships for ordinary portland cement concrete and alkali-activated slag cement concrete, it can be seen that the modulus of elasticity of alkali-activated slag cement concrete is always lower than that of ordinary portland cement concrete.

Thus, in alkali-activated slag cements, it is very easy to add adjustment components to the system to increase the fire-resistance of the cement. For example, a slag with a composition close to melilite was combined with 10% sodium metasilicate to manufacture a cement. In order to increase the fire resistance of the cement, different amounts of slag were replaced by calcined serpentinite rock to modify the composition of the hydration products. The effect of the replacement on strength of the cement mortars with a cement-to-sand ratio of 1:3 is shown in Figure 7.22. It can be seen that the replacement of the slag with calcined serpentinite rock decreased the strength of the mortars after steam curing and drying at 100°C, but increased the strength of the mortars after heating at 950°C. The optimum replacement is around 30 to 40%.

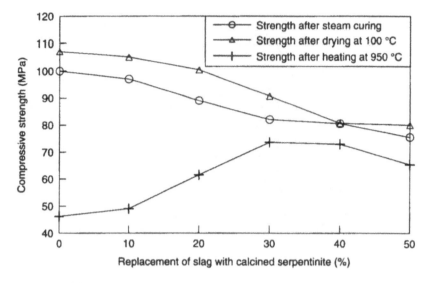

Figure 7.22 Effect of slag replacement with calcined serpentinite on strength of alkali-activated cement at high temperatures (data from Krivenko 1994b).

Because of the low capillary porosity, there is a higher risk of explosive spalling for alkali-activated slag cement concrete when exposed to high temperatures. The use of organic fibres may be able to eliminate the spalling problem (Sarvaranta and Mikkola 1994, Kalifa *et al.* 2001). Actually, polypropylene fibre is the most efficient one among various organic fibres. A recent study (Kalifa *et al.* 2001) has indicated that the addition of about 2 kg of polypropylene fibre in each cubic meter of concrete can solve the spalling problem. Detailed microstructure investigation indicated that polypropylene fibre will be molten at around 200 °C, which results in the formation of a continuous pore network within the concrete and which can accumulate the high pressure due to the water vapour from the decomposition of hydration products and enclosed air.

7.13 Wearing resistance

Wearing can be classified into three categories: abrasion, erosion and cavitation (Mindess and Young 1981). Wearing refers to repeated rubbing or frictional process, which is usually in connection with traffic wear on pavements and industrial floors. Erosion is the abrasive action of fluids and suspended solids. It is a special case of abrasion and occurs in water-supply installations such as canals, conduits, pipes and spillways. Cavitation is the impact damage caused by high velocity disturbed liquid flow, which happens at spillways and sluiceways in dams and irrigation installation. Mechanical abrasion is the dominant abrasion for pavements and bridge decks exposed to studded tires. Pu *et al.* (1989) reported that alkali-activated slag cement concrete showed better abrasion resistance that conventional portland cement concrete.

7.14 Wetting–drying resistance

Wetting and drying cycles increase irreversible shrinkage and may destroy the integrity of concrete structure (Sheikin 1974). Malolepszy and Deja (1988) found that wetting and drying of alkali-activated slag concrete did not virtually affect its compressive strength. However, it can have significant adverse effects on flexural strength (Kutti and Malinowski 1982), like on portland cement concrete (Craf 1960).

Zeolites calcined at 600–900 °C and petrolatum, which is a blend of high-molecular solid hydrocarbons (paraffins and ceresins) produced in petroleum oil processing, were added to improve the weathering resistance of alkali-activated slag cement concrete. Petrolatum was added in a form of emulsion by blending with fatty acids with general formula $CH_3(CH_2)nCOOH$ ($n = 14$–16). Waterglasses with moduli of 1, 1.5 and 2 were used as alkaline activators. The effects were evaluated by measuring the compressive and flexural strength of $4 \times 4 \times 16\,cm$ mortars with slag:sand $= 1 : 3$. The mortar specimens were steam-cured with a curing regime of $3+6+3$ hours with the highest temperature of 85 °C, then tested at specific ages after continued curing in water, in air and wetting–drying cycles. One cycle consisted of seven days in water and seven days in dry air at 18–20 °C and relative humidity of 60%. Figure 7.23 shows the effect of different additives on flexural and compressive strength of alkali-activated cement mortars under different conditions after steam curing. The results indicated that the use of those additives increases the compressive strength and those different conditions did not show any obvious effect on the compressive strength of the specimens. Under normal curing conditions, those additives and curing conditions did not show a significant effect on flexural strength of the mortars. When cured in water, air or wetting–drying cycles, the flexural strength decreased during the first 60 days then increased gradually. The results in Figure 7.23-II-a indicated that the first two cycles had the most obvious effects on flexural strength. The flexural strength lost 6–8% of the initial value after steam curing after the first cycle, but lost almost 40% after the second cycle. The use of those additives obviously decreased the loss of flexural strength in air and wetting–drying cycles. The use of 10% calcined zeolite and 2% petrolatum emulsion was most effective for decreasing the loss of flexural strength when cured in water, air or wetting–drying cycles.

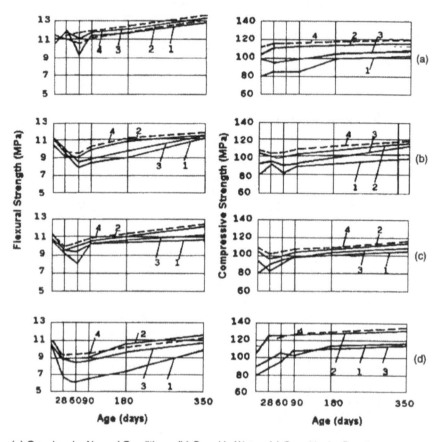

(a) Cured under Normal Conditions; (b) Cured in Water, (c) Cured in the Dry Air,
and (d) Cured in Wetting–Drying Cycles

1-Slag Alkali-activated Cement (SAC); 2-SAC + 10% (by mass) dehydrated zeolite rock;
3-SAC + 2% (mass) Petrolatum Emulsion; 4-SAC + 10% (by mass)
Dehydrated Zeolite Rock + 2% (by mass) Petrolatum Emulsion.

Figure 7.23 Changes in compressive (I) and flexural (II) strength of the alkali-activated
slag cement-based specimens.

Differential thermal analysis and infrared spectroscopy analysis on the
binder containing petrolatum emulsion indicated absorption of a polymer
on the surface of slag grains and the formation of new products. This interaction caused the formation of hydration products with a lower crystallinity
and less porous structure, which resulted in better physical-mechanical
properties and less deformations of the material.

7.15   Summary

This chapter has discussed most durability aspects of alkali-activated slag cement and concrete and can be summarized as follows:

1   The performance of alkali-activated slag cement and concrete is predominately controlled by the nature of the slag, the nature and dosage of the activator(s) used, and water to slag ratio.

2   In moist conditions, alkali-activated slag cements can have lower water and chloride permeability, and better resistance to corrosive media such as acid, sulphate, chlorides than conventional cement and concrete.

3   Under both saturated and moisture-controlled conditions, alkali-activated cements can be carbonated faster than conventional cement and concrete.

4   When alkali-reactive aggregate is used, alkali-activated cement concrete may show high expansion. However, the expansion depends on the nature of the cementing components, the nature and dosage of activator and water-to-cement ratio. However, the alkali–aggregate reaction expansion may be reduced or eliminated by introduction of pozzolans such as low calcium fly ash, silica fume and metakaolin.

5   Air-entrainment agents for portland cement concrete may not work with alkali-activated slag cement concrete. For given air void content and spacing, alkali-activated slag concrete exhibits as good as or even better resistance to freezing–thawing cycles than portland cement concrete.

6   Since hardened alkali-activated slag pastes and concrete contain no free $Ca(OH)_2$, it shows better fire resistance than portland cement concrete.

Chapter 8

Mix design of alkali-activated slag cement concrete

8.1 Introduction

A mix design method for alkali-activated slag cement concrete was first developed in 1967 (Sikorsky 1967). Many methods have been proposed since then (Frenkel and Shakhmuratyan 1974, Astapov 1976, Glukhovsky and Pakhomov 1978, Batalin *et al.* 1979, Belitsky 1994, Firsov 1984, Akimov *et al.* 1984, Latina *et al.* 1984). Generally speaking, these mix design methods can be classified into two categories: empirical methods and experimental methods. The disadvantages of these mix design methods are that they require a large amount of experiments and use statistics to establish relationships between the properties of concrete and proportions of the concrete ingredients. This chapter will describe the general procedures of these two mix design methods.

8.2 Empirical mix design method

In the empirical mix design method, relationships between strength and slag-to-alkaline activator solution ratio – S/W_{AAS} (for a given density of the alkaline activator solution) need to be established first, as shown in Figure 8.1. The following factors should be considered: (1) the strength of alkali-activated slag cement paste determines strength of the concrete; (2) the amount of alkali-activated slag cement paste controls the workability of the concrete mixture; and (3) the workability of the alkali-activated slag cement concrete mix is evaluated using the Vebe test due to its high viscosity.

Step 1: Selection of slag-to-alkaline activator solution ratio and density of alkaline activator

Based on design strength requirement, the relationships between the concrete strength and S/W_{AAS} are established, as shown in Figure 8.1. The approximate concrete mix proportions for trial batches are recommended in Tables 8.1 and 8.2. Then, the S/W_{AAS} is further optimized based on the nature and concentration of activators, as shown in Table 8.3.

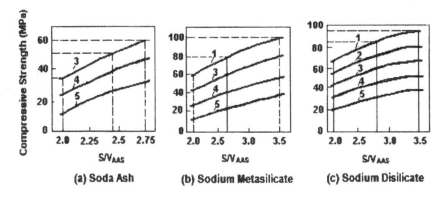

Figure 8.1 Relationships between concrete strength and slag-to-alkaline activator solution ratio (RSN 336-84 1984).

Table 8.1 Recommended trial proportions for non-prestressed reinforced concrete (RSN 336-84 1984)

No.	S/V_{AAS}	Slag (kg/m³)	Sand (kg/m³)	Crushed stone (kg/m³)	Alkaline activator solution (L/m³)
1	2.0	300	820	1180	150
2	2.5	425	690	1155	170
3	3.0	550	590	1100	183

Note: S – Slag content (kg/m³); V_{AAS} – volume of alkaline activator solution (L/m³).

Table 8.2 Variant II: recommended trial proportions for prestressed reinforced concrete (RSN 336-84 1984)

No.	S/V_{AAS}	Slag (kg/m³)	Sand (kg/m³)	Crushed stone (kg/m³)	Alkaline activator solution (L/m³)
1	2.3	300	600	1455	130
2	2.93	425	515	1365	145
3	3.43	550	420	1285	160

Note: S – Slag content (kg/m³); V_{AAS} – volume of alkaline activator solution (L/m³).

Step 2: Determination of workability of the concrete mixture

Based on the type of concrete product and casting process, the workability of the concrete mixture is selected using the guidelines in Table 8.4. Then, the quantities of the alkaline activator solution are determined according to the aggregate size and workability requirement of the concrete mixture as shown in Table 8.5.

Table 8.3 Approximate slag-to-alkaline activator solution ratio of concretes vs alkaline activator (RSN 336-84 1984)

Alkaline activator	Optimal ratio (kg/L)			Normal consistency of the paste (%)	
Type	Density (kg/m^3)	S/V$_{AAS}$	S/W	Alkaline activator solution	Water
Blast furnace slag (S = 330 m^2/kg)					
Sodium metasilicate, Ms = 1	1100	3.13	3.45	24.2	21.9
	1150	3.03	3.57	24.5	21.0
	1200	2.94	3.70	25.0	20.4
	1100	3.23	3.57	22.8	20.6
	1150	3.23	3.85	22.6	19.5
	1200	3.23	4.00	22.7	18.5
	1250	3.23	4.17	23.0	17.8
Sodium disilicate, Ms = 2	1100	3.23	3.57	23.5	21.0
	1150	3.23	3.70	23.3	19.8
	1200	3.23	3.85	23.5	18.9
	1250	3.13	4.17	23.8	18.1
	1300	3.13	4.17	24.5	17.7
Sodium silicate 2 < Ms < / = 3	1100	3.13	3.57	24.1	21.2
	1150	3.03	3.70	24.3	20.2
	1200	3.03	3.85	24.7	19.4
	1250	2.94	3.85	25.6	19.0
	1300	2.78	4.00	26.7	18.8

Table 8.4 Workability of concrete mixtures for different products and casting processes (RSN 336-84 1984)

Product	Casting process	Workability (Vebe sec)
Narrow hollow-core blocks, Sewage water collector Rings, pipes, etc. of height less than 1.2 m	Immediate demolding (Full or Partial)	30–50
Hollow core floor slabs	Cast horizontally with loading	20–50
Columns, girders, beams, slabs, foundation shoes, curbs, etc.	Cast without loading	15–30
Flat or ribbed floor panels, structures with more than 10% reinforcement	Cast horizontally with loading	10–25
Heavily reinforced thin-wall structures	Cast in cassettes	5–15

Table 8.5 Approximate amount of water/alkaline activator solution for required workability of concrete mixture (RSN 336-84 1984)

| Workability (Vebe sec) | Water/alkaline consumption (L/m³) at different aggregate size (mm) | | | | | |
| | Gravel | | | Crushed stone | | |
	10	20	40	10	20	40
40–50	112/140	97/125	87/115	122/150	112/140	97/125
25–35	122/150	107/135	92/120	132/160	122/150	107/135
15–20	126/155	112/140	97/125	137/165	126/155	112/140
10–15	137/185	122/150	107/135	148/175	137/165	122/150
6–10	152/180	137/165	122/150	162/190	152/180	137/165
4–6	162/190	147/175	132/160	172/200	162/190	147/175
2–4	167/195	152/180	137/165	177/205	167/195	152/180

Notes: For a concrete mixture containing granulated slag with of basicity of 1, sodium disilicate with $\rho = 1300\,kg/m^3$ and river sand (paste has a normal consistency of 17.5%); With 1% decrease/increase in a normal plasticity of the alkali-activated slag cement paste, the consumption of the alkaline activator solution decreases/increases by 2–3 L/m³.

After the workability is selected, some trial batches are prepared and tested as follows:

1   Prepare the first batch (10–15 L) of concrete mixture with the recommended proportions, and then measure consistency and density of the concrete mixture.
2   If the consistency is lower than the requirement, 5–10% of slag and alkaline activator solution is added to the trial batch, while maintaining accepted alkaline activator solution-to-slag ratio.
3   If the consistency is higher than the requirement, 5–10% of sand and coarse aggregate in the accepted ratio is added to the trial batch.
4   After the required consistency of the mix is achieved, the concrete mix composition is re-calculated and a new batch for three control specimens is made.

Step 3: Determination of fine and coarse aggregate contents
First of all, the slag content in one cubic meter of concrete mix is:

$$S = V_{AAS} \times S/V_{AAS} \tag{8.1}$$

where
   V_{AAS} - volume of the alkaline activator solution (L);
   S - slag content (kg).

The total mass of aggregates (sand and crushed stone) (C_{AGG}) is calculated as follows:

$$C_{AGG} = M_{CC} - S - V_{AAS} \times \rho_{AAS} \tag{8.2}$$

where
   M_{CC} – mass of one cubic meter of compacted concrete mix (kg);
   ρ_{AAS} – density of alkaline activator solution (kg/m³).
Then, the fine aggregate content is:

$$C_{FA} = (C_{AGGR}) \times r \tag{8.3}$$

where
   r – ratio of fine aggregate to total aggregate in the mixture (Table 8.6)
The coarse aggregate content is:

$$C_{CA} = (C_{AGGR}) - C_{FA} \tag{8.4}$$

Step 4: Adjustment of concrete mix proportions
After the initial determination of concrete proportion for workability, the mix proportion will be further adjusted for strength purpose. For this reason, three batches of concrete should be made: the one determined based on workability and another two batches with the same mixing proportions, but with the slag-to-alkaline activator solution ratios of ±0.14. After the slag-to-alkaline activator solution ratio is changed, the slag and sand contents are adjusted to get the same volume of concrete mixture, but keep the mortar and coarse aggregate contents unchanged. The final mixing proportion is selected with required strength.

Table 8.6 Optimum fine aggregate to total aggregate ratios in concrete mixtures (Belitsky 1994)

Slag content (kg/m³)	Fine aggregate to total aggregate ratios at different maximum size of coarse aggregate					
	Gravel (mm)			Crushed stone (mm)		
	10	20	40	10	20	40
300	0.4	0.38	0.36	0.43	0.417	0.4
400	0.38	0.36	0.35	0.4	0.38	0.37
450	0.37	0.35	0.34	0.39	0.37	0.36
500	0.36	0.35	0.34	0.38	0.36	0.35
550	0.35	0.34	0.33	0.378	0.35	0.34

8.3 Experimental design method

The experimental design method uses previous experience and some empirical data to design an initial mixing proportion. The mixing proportion is then adjusted by trial and error method. Any admixtures that will be used should be considered during the mixture design. This method can be used to design alkali-activated slag cement concrete with strength from 10 to 110 MPa with a wide range of workability/consistency requirements.

This design method is applicable to both silicate alkaline activators (sodium and potassium silicates) and non-silicate alkaline activators (sodium and potassium hydroxide or carbonate). When high-modulus sodium silicate (>2.5) is used, the mix design procedure is correct only if special additives are introduced into the mixture to retard quick initial setting (Skurchinskaya and Belitsky 1989).

Step 1: Property information of concrete ingredients
Some property information of the raw materials should be known for concrete mixture design purpose, which include maximum size of coarse aggregates, percentage of void space in the compacted coarse aggregate, bulk density and moisture content of the aggregates, water content of fine aggregate, density of the slag and alkaline solution, amount of solution required for alkali-activated slag cement paste of normal consistency, and compressive strength of alkali-activated slag cement mortars based on 40 mm × 40 mm × 160 mm prisms with sand/slag ratio of 3 and specified flowability in accordance with the Ukrainian Standard RST 5024-83 (1983).

Step 2: Calculation of slag content
The slag content is primarily dependent upon strength and workability requirements of the designed concrete, compressive strength of the alkali-activated slag cement and nature of the alkaline activator used. It is expressed as amount of slag per unit volume of concrete and can be determined based on the past experience with the materials or can be obtained from Table 8.7. The strength of concrete in Table 8.7 is based on 150 mm × 150 mm × 150 mm cubes in accordance with the Ukrainian Standard Technical Specification RST 5025-84 (1984). If there are severe exposure conditions, such as freezing and thawing, or special requirements as to water permeability, the slag content may need to be adjusted in accordance with the Ukrainian Building Norms RSN 344-87 (1987).

Step 3: Determination of alkaline activator solution content
The alkaline activator solution content is primarily dependent upon workability of the designed concrete mixture, maximum size of coarse aggregate used, water requirement of fine aggregate and quantity of alkaline activator solution required for obtaining normal consistency of the alkali-activated slag cement paste. It is expressed as a unit volume of alkaline activator

Table 8.7 Recommended slag content (kg/m^3) for alkali-activated slag cement concrete (Belitsky 1994)

Strength of concrete (MPa)	Compressive strength of mortar (MPa)	Levels of consistence (by a Vebe test), sec.				
		<2	2–4	5–10	11–20	>21
For non-silicate alkaline activators						
10	30	300	300	275	275	275
15	30	375	375	350	350	350
	40	325	300	300	300	300
20	30	475	475	450	450	450
	40	425	425	400	400	400
25	40	550	525	525	500	500
	50	425	400	400	375	375
30	50	475	450	450	425	425
	60	325	300	300	300	300
40	50	550	525	525	500	500
	60	450	425	425	400	400
50	60	525	500	500	475	475
55	60	600	575	575	550	550
Silicate activators						
10	30	250	250	250	250	250
15	30	325	300	300	300	300
	40	275	275	275	250	250
20	30	375	375	350	350	350
	40	350	325	325	300	300
25	30	450	425	425	400	400
	40	400	375	375	350	350
30	30	500	500	475	475	450
	40	425	400	400	375	375
	50	350	325	325	300	300
40	40	500	500	450	450	425
	50	400	400	375	375	350
	60	350	350	325	325	300
50	50	500	500	475	475	450
	60	450	450	425	425	400
	70	400	400	375	375	350
55	60	475	475	450	450	425
	70	350	350	325	325	300
60	60	550	550	550	525	525
	70	450	450	425	425	400
	80	350	350	325	325	300
70	70	550	550	550	525	525
	80	450	450	450	425	425
	90	350	350	350	325	325
80	80	550	550	550	525	525
	90	475	475	475	450	450
	100	400	400	400	375	375

90	90	550	550	550	525	525
	100	475	475	475	450	450
	110	400	400	400	375	375
95	100	525	525	525	525	500
	110	475	475	475	450	450
	120	425	425	400	400	400
100	110	525	525	525	525	500
	120	475	475	475	475	450
110	120	550	550	550	550	525

Table 8.8 Approximate contents of alkaline activator solution (L/m^3) to obtain various levels of workability (Belitsky 1994)

Workability		Approximate contents of alkaline activator solution (L/m^3)							
		Crushed				Uncrushed			
Slump (mm)	Vebe test (sec.)	Maximum coarse aggregate size (mm)							
		10	20	40	70	10	20	40	70
Non-silicate alkaline activator									
>16	–	235	230	215	205	225	220	205	195
10–15	–	225	215	200	190	215	205	190	180
5–9	<2	210	200.	185	180	200	185	170	165
1–4	2–4	200	190	175	170	190	175'	160	155
–	5–10	185	175	165	160	175	160	150	145
–	11–20	175	165	155	150	165	155	145	140
–	21–30	170	160	150	145	160	150	140	135
	>31	165	155	145	140	155	145	135	130
Silicate alkaline activator									
>16	–	185	180	175	170	175	170	165	160
10–15	–	180	175	170	165	170	165	160	155
5–9	<2	175	170	165	160	165	160	155	150
1–4	2–4	170	165	160	155	160	155	150	145
–	5–10	165	160	155	150	155	150	145	140
–	11–20	160	155	150	145	150	145	140	135
–	21–30	155	150	145	140	145	140	135	130
	>31	150	145	140	135	140	135	130	125

solution per unit volume of concrete and can be determined from past experience with the materials, or can be obtained from Table 8.8. The following recommendations should be considered during the mix design process:

• If the water content of fine aggregate is decreased by 7%, the content of alkaline activator solution may be increased by 5 L/m^3 (or by 2.5 L/m^3 – in case of alkali metal silicate) with each one percent increase in water requirement and may be reduced by the same value with decrease in the water content.

- If the amount of alkaline activator solution required for the paste of normal consistency is different from 0.26 L/kg, the content of the alkaline solution may be increased by 5 L/m³ or (or by 2.5 L/m³ – in case of alkali metal silicate) for each 0.01 L/kg increase in the quantity of the alkaline activator solution required for the paste of normal consistency and may be reduced by the same value with decrease in plasticity of the paste.
- If the slag content exceeds 400 kg/m³, the content of alkaline activator solution may be increased by 10 L/m³ (or by 5 L/m³ – in case of alkali metal silicate) with each 100 kg/m³ increase in the slag content.

Step 4: Determination of alkaline activator solution-to-slag ratio
After the slag and alkaline solution contents are determined, the ratio of alkaline activator solution to slag (W_{AAS}/S) is determined very easily and is expressed as a volume unit of alkaline activator solution to mass unit of slag.

Step 5: Calculation of increase in cement paste content when taking into account aggregate distribution in concrete.
The loosening coefficient of coarse aggregate by mortar (α) is a ratio of the volume of mortar in the concrete mixture to the volume of void content in the compacted coarse aggregate. This coefficient can be selected from Table 8.9, depending upon slag content, alkaline activator solution-to-slag ratio and water requirement of the fine aggregate. The selection of the appropriate coefficient (α) may be governed by special technological requirements, depending upon casting technique and peculiarities of structures.

Step 6: Calculation of coarse aggregate content
The coarse aggregate content (C_{CA}) in kg/m³ is calculated using the following equation:

$$C_{CA} = (1 - \alpha C_V/100) \times \rho_{CA} \tag{8.5}$$

where
   C_V = percentage of voids content of compacted coarse aggregate (%);
   ρ_{CA} = bulk density of coarse aggregate (kg/m³).

Step 7: Calculation of fine aggregate content
The fine aggregate content (C_{FA}) in kg/m³ is calculated using a "volumetric" method:

$$C_{FA} = (1 - S/\rho_S - V_{AAS}/\rho_W - C_{CA}/\rho_{CA})\rho_{FA} \tag{8.6}$$

Table 8.9 Coefficient of coarse aggregate arrangement by mortar (Belitsky 1994)

$V_{AAS}/S(L/kg)$	Slag content (kg/m³)						
	250	300	350	400	450	500	550
0.25						1.16	1.20
0.30					1.15	1.20	1.25
0.35				1.15	1.19	1.26	1.32
0.40				1.19	1.24	1.31	
0.45			1.17	1.22	1.30		
0.50		1.17	1.22	1.29			
0.55		1.22	1.28				
0.60	1.16	1.27					
0.65	1.20	1.31					
0.70	1.24						

Notes: If the water demand of fine aggregate is not 7%, the coefficient of loosening of coarse aggregate by mortar may be decreased by 0.03 for increase of each percent of water demand and it may be reduced by the same value when water demand increases. However, it cannot be lower than 1.1.

where

S = slag content (kg/m³);

V_{AAS} = volume of alkaline activator solution (L/m³);

ρ_S = density of slag (kg/m³);

ρ_W = density of water (kg/m³);

ρ_{FA} = bulk density of fine aggregate (kg/m³).

If the calculated fine aggregate content is less than 40% of the total aggregate, the proportions of all the ingredients (excluding the coarse aggregate content) should be re-calculated using the following equations:

(a) The re-calculated slag content is

$$S' = 2.5 \frac{1 - \frac{C_{CA}}{\rho_{CA}}}{\frac{1}{\rho_{FA}} + \frac{2.5}{\rho_C} + \frac{V_{AAS}}{400S}} \tag{8.7}$$

(b) The re-calculated alkaline activator solution content is

$$V'_{AAS} = \frac{V_{AAS}}{S} S' \tag{8.8}$$

(c) The re-calculated fine aggregate content is

$$C'_{FA} = 0.4S' \tag{8.9}$$

Step 8: Adjustment of concrete mixing proportion when taking into account water absorption of aggregates

If the aggregates are wet, they will increase the workability of a fresh concrete and decrease concentration of the alkaline activator solution. Therefore, the alkaline activator solution content may be reduced; taking into account the water absorption of aggregates, the concentration of the alkaline activator solution may be increased, but the alkaline activator solution content will be reduced.

However, if the recalculated concentration of the alkaline activator solution is more than the upper limit of saturation, it is better to use aggregates with lower moisture contents.

Step 9: Trial batch

A trial batch is prepared using the calculated proportions. The produced concrete mixture should be tested for workability. If the workability is incorrect, a new alkaline activator solution content may be estimated and then, the fine aggregate content should be recalculated using Eqn (8.2). After the required workability is achieved, the unit weight of fresh concrete should be determined. Finally, the mix is adjusted in terms of a difference between the required and estimated unit weight of the trial batch.

8.4 Summary

This chapter summarizes two design methods for alkali-activated slag cement concrete: empirical mix design method and experimental mix design method. Since the nature of the slag and activators, and the dosage of activators have significant effects on strength, both mix design methods require establishment of relationships between strength and ingredients of the concrete mixtures.

Chapter 9

Alkali-activated portland cement-based blended cements

9.1 Introduction

Blast furnace slag, phosphorus slag, coal fly ashes, natural pozzolans and silica fume are widely used for the production of blended cement or as a cement replacement in concrete. The use of these materials usually increases setting time and decreases early strengths of the cement and concrete. Many researches have indicated that the addition of alkaline activators can activate potential pozzolanic or cementitious properties of those materials, especially at early ages. This chapter discusses the alkaline activation of portland cement-based blended cements.

9.2 Portland blast furnace slag cement

9.2.1 Introduction

In Europe, China and Japan, more than 80% of the production of blast furnace slag is used for the production of blended cement or as a cement replacement in concrete. In most cases, the replacement of portland cement with blast furnace slag increases setting time and decreases early strengths of the cement and concrete. In order to overcome these disadvantages, attempts have been made to use alkaline activators in portland blast furnace slag cement.

9.2.2 Rheological properties of cement pastes

The replacement of portland cement with supplementary cementing materials or the addition of alkaline activator changes the rheological properties of cement pastes. For a given water-to-cementing material ratio, the apparent viscosity and yield stress decrease as the slag content increases. However, the addition of NaOH significantly increases the apparent viscosity and yield stress of the system, as shown in Table 9.1.

Table 9.1 Effect of alkaline activator on apparent viscosity and yield stress of portland slag cement paste (Jiang 1997)

Material	Water/(slag + cement) ratio	Apparent viscosity (MPa · s)	Yield stress (MPa)
100% Slag	0.75	6.2	32
	1.5	3.2	21
100% Portland cement (PC)	0.75	676	5270
	1.5	43	710
70% PC + 30% Slag	0.75	405	4300
	1.5	92	980
70% PC + 30% Slag +1.5% NaOH	0.75	860	6700
	1.5	120	1100
50% PC + 50% Slag	0.75	315	4700
	1.5	80	820
50% PC + 50% Slag +2.5% NaOH	0.75	590	5800
	1.5	98	940
30% PC + 70% Slag	0.75	205	3050
	1.5	45	340
30% PC + 70% Slag +3.5% NaOH	0.75	445	4700
	1.5	59	430

9.2.3 Setting time

One significant disadvantage of the replacement of portland cement with ground granulated blast furnace slag is that it increases setting time of the cement. The use of alkaline activators definitely decreases setting time of the cement. Figure 9.1 shows the effect of NaOH dosage on setting times of cement consisting of 50% portland cement and 50% ground granulated

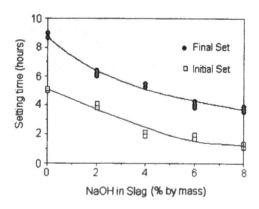

Figure 9.1 Effect of NaOH dosage on setting times of portland blast furnace slag cement consisting of 50% cement and 50% of slag (Jiang 1997).

blast furnace slag. It can be seen that both the initial and final setting times decreased with the increase of NaOH dosage.

9.2.4 Strength development

The use of alkaline activators can significantly increase the early strength of portland slag cement. Of course, the increase in strength will be dependent on the nature and dosage of the activator, and the nature and content of slag in the cement. Figure 9.2 shows the effect of different activators on the 14-day compressive strength of portland blast furnace slag cement containing 50% different blast furnace slags. Among the four activators – Na_2SiO_3, Na_2CO_3, NaOH and Na_2SO_4, Na_2SiO_3 is the most and Na_2SO_4 is the least effective activator. The strength of Na_2SiO_3-activated cements was around 50 MPa, while the strength of Na_2SO_4-activated cements was around 10 MPa. There is no much obvious difference observed between the four slags used. In another study (Singh *et al.* 2001), it was found that the addition of 4% Na_2SO_4 into a cement containing 50% blast furnace slag could significantly increase the strength of the cement. Actually, industrial wastewater containing alkalis can also be a good activator for portland slag cement (Sersale and Frigione 2004).

High volume replacement of cement with slag can drastically reduce the strength of cement mortars at early ages. The addition of alkali silicate can be very effective in increasing the early strength of high volume slag cements. Zivica (1993) found that a replacement of cement with 70% or 90% slag

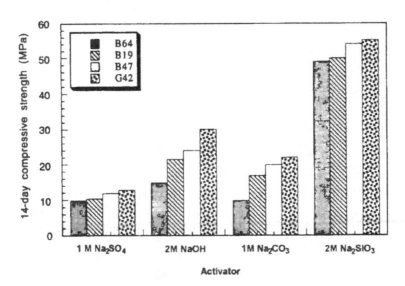

Figure 9.2 Effect of different activators on strength of portland blast furnace slag cement containing 50% slag (Jiang 1997).

decreased the strength of the cement very significantly from 1 to 90 days. However, the addition of sodium silicate into cement mortars containing 70% slag increased the strength very significantly, which was higher at 1 day but lower at 28 and 90 days than the portland cement mortars. The sodium silicate-activated cement mortars containing 90% slag had a higher strength than the portland cement mortars at 1 and 28 days, and they showed a similar strength at 90 days. The other feature was that the strength of sodium silicate-activated high volume slag cement mortars did not show an observable increase in strength from 28 to 90 days, while that of the portland cement mortars continued to increase from 28 to 90 days.

A measurement of strength development over a longer time period indicated that the use of alkaline activator in portland slag cement increases the strength of the cement during the first 90 days, and does not show any effect thereafter (Figure 9.3).

9.2.5 Distribution of silicate anions

The presence of alkaline activators accelerates the early hydration of slag and affects the silicate anion distribution in the cement pastes (Figure 9.4). TMS studies on alkali-activated portland slag cement have indicated that the content of polymer increases and the content of low-molecular-weight

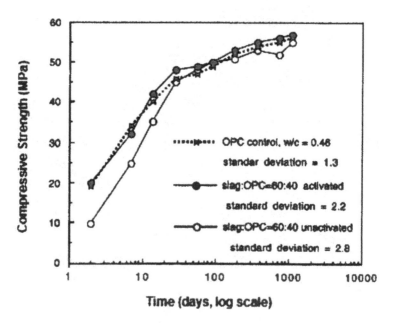

Figure 9.3 Long-term strength development of OPC, portland slag cement and activated portland slag cement (with 2M NaOH solution) (Jiang 1997).

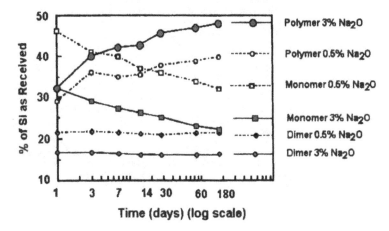

Figure 9.4 Distribution of silicate anions in alkali-activated portland slag cement (Roy and Jiang 1994).

polymers decreases as the dosage of alkaline activator increases (Roy and Jiang 1994).

9.3   Portland phosphorus slag cement

9.3.1   Introduction

As discussed in Chapter 3, granulated phosphorus slag is less reactive than blast furnace slag. It can be expected that the replacement of portland cement with ground phosphorus slag will result in longer setting time and lower early strengths. This section discusses the effect of alkaline activators on setting time, strength development and microstructure of portland phosphorus slag cement.

9.3.2   Setting time

The use of alkali activators can decrease the setting time of portland phosphorus slag cements, but the effectiveness will be dependent on the nature and dosage of the activator used (Tang 1986). Table 9.2 shows the effect of different activators, and phosphate and fluoride on the setting time of alkali-activated portland phosphorus slag cement. Comparing the results of Batch No. 1 and No. 5, it can be seen that the setting times of $Na_2O \cdot 2SiO_2$-activated portland phosphorus slag cement are obviously shorter than Na_2SO_4-activated portland phosphorus slag cement although the former has a lower clinker content than the latter. The addition of

Table 9.2 Setting times of alkali-activated portland phosphorus slag cements (Tang 1986)

No.	Activator (%Na$_2$O by mass)	Clinker content (%)	Water demand (%)	Na$_3$PO$_4$ (% P$_2$O$_5$)	Setting time (h:m)	
					Initial	Final
1	1.7% Na$_2$SO$_4$	30	26.0	0	4:10	7:30
2			26.0	0.2	7:45	13:45
3			25.5	0.5	11:05	20:00
4			25.0	0.8	15:30	26:35
5	2% Na$_2$O·2SiO$_2$	20	24.0	0	2:00	3:00
6			24.0	0.2	3:20	6:30
7			24.0	0.5	6:00	11:45
				NaF (% F)		
8	1.7% Na$_2$SO$_4$	30	25.5	0.3	3:45	6:55
9			25.5	0.5	3:25	6:30

Na$_3$PO$_4$ increases the setting time of the cement, while the addition of NaF even decreased the setting time of the cement.

9.3.3 Strength development

9.3.3.1 Effect of phosphorus slag content on strength development

As discussed in Chapter 3, granulated phosphorus slag is usually less reactive than granulated slag due to its lower content of Al$_2$O$_3$. Also, the presence of P$_2$O$_5$ in phosphorus slag retards the early hydration of cement. Thus, the replacement of portland cement with phosphorus slag can result in a significant drop in strength, especially at early ages and high replacement levels. Several studies have confirmed that the use of alkaline activators such as Na$_2$SO$_4$ is very effective in increasing early strength of portland phosphorus slag cement (Tang 1986, Shi *et al.* 1989b, Wang and Zhao 1990). The effect of phosphorus slag content on strength development of portland phosphorus slag cement is shown in Figure 9.5. At three and seven days, the strength of the cement decreases linearly with slag content. However, at 28 days, slag content does not show an observable effect on strength when slag content increases from 0 to 75%, and the further increase of slag content decreases strength significantly.

9.3.3.2 Nature and dosage of activators

A wide variety of chemicals can be used as the activator of portland phosphorus slag cement. However, based on the availability and cost, Na$_2$SO$_4$ and waterglass are the most promising ones (Tang 1986). Figure 9.6 shows the effect of Na$_2$SO$_4$ dosage on strength of portland phosphorus slag cement. It can be seen that the optimum Na$_2$SO$_4$ dosage is about 3–5%.

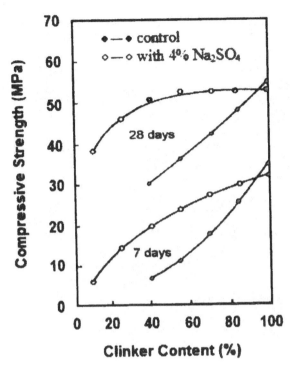

Figure 9.5 Effect of the addition of 4% Na₂SO₄ on compressive strength of portland phosphorus slag cement with different amounts of clinker (Shi *et al.* 1989b).

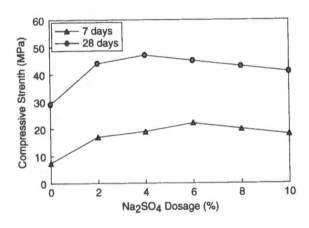

Figure 9.6 Effect of Na₂SO₄ dosage on strength of portland phosphorus slag cement consisting of 20% portland cement clinker and 80% phosphorus slag (data from Tang 1986).

Experimental results also indicate that by-product Na_2SO_4 from HCOOH production is more effective than natural sodium sulphate due to the presence of other sodium compounds such as NaCl and NaCOOH in the by-product.

The modulus of waterglass has a significant effect on strength development of high volume phosphorus slag cement. It has been found that the optimum modulus is between 1.5 and 2.5. The strength of the cement increases linearly with activator dosage and reaches a plateau or a peak at 4% dosage. However, the cement activated with optimum waterglass has a lower clinker content but higher strength than Na_2SO_4-activated portland phosphorus slag cement

9.3.3.3   Soluble phosphates and fluorides

As discussed above, soluble phosphates retard the setting of portland phosphorus slag cement. They decrease the strength of waterglass-activated portland phosphorus slag cement at 7 days, but do not show any effect on strength at 28 days, and decrease the strength of Na_2SO_4-activated portland phosphorus slag cement at both 7 and 28 days very significantly, especially when the P_2O_5 content is above 0.2%. The addition of NaF increases the strength of Na_2SO_4-activated portland phosphorus slag cement by more than 10% at both 7 and 28 days.

9.3.3.4   Curing conditions

Standard mortar strength testing methods, such as ASTM C 109 (2003), require that the mortar specimens be cured in $Ca(OH)_2$ saturated water after demoulding. Alkaline activators in alkali-activated slag cements are readily water-soluble compounds and can be leached out easily once it is immersed in water. If specimens are exposed to water too early, the activator may not have completed the activation action on the slag, and it can seriously affect the hydration of the slag and properties of the cement, especially the surface layer. Experimental results indicated that the 28-day strength of Na_2SO_4-activated portland phosphorus slag cement is decreased significantly if the specimens are immersed into water within five days after the preparation of the specimens (Tang 1986).

9.3.4   Hydration kinetics

The discussions above indicate that the use of Na_2SO_4 accelerates the setting and improves the strength development of portland phosphorus slag cements. Table 9.3 shows the effect of Na_2SO_4 on non-evaporable water, reacted phosphorus slag and compressive strength. It can be seen that the

Table 9.3 Effect of Na_2SO_4 on hydration kinetics of portland phosphorus slag cement (Wang and Zhao 1990)

Age (days)	67% Clinker + 30% Phosphorus slag + 3% Na_2SO_4			67% Clinker + 30% Phosphorus slag + 3% Gypsum		
	Compressive strength (MPa)	Non-evaporable water (%)	Reacted phosphorus slag (%)	Compressive strength (MPa)	Non-evaporable water (%)	Reacted phosphorus slag (%)
0.25	–	9.2	1.4	–	2.2	0.6
1	20.4	11.9	5.2	4.3	5.6	1.1
3	33.5	13.2	9.3	13.1	7.9	4.3
7	43	14.3	19.8	25.3	12	11
28	71.2	17.8	35	50	15.4	19.5

replacement of gypsum with Na_2SO_4 accelerates the hydration of phosphorus slag very significantly. After six hours of hydration, Na_2SO_4-activated portland phosphorus slag cement shows a non-evaporable water content of 9.2%, while gypsum-activated portland phosphorus slag cement shows a non-evaporable water content of only 2.2% at that time. Although the difference in non-evaporable water content decreases with time, Na_2SO_4-activated portland phosphorus slag cement showed significantly higher non-evaporable water content than gypsum-activated portland phosphorus slag cement from 1 to 28 days.

The differences in hydration degree of phosphorus slag between Na_2SO_4-activated and gypsum-activated portland phosphorus slag cements are also very significant, as shown in Table 9.3. The percentage of hydrated phosphorus slag in Na_2SO_4-activated portland phosphorus slag cement is almost twice as that in gypsum-activated portland phosphorus slag from 1 to 28 days.

9.3.5 Hydration products and microstructure

Microstructural examinations indicated that if phosphorus slag content is low, C–S–H, $Ca(OH)_2$ and AFt are the main hydration products; if phosphorus slag content is high, C–S–H is the only detected hydration product of Na_2SO_4-activated portland phosphorus slag cement (Wang and Zhao 1990).

9.4 Portland silica fume cement

Silica fume is a by-product in the production of metallic silicon or siliconalloy. The SiO_2 content of silica fume may exhibit a wide variation range due to different sources, form 25% from SiMn alloy production to 94% from Si production (Malhotra *et al.* 1987). Because of its high specific surface area, silica fume with a high content of SiO_2 shows high pozzolanic reactivity, as discussed in Chapter 3.

The replacement of portland cement with small amount of silica fume will increase, but a high volume replacement will reduce, the strength of cement. The replacement of portland cement with 30 and 50% silica fume decreases the early strength significantly, as shown in Figure 9.7 (Zivica 1993). The addition of sodium silicate increases the early and later strength of 30% silica fume replacement mortars, increasing the one-day strength from 4 MPa to 19.7 MPa, compared with 5.6 MPa for portland cement mortars. The activated mortars with 30% silica fume replacement also show significantly higher strength than portland cement mortars at 28 and 90 days. The addition of sodium silicate to 50% silica fume replacement cement mortars slightly increases the strength at 1 and 28 days but significantly increases the strength at 90 days. However, the strength of activated cement

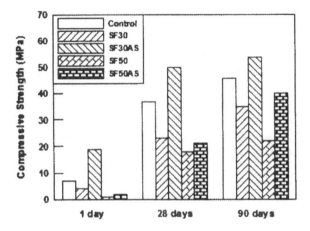

Figure 9.7 Strength development of high volume silica fume cement mortars (data from Zivica 1993).

mortars with 50% silica fume replacement is still lower than that of portland cement mortars.

Figure 9.8 is a comparison of cumulative pore volume of non-activated and activated cement mortars with 50% silica fume replacement. It can be seen that the addition of sodium silicate decreases the pore volume in excess of 20 nm.

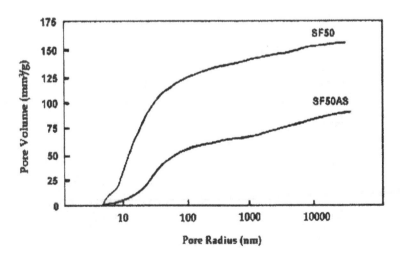

Figure 9.8 Cumulative pore volume of control (SF50) and sodium silicate activated (SF50AS) cement mortars with 50% silica fume (Zivica 1993).

9.5   Portland steel slag-iron slag cement

9.5.1   Introduction

Portland steel slag-iron slag cement, which is composed of 30–45% steel slag, 30–45% blast furnace slag, 20–35% portland cement clinker and 5–7% gypsum, has been commercially marketed in China since the 1970s. This type of cement has the advantages of lower energy cost, higher abrasion resistance, lower hydration heat evolution and higher later strength development, but the disadvantage of longer setting time and lower early strength when compared with portland cement (Sun 1983, Sun *et al.* 1993, Li *et al.* 1997, Zhu and Sun 1999). This section discusses how to use alkaline activators to overcome the disadvantages of portland steel slag-iron slag cement.

9.5.2   Setting time

Some attempts have been made to shorten the setting time, to increase the early strength and to decrease the cement clinker content of portland steel slag-iron slag cement using alkaline activators. Table 9.4 shows the addition of 4% Na_2SO_4 on the setting times of portland steel slag-iron slag cement. The results indicate that the addition of activators can decrease both the initial and final setting time of Grade-325 cement, and there is no obvious effect on Grade-425 cement.

9.5.3   Strength development

Figure 9.9 shows that the addition of 4% Na_2SO_4 to Grade-325 cement can increase its strength greatly; both its compressive and flexural strengths are even higher than those of Grade-425 cement (Shi *et al.* 1993). The addition of 4% Na_2SO_4 to Grade-425 can also increase its strength, but not as effectively as on Grade-325 cement. The seven-day and 28-day strengths of the activated Grade-425 cement are lower than those of the activated Grade-325 cement.

Table 9.4 Effect of Na_2SO_4 on setting times of portland steel slag-iron slag cements (Shi *et al.* 1993)

No.	Mixing proportions (%)				Setting time (h:m)	
	Steel slag	BFS	Clinker	Gypsum	Initial	Final
1 (Grade-325, control)	30	35	30	5	8:22	13:20
2	Control (Grade-325) + 4% Na_2SO_4				4:37	9:38
3 (Grade 425, control)	18	18	60	4	4:06	7:54
4	Control (Grade-425) + 4% Na_2SO_4				3:49	8:11

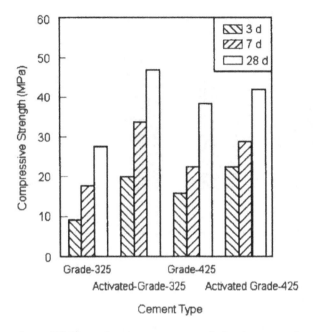

Figure 9.9 Effect of activator on strength development of portland steel slag-iron slag cements (Shi *et al.* 1993).

9.6   Portland steel slag-fly ash cement

In some countries, ground granulated blast furnace slag is widely used as a cement replacement and its price is much higher than coal fly ash. Researches have indicated that fly ash can be used together with steel slag to eliminate the potential soundness problem with satisfactory strengths. Figure 9.10 shows the strength development of Portland steel slag-fly ash cement (PSSFAC) mortars (Wu *et al.* 1999). The control cement consists of 93% cement clinker and 7% gypsum, PSSFAC1 consists of 45% cement clinker, 5% gypsum, 30% steel slag and 20% fly ash, and PSSFAC2 consists of 45% cement clinker, 3% gypsum, 2% high alumina cement, 30% steel slag and 20% fly ash. PSSFAC1 exhibited slightly lower strengths than the control cement mortars from 3 to 28 days. However, if 2% gypsum was replaced by high alumina cement as used in PSSFAC2, it gave even higher strength than the control cement mortars at three days. No difference in strength was observed at seven days, but PSSFAC2 also showed higher strength than the control cement at 28 days. It is also reported that PSSFACs showed lower drying shrinkage and much better corrosion resistance than the control cement (Wu *et al.* 1999).

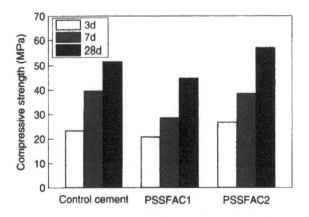

Figure 9.10 Strength development of portland-steel slag-fly ash cement (data from Wu *et al.* 1999).

9.7 Portland fly ash cement

It is well known that the strength of cement decreases as the fly ash content increases. Alkali sulphates are very effective activators for portland fly ash cement when its portland cement content is greater than 20%. Usually, the higher the fly ash content, the more effective is the alkaline activation. Figure 9.11 shows the effects of the addition of 3% Na_2SO_4 on the strength of the blended cement mortars containing 30 and 70% coal fly ash (Qian *et al.* 2001). When 30% cement is replaced with coal fly ash, the addition of 3% Na_2SO_4 increases strength by approximately 40% from 3 to 28 days. However, as fly ash replacement level is increased from 30 to 70%, the addition of 3% Na_2SO_4 increases strength by approximately 80% from 3 to 28 days. This means that the use of Na_2SO_4 is particularly effective for the cement with high fly ash replacements. A recent study compared the strength and microstructure of portland fly ash cements, which contained 40% fly ash and is activated by both Na_2SO_4 and K_2SO_4 (Lee *et al.* 2003). Both Na_2SO_4 and K_2SO_4 could increase the strength of the cement significantly, but K_2SO_4 was more effective in increasing the strength of the cement than Na_2SO_4. Na_2SO_4 was found very effective in improving the strength of a blended cement containing coal fly ash and spray dry ash from flue gas desulphurization collection system (Wu and Naik 2003).

When Na_2SO_4 is added, Na_2SO_4 reacts with $Ca(OH)_2$ first as follows:

$$Na_2SO_4 + Ca(OH)_2 + 2H_2O \rightarrow CaSO_4 \cdot 2H_2O \downarrow + 2NaOH \qquad (9.1)$$

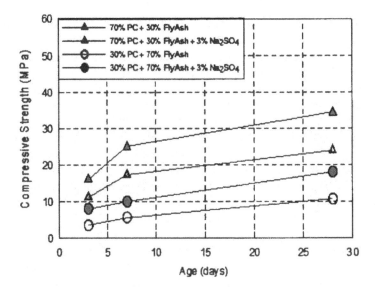

Figure 9.11 Effect of Na₂SO₄ on strength development of portland fly ash cement (data from Qian *et al.* 2001).

The reaction increases the pH of the solution, accelerates the dissolution of fly ashes and speeds up the pozzolanic reaction between $Ca(OH)_2$ and fly ashes. At the same time, the introduction of Na_2SO_4 increases the concentration of SO_4^{2-} and results in the formation of more ettringite (AFt). The formation of AFt increases the solid volume by 164%, but the formation of C–S–H increases the solid volume only by 17.5%. Thus, the generation of AFt densifies the structure and increases the early strength of hardened cement pastes very significantly (Shi and Day 2000b, Lee *et al.* 2003, Wu and Naik 2003). Microstructural examinations (Lee *et al.* 2003) have indicated that the use of Na_2SO_4 or K_2SO_4 as activator obviously increased the consumption of $Ca(OH)_2$ and the formation of ettringite. At the same, it also decreased the total porosity and increased the portion of small pores. Actually, it was noticed that K_2SO_4 was more effective than Na_2SO_4 in reducing the total porosity and pore size. Thus, the high early strength of the alkali sulphate-activated pastes is attributed to two aspects: acceleration of early pozzolanic reactions and the formation of more AFt.

When cement content is less than 20%, waterglass can be a very effective activator for the cement (Zhang *et al.* 1997). The addition of certain amount of gypsum (less than 2% by mass of SO_3) can enhance the activation effect, excessive gypsum (greater than 2% by mass of SO_3) will decrease the strength of the cement drastically.

9.8   Multiple-components blended cements

Dense-packing is an essential feature of DSP materials. It has been demonstrated that a small portion of fine particles can have a significant effect on the pore structure and properties of hardened cement materials (Lu and Young, 1993). Based on the particle-packing theories, it has been found that the effective packing of various size grain mixtures also plays an important role in the performance of alkali-activated cementing systems (Roy and Malek 1993, Roy and Silsbee 1994, 1995).

After fairly extended studies, it was found that rather optimal performance was achieved in materials containing about 50% portland cement, with the remainder 50% solids distributed among several components (Roy and Silsbee 1992). Figure 9.12 shows the chemical composition of alkali-activated cementing materials and how the addition of a small amount of silica fume (4% by mass) affects the strength of the cement.

The composition with silica fume had developed a compressive strength of approximately 28 MPa by five hours after mixing in contrast to the formulation without silica fume which was just beginning to pass through the final set at this time. This indicates the importance of the silica fume in the cement chemistry and microstructurual development even at these early ages.

A study (Roy and Silsbee 1992) compared two nearly identical formulations except for different types of fly ash used. The one with Class C (high calcium) fly ash containing paste had developed strength of approximately 28 MPa by five hours whereas the Class F (low calcium) fly ash containing

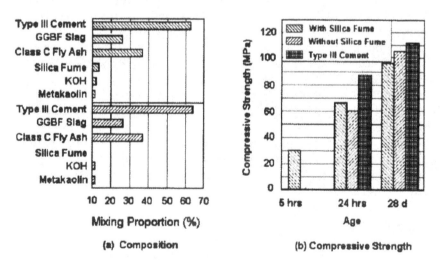

Figure 9.12 Composition and strength of alkali-activated multiple component cementing systems (Roy and Silsbee 1992).

paste had not passed the final set at this time. The one-day compressive strengths indicate that the high Ca fly ash, being more reactive, has contributed more to the strength development at this time. However, by 28 days the products of the less reactive low Ca fly ash containing paste have reached similar strengths.

X-ray diffraction analyses were conducted on samples of alkali-activated blended cement pastes containing Class C and Class F fly ashes after nine months of curing at room temperature in a moist atmosphere. The patterns indicate that at this time the principal crystalline phases present are $Ca(OH)_2$ and residual cement phases, with smaller amounts of ettringite being present in both cases. The patterns give evidence that the low calcium (Class F) fly ash was more effective in reducing the content of calcium hydroxide (Roy and Silsbee 1992).

Malek and Roy (1997a) prepared a broad range of alkali-activated aluminosilicates with target composition of variable n and m in the formula:

$$\frac{n}{x}M^{x+}((SiO_2)_m(AlO_2)_n)\cdot yH_2O$$

where M is an alkali or alkaline earth metal cation. Two systems were studied, namely, low-calcium and high-calcium systems. The low-calcium system is composed of metakaolinite, Class F fly ash, potassium silicate and potassium hydroxide. The high-calcium system is composed of metakaolinite, Class C fly ash, ground granulated blast-furnace slag, potassium silicate and potassium hydroxide. Two- and three-dimensional composition diagrams were used to design the target compositions. Figure 9.13 represents the composition diagrams for aluminosilicate mixtures presented in the present paper, where $x = 1$, $n = 1$, and $m = 4$, for both low-calcium and high-calcium formulations. Blends with cement were prepared as follows:

a   Portland cement 80% + high-calcium cement 20%
b   Portland cement 50% + high-calcium cement 50%
c   Portland cement 80% + low-calcium cement 20%
d   Portland cement 50% + low-calcium cement 50%

Compressive strength of samples cured for 7 and 14 days at 38 °C are presented in Figure 9.13. In these figures, the compressive strength of the blended cements containing low-calcium aluminosilicate cement showed slightly lower strength, but the blended cements containing high-calcium aluminosilicate cement showed slightly higher strength than portland cement pastes and a commercially available aluminosilicate binder (Pyrament) at the same age and temperature. It is evident that the compressive strength values of cement pastes containing 20% and 50% of high-calcium materials are equivalent to those of neat cement pastes and slightly exceed those of the Pyrament. The corresponding blends made with low-calcium

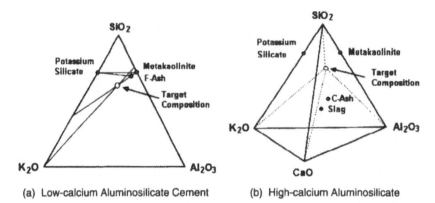

(a) Low-calcium Aluminosilicate Cement   (b) High-calcium Aluminosilicate

Figure 9.13 Target compositions for low- and high-calcium aluminosilicate cements (Malek and Roy 1997a).

materials are lower in strength but their strength seems to increase with time. The pore structure evolution has been studied by mercury intrusion porosimetry. Results of pore size distribution measurement indicated that the alkali-activated systems develop a finer pore structure compared to neat cement pastes. In addition, the blends with high-calcium materials seem to develop the finest pore structure, possibly because they are more fully reacted. The 28-day mortar bar expansions using ASTM C 227 to test potential alkali–silica reaction clearly displayed that the alkali-activated aluminosilicate materials gave the lowest reaction rate with reactive silica.

9.9 Summary

This chapter has discussed the alkaline activation of portland-based cements containing different supplementary cementing materials such as blast furnace, phosphorus slag, steel slag, coal fly ash, silica fume, metakaolin or a combination of two or more of them. When portland cement content is high, Na_2SO_4 can be a very effective activator. When portland cement content is low, sodium silicate can be an effective activator.

Chapter 10

Alkali-activated lime-pozzolan cements

10.1 Introduction

Lime-pozzolan mortar was one of the earliest building materials. It was widely used in the masonry construction of aqueducts, arch bridges, retaining walls and buildings during Roman times (Hazra and Krishnaswamy 1987). Many Roman monuments such as the bridges of Fabricus, Aemilius, Elius and Milvius, the arches of Claudius and Trajan at Ostia and Nero at Antium, together with many maritime works, some of which, such as those erected by Trajan at Ancona and Civitavecchia, are still in use today, and stand as a tribute to the permanence of lime-pozzolan mortars (Bogue 1955, Lea 1974). Lime–Surkhi mixtures (Surkhi is pulverized fire clay or brick-earth) have commonly been used since Greece and Roman times and are still being widely used in India (Spence 1974).

One obvious disadvantage of lime-pozzolan cement is its slow strength development during ambient temperature curing. The invention of portland cement in the 19th century resulted in a drastic reduction in the use of lime-pozzolan cement because of the faster setting and higher early strength that could be realized. Nevertheless, the low cost and excellent durability of lime-pozzolan cement make them attractive for some applications. In the past 50 years, lime-pozzolan cement has again been used for manufacturing construction products and other uses, especially in some developing countries, because of their low cost (Coad 1974, Dave 1981, Spence and Sakela 1982, Apers and Pletinck 1985). This chapter discusses the activation of lime-pozzolan and lime-slag cements.

10.2 Activation of lime-natural pozzolan cement

10.2.1 Effect of activator nature and dosage on activation of lime-pozzolan cements

It can be expected that most alkaline compounds can be used as activators of lime-pozzolan cement. However, different activators may show different

Table 10.1 Effect of alkaline activators on strength of lime-pozzolan cement pastes (W/S = 0.45) (Shi et al. 1994)

No.	Composition	Compressive strength (MPa)	Relative strength (%)	Non-evaporable water content (%)
1	80% NP + 20% Lime	3.5	100	7.62
2	(80% NP + 20% Lime) + 4% Na_2SO_4	5.5	157	6.28
3	(80% NP + 20% Lime) + 4% Na_2CO_3	4.0	114	7.09
4	(80% NP + 20% Lime) + 4% NaOH	4.8	137	8.05
5	(80% NP + 20% Lime) + 6% $Na_2SiO_3 \cdot 5H_2O$	3.7	106	7.16

effects. Table 10.1 shows the strength of lime-pozzolan cements activated by 4% Na_2SO_4, NaOH, Na_2CO_3 and 6% $Na_2SiO_3 \cdot 5H_2O$ after three days at 25 °C and four days at 50 °C. These alkaline activators increase the compressive strength from 5 to 60% compared with that of the control, and the batch added with Na_2SO_4 exhibits the highest strength. Since Na_2SO_4 is a widely available chemical at reasonable costs, it can be regarded as the most effective activator for lime-pozzolan cement.

Table 10.1 also lists the non-evaporable water of these cement pastes. All alkali activated lime-pozzolan pastes show lower non-evaporable water contents than the control paste, except Na_2CO_3. Examination of TGA curves indicated that Na_2CO_3-activated lime-pozzolan pastes have two weight loss peaks that other pastes did not have. One corresponds to the decomposition of $CaCO_3$, and the other one appeared between 500 and 600 °C, which may be attributed to $C_3A \cdot CaCO_3 \cdot 12H_2O$ (Shi et al. 1994).

Since Na_2SO_4 is the most efficient activator for lime-pozzolan cement, a further detailed study has indicated that the strength of the lime-pozzolan cement pastes increases linearly with the Na_2SO_4 dosage at early ages, as shown in Figure 10.1 (Shi and Day 1993b). At 28 and 90 days, the strength of the pastes increases noticeably when the Na_2SO_4 dosage is increased from below 1% to within the range 2–4%. At three days, the strength of the pastes with 4% Na_2SO_4 is about 3 times as high as that of the control pastes, and about 1.5 times that of the control at 7, 28 and 90 days. Although 4% dosage is the optimum amount, the marginal strength gain achieved in going from 2 to 4% dosage must be balanced against the concomitant increase in cost.

10.2.2 Activation of lime-pozzolan cements at different temperatures

One obvious disadvantage of lime-pozzolan cement is its slow strength development during room temperature curing. Several studies have

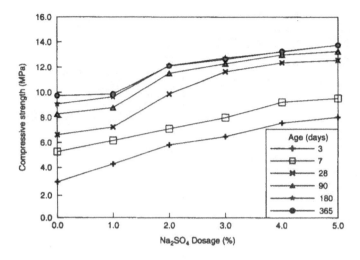

Figure 10.1 Effect of Na$_2$SO$_4$ dosage on strength development of lime-natural pozzolan cement pastes cured at 50 °C (Shi and Day 1993b).

confirmed that elevated temperature curing is very helpful to increase the early strength of lime-pozzolan cements. For example, mortars made with Santorin Earth did not show any strength at 7 and 28 days at 0 °C, but they did show high strength when cured at 35 °C (Lea 1974). In another study, lime-pozzolan pastes cured at ambient temperatures and tested at seven days gave strengths of about 0.8 MPa, but gave 6 MPa when the pastes were cured at 52 °C (Day 1992). For practical use of lime-pozzolan cements it is essential that they obtain satisfactory strength in an acceptable time period. Figure 10.2 shows the effect of curing temperature on strength development of the lime-pozzolan cement pastes without any activator. A rise in curing temperature accelerates early age strength development greatly. The pastes do not show measurable strength until about four days at 23 °C, but have strength of about 2.5 MPa after one day at 65 °C. Strength levels off to a shallow slope (plateau) at about 75 days at 23 °C, about 20 days at 35 °C, 15 days at 50 °C and 10 days at 65 °C. The rate of initial strength development increases as curing temperature increases.

The addition of Na$_2$SO$_4$ does not change the trends just noted (as shown in Figure 10.3), but the pastes with 4% Na$_2$SO$_4$ show higher strength than the control pastes at a given curing temperature and age. Characteristics of the strength plateau are also similar to that of control pastes at the same curing temperature, but the chemically activated cement pastes exhibit much higher strength than the control pastes.

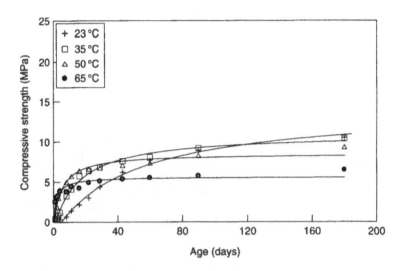

Figure 10.2 Effect of curing temperature on strength development of control pastes (Shi and Day 1993a).

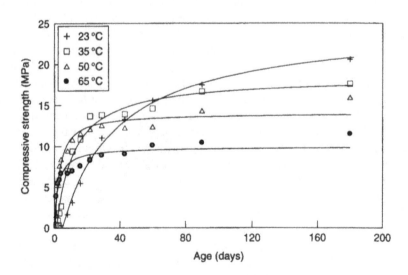

Figure 10.3 Effect of curing temperature on strength development of lime-pozzolan pastes with 4% Na_2SO_4 (Shi and Day 1993a).

The strength–age relationship of cement pastes, mortars and concrete can be described by the following equation (Knudsen 1980, Carino 1984, Roy and Idorn 1985):

$$S = S_o \frac{K_T(t - t_o)}{1 + K_T(t - t_o)} \tag{10.1}$$

where
  S = compressive strength (MPa);
  S_o = ultimate compressive strength at infinite age (MPa);
  K_T = strength development constant at curing temperature, T (day^{-1});
  t = actual curing age at temperature, T (days);
  t_o = theoretical initial hardening time (days).

Based on the data in Figures 10.2 and 10.3, the theoretical best-fit values of S_o, K_T and t_o are outlined in Table 10.2. The curves in Figures 10.2 and 10.3 are derived from Eqn (10.1). Some features of these results in Table 10.2 can be summarized as follows (Shi and Day 1993a):

- Initial hardening time t_o decreases as curing temperature increases.
- Strength development rate constants of Na_2SO_4-activated cements appear to be less efficient above 50°C, which may be caused by different pozzolanic reactions and different reaction products occurring above 50°C.
- The values of the theoretical ultimate strength S_o of control pastes decreases linearly with the curing temperatures over temperature between 23 and 65°C; this is valid for Na_2SO_4-activated pastes between 35 and 65°C.

Table 10.2 Strength development characteristics of lime-pozzolan cement pastes (Shi and Day 1993a)

Sample	Curing temperature (°C)	S_o (MPa)	σ_1^*	K_T(day^{-1})	σ_2	t_o (days)	σ_3
	23	14.3	0.87	0.019	0.003	4.65	1.01
Control	35	11.0	0.36	0.060	0.007	1.12	0.35
	50	8.5	0.31	0.197	0.032	0.50	0.35
	65	5.6	0.27	0.520	0.226	Nil	Nil
	23	24.8	0.99	0.029	0.003	4.71	0.63
Control +	35	18.4	0.67	0.096	0.014	1.34	0.30
4% Na_2SO_4	50	14.0	0.51	0.377	0.094	0.50	0.31
	65	9.8	0.52	0.456	0.205	Nil	Nil

* Estimated standard error of the regression parameters; $1 = S_o, 2 = K_T, 3 = t_o$.

The Arrhenius equation can be employed to depict the effect of curing temperature on cement hydration or the strength development of cement pastes (Knudsen 1980, Roy and Idorn 1982, Wu et al. 1983):

$$K_T = A \exp\left(\frac{-E_a}{RT}\right)$$ (10.2)

where
  A = frequency constant (day^{-1});
  E_a = apparent activation energy (J/mol);
  R = gas constant (8.314 J/K.mol);
  T = Kelvin temperature (K).

If K_T vs $1/T$ (K^{-1}) is plotted in log scale, the relationships demonstrate excellent linearity within the studied temperature range except for the Na$_2$SO$_4$-activated pastes at 65°C (Shi and Day 1993a). Table 10.3 outlines the regression results of these straight lines within the studied temperature range. Compared with the hydration activation energy of portland cement (about 45 KJ/mol) (Wu et al. 1983), the hydration activation energies of lime-pozzolan cement pastes are high (66 KJ/mol); the addition of 4% Na$_2$SO$_4$ influences hydration by increasing the activation energy by about 15%. Thus, curing temperature will have more effect on the strength development of the Na$_2$SO$_4$-activated pastes than on the control pastes. Compared to the control pastes, Na$_2$SO$_4$-activated pastes achieve a larger percentage of ultimate strength at early age and then approach the ultimate strength more gradually.

A detailed comparison based on technical and economical analyses indicated that chemical activation is much more effective and efficient than either thermal activation or mechanical activation (Shi and Day 2001).

10.2.3 Effect of Na$_2$SO$_4$ on pozzolanic reaction in lime-pozzolan cement

The tendencies for the consumption of Ca(OH)$_2$ in Na$_2$SO$_4$-activated lime-pozzolan pastes from 23 to 65°C are the same as the control pastes (Shi and

Table 10.3 Regression results for strength rate constant–age relationship (Shi and Day 1993a)

Pastes	A(day^{-1})	Correlation coefficients	Activation energy Ea (KJ/mol)
Control	8.71×10^9	0.999	66
Na$_2$SO$_4$-activated Pastes*	5.74×10^{11}	1.000	75

* Between 23 and 50°C.

Day 2000a); however, $Ca(OH)_2$ in Na_2SO_4-activated pastes was consumed more quickly than that in control pastes during the initial rapid reaction stage. Thus, the addition of Na_2SO_4 accelerated the early pozzolanic reaction. But a small amount of $Ca(OH)_2$ could still be detected in Na_2SO_4-activated pastes at later stages regardless of curing temperatures, but the remaining $Ca(OH)_2$ in Na_2SO_4-activated pastes was less than that in the control pastes.

The modified Jander's Equation is often used to describe the hydration kinetics of cementing materials (Kondo *et al.* 1976):

$$(1 \,\sqrt[-3]{1-\alpha})^N = Kt \tag{10.3}$$

where
  α = reaction degree;
  K = reaction constant;
  t = reaction time; and
  N = reaction grade.

1 If the reaction is controlled by the reaction happening on the surface of grains, or by the dissolution of reactants or the precipitation of reaction products, then $N < 1$.
2 If the reaction is controlled by the diffusion of reactants through a layer of porous reaction products, then $1 < N < 2$.
3 If the total reaction is controlled by the diffusion of reactants through a layer of dense reaction products, then $N > 2$.

Figure 10.4 is the plot of the degree of reacted $Ca(OH)_2$ in lime-pozzolan cements at different times using Eqn (10.3) in log-scale. The slopes of these straight lines are the reaction grade N. Table 10.4 lists the grade N and phase period of pozzolanic reactions, which took place in the control and Na_2SO_4-activated pastes at different temperatures. For the control pastes, Phase I was not detected regardless of curing temperatures. This means that Phase I completed before the determination of free lime in this study. From 23 to 50 °C, pozzolanic reactions in the control pastes behaved in the same way: Phase II ended after approximately three days of curing and Phase III started thereafter. As curing temperature increased to 65 °C, only Phase III could be identified.

From 23 to 65 °C, the trends of pozzolanic reactions in the Na_2SO_4-activated pastes are very similar to that in the control pastes, except that some offset of the reaction degree occurred. This means that the addition of Na_2SO_4 accelerates the consumption of lime or the pozzolanic reaction rate mainly during the first day.

The measurable compressive strength and the degree of reaction of $Ca(OH)_2$ for the control and Na_2SO_4-activated lime-pozzolan pastes at different temperatures can be correlated with fair linear relationships. However, the degrees of reaction of $Ca(OH)_2$ corresponding to the initial measurable strength and the slopes of these linear relationships are very different,

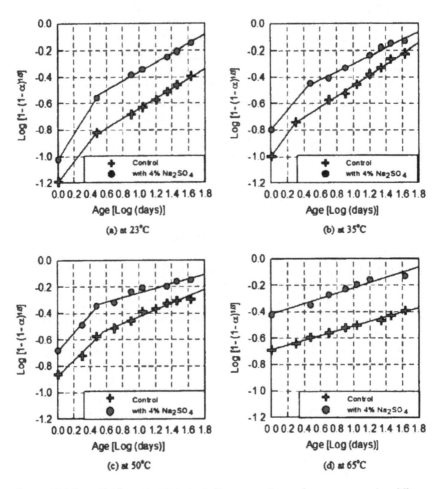

Figure 10.4 Pozzolanic reaction kinetics in lime-natural pozzolan pastes cured at different temperatures (Shi and Day 2000a).

which might be attributed to the formation of different hydration products (Shi and Day 2000a).

10.2.4 Effect of fineness of natural pozzolan on strength of lime-pozzolan cements

Increasing the fineness of cement within a proper range can improve its early strength, especially as described for the cement containing mineral admixtures, because lime-pozzolan cement pastes have much higher hydration activation energy than portland cement pastes. It can be expected that the fineness of a lime-pozzolan cement, mainly the fineness of the natural

Table 10.4 Reaction grade N and phase period of the two pastes at different temperatures (Shi and Day 2000a)

Curing temperature	Paste	Phase I (N < I)		Phase II (2 > N > I)		Phase III (N > 2)	
		N	Period	N	Period	N	Period
	Control	N/A	N/A	1.28	0–3 d	2.90	>3 d
23 °C	+ 4% Na₂SO₄	N/A	N/A	1.02	0–3 d	2.68	>3 d
	Control	N/A	N/A	1.16	0–3 d	2.53	>3 d
35 °C	+ 4% Na₂SO₄	N/A	N/A	1.36	0–3 d	3.40	>3 d
	Control	N/A	N/A	1.69	0–3 d	3.87	>3 d
50 °C	+ 4% Na₂SO₄	N/A	N/A	1.43	0–3 d	5.81	>3 d
	Control	N/A	N/A	N/A	N/A	5.68	>1 d
	+4 % Na₂SO₄	N/A	N/A	N/A	N/A	5.11	>1 d

N/A – Not applicable.

pozzolan used, shows a more obvious effect on the early strength of its pastes than on that of portland cement pastes.

Some researches have been conducted on how the fineness of pozzolan influences pozzolanic reaction between lime and pozzolan or the strength of lime-pozzolan cement pastes. In an early investigation of the reactivity of 6 siliceous rocks, Alexander (1960) found that siliceous materials such as quartz and basic or devitrified volcanic rocks, which were not regarded as likely sources of active pozzolan, became highly reactive when ground to produce ultrafine powders. He attributed it to the presence of a disturbed layer of highly reactive material, which was formed on the surface of siliceous mineral particles as a result of prolonged grinding. Greenberg (1961) found that for different siliceous materials with different surface areas, the reaction rate with $Ca(OH)_2$ solution did not show any correlation with the surface areas of the siliceous materials.

Costa and Massazza (1974) studied 6 commercial Italian pozzolans with different surface areas. They observed that the correlation between combined lime and surface area was valid only before seven days, and the activity of the pozzolans depended on the reactive content $(SiO_2 + Al_2O_3)$. But no correlation was found between the fixed $Ca(OH)_2$ and the surface areas for different pozzolans in another research (Mortureux, B. *et al.* 1980). Rossi and Forchielli (1976) also did not find any proportionality between surface areas and reactivity with lime for the same material. A strength test on 22 pozzolans (Chatterjee and Lahiri 1967) indicated that there was no general correlation between the compressive strength at 28 or 60 days and surface area (either by Blaine or BET method) of different materials. However, the strength increased as fineness increased for a single material.

Most of the above studies used different pozzolans with different adsorption surface areas. Fineness is not the only factor affecting the reactivity of the pozzolan; the chemical compositions, mineral composition, glass

content of the pozzolan and the cooling rate of the fused magma are also
very important for the reactivity of the pozzolan. Day and Shi (1994) found
that the compressive strength of lime natural-pozzolan cement pastes cured
at 50 °C correlates linearly with the Blaine fineness of the natural pozzolan
in all cases, and the slope of the linear relationship does not change significantly between 3 and 90 days. It seems that the fineness has its greatest
influence before three days under such curing conditions. For the pastes
with 4% Na_2SO_4, the Blaine fineness of the natural pozzolan shows some
effect on strength at three and seven days, and the lines are parallel after
seven days.

10.2.5 Microstructure

Extensive XRD analyses on hardened control and Na_2SO_4-activated lime-
pozzolan pastes indicated that the addition of Na_2SO_4 results in or enhances
the formation of AFt (Shi 1992). The diffraction peaks of AFt are intensified
with time. The diffuse band of C–S–H at $d = 3.04$ Å in Na_2SO_4-activated
lime-pozzolan pastes is sharper than that in control pastes. The diffraction
peaks of $Ca(OH)_2$ of the former diminish more quickly at early ages but
are much more similar to those in the control pastes between 90 and 180
days. As the curing temperature is elevated from 23 to 65 °C, the XRD
patterns of Na_2SO_4-activated pastes at 35 °C were the same as those at
23 °C; however, AFm appeared at 50 and 65 °C in addition to those peaks
appearing at 23 °C and 35 °C.

Figures 10.5 and 10.6 show SEM pictures of control and Na_2SO_4-activated
lime-natural pozzolan pastes cured at 23 °C for different ages (Shi 1992).
At a given age, the structure of Na_2SO_4-activated lime-pozzolan cement pastes
is clearly less porous than that of the control pastes. The microstructure of
Na_2SO_4-activated lime-pozzolan cement pastes at three days is similar to that
of the control pastes: all pozzolan particles are covered by a layer of grape-
like C–S–H, and most C–S–H-covered pozzolan particles are also isolated,
but the C–S–H gel surrounding pozzolan particles in the Na_2SO_4-activated
lime-pozzolan cement pastes is thicker than that in the control pastes. Many
fine needles, which are attributed to AFt, can be observed between the C–S–H-
covered pozzolan particles. Figure 10.6b shows that a network made with foil-
like C–S–H can be observed as the age proceeds to seven days. At the same time,
coarse AFt needles can be observed under SEM in most locations. At 90 days,
some needle clusters can be observed in air voids (Figure 10.6c). Figure 10.6d
shows AFt in the voids of Na_2SO_4-activated lime-pozzolan cement pastes at
180 days; needle clusters in the voids become bigger and coarser compared
with those at 90 days.

Elevation of curing temperatures accelerates the pozzolanic reaction, especially at early ages (Shi and Day 1993a). The microstructure of Na_2SO_4-
activated lime-pozzolan cement pastes at 35 °C is similar to that at 23 °C.
However, in addition to needle-like crystals, there are some hexagonal

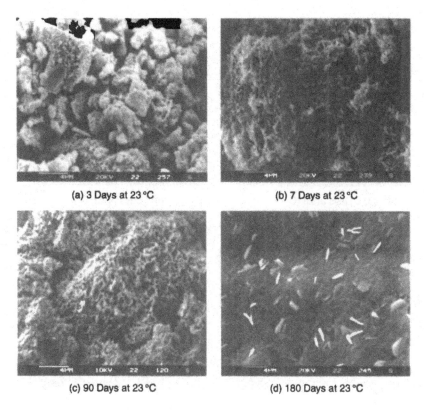

| (a) 3 Days at 23 °C | (b) 7 Days at 23 °C |
| (c) 90 Days at 23 °C | (d) 180 Days at 23 °C |

Figure 10.5 SEM pictures of hardened control lime-pozzolan cement pastes (Shi 1992).

plates, which are always accompanied with needles at 50 °C. It can be deduced that these hexagonal plates are AFm. The amount of hexagonal plates decreases with time, and no more hexagonal plates but coarse needles could be observed at 90 days and beyond. At 65 °C, some coarse gypsum exists in some voids in addition to AFt and AFm in Na_2SO_4-activated pastes at one day, and gypsum disappears, and coarse AFt and AFm co-exist at three days and thereafter.

10.2.6 Pozzolanic reaction mechanism in the presence of Na_2SO_4

10.2.6.1 Lime-pozzolan cement

In the absence of an activator, when a lime-pozzolan blend is mixed with water, $Ca(OH)_2$ in the blend hydrolyses first and the solution reaches a high pH value (approximately 12.5 at 20 °C) very quickly:

$$Ca(OH)_2 \rightarrow Ca^{2+} + 2OH^-$$

$$(10.4)$$

(a) 3 Days at 23 °C   (b) 7 Days at 23 °C

(c) 90 Days at 23 °C   (d) 90 Days at 23 °C

(e) 90 Days at 50 °C   (f) 28 Days at 65 °C

Figure 10.6 SEM pictures of hardened Na_2SO_4-activated lime-pozzolan cement pastes (Shi 1992).

Under the attack of OH^- in such a high pH solution, network modifiers, such as Ca^{2+}, K^+, Na^+, etc., in the pozzolan are dissolved into the solution very quickly. Silicate or aluminosilicate network formers in the pozzolan are also depolymerized and dissolved into the solution. However, a larger fraction of network modifiers rather than network formers is dissolved into the solution due to the lower bond energy of network modifiers. Depolymerized monosilicate and aluminate species enter the solution as follows:

$$\equiv Si-O-Si \equiv + 3OH^- \rightarrow (SiO(OH)_3)^- \tag{10.5}$$

$$\equiv Si-O-Al \equiv + 7OH^- \rightarrow (SiO(OH)_3)^- + (Al(OH)_4)^- \tag{10.6}$$

When Ca^{2+} ions contact these dissolved monosilicate and aluminate species, calcium silicate hydrate (C–S–H) and calcium aluminate hydrate (C_4AH_{13}) form:

$$Y(SiO(OH)_3)^- + XCa^{2+} + (Z-X-Y)H_2O$$
$$+(2X-Y)OH^- \rightarrow C_X-S_Y-H_Z \tag{10.7}$$

$$2(Al(OH)_4)^- + 4Ca^{2+} + 6H_2O + 6OH^- \rightarrow C_4AH_{13} \tag{10.8}$$

Although lime-pozzolan pastes do not show a significant strength after three days at 23 °C, SEM observation shows that all pozzolan particles are coated by a layer of grape-like C–S–H gel (Shi 1992). The chemical composition of C–S–H varies with the local concentration of reactants and reaction conditions (Takemoto and Uchikawa 1980).

Because dissolved monosilicate species diffuse much more quickly than dissolved aluminate species, and a higher concentration of Ca^{2+} is needed for the formation of calcium aluminate hydrates than for hydrated calcium silicates, hydrated calcium aluminates precipitate away from pozzolan particles and hydrated calcium silicates precipitate around the pozzolan particles (Takemoto and Uchikawa 1980).

In the presence of sulphate ions, AFt forms first. Because the SO_3 content of the natural pozzolan is low, AFt transforms to AFm very quickly. These reactions can be expressed as follows:

$$2(Al(OH)_4)^- + 3SO_4^{2-} + 6Ca^{2+} + 4OH^-$$
$$+26H_2O \rightarrow C_3A \cdot 3CaSO_4 \cdot 32H_2O \tag{10.9}$$

$$6Ca^{2+} + 2(Al(OH)_4)^- + 3CaO \cdot Al_2O_3 \cdot 3CaSO_4 \cdot 32H_2O$$
$$+10OH^- \rightarrow 3(3CaO \cdot Al_2O_3 \cdot CaSO_4 \cdot 12H_2O) + 5H_2O \tag{10.10}$$

Since the dissolution of aluminosilicate glass is the slowest process during the initial pozzolanic reaction, it determines the total pozzolanic reaction rate. After a certain period, the surface of pozzolan particles is covered by precipitated hydration products; then further reaction is controlled by the diffusion of OH^- and Ca^{2+} through the precipitated products and into the inner side of precipitated products. The later hydration is no longer a solution–precipitation reaction but a topochemical reaction. Based on the analysis of reaction kinetics and hydration products in previous sections, it seems that a rise in curing temperature does not affect the hydration process, but the hydration rate.

10.2.6.2 Na_2SO_4-activated lime-pozzolan cement

When Na_2SO_4 is added, the reaction between $Ca(OH)_2$ and Na_2SO_4, as expressed by Eqn (10.11), happens first (Roy 1986):

$$Na_2SO_4 + Ca(OH)_2 + 2H_2O \rightarrow CaSO_4 \cdot 2H_2O\downarrow + 2NaOH \qquad (10.11)$$

This reaction results in a higher alkaline solution than the $Ca(OH)_2$-saturated solution. Laboratory measurements (Shi and Day 2000a) indicated that the addition of 4% Na_2SO_4 raises the pH value from 12.50 to 12.75. The dissolution rate and solubility of $Ca(OH)_2$ may be decreased due to the increase in alkalinity of the solution.

Usually, natural pozzolans consist of siliceous or aluminosilicate glass. According to their structure, it can be inferred that pH will have a similar effect on the dissolution of natural pozzolans as that on the amorphous SiO_2. The presence of Al increases the dissolution of natural pozzolans due to the lower bonding energy of the Al–O bonds compared to that of the Si–O bonds. It is well known that there is an abrupt rise in dissolution of amorphous SiO_2 at about $pH = 12.5$ at room temperature; after this point, the dissolution of amorphous SiO_2 increases very steeply with pH (Iler 1979, Tang and Han 1981). This means that the addition of Na_2SO_4 accelerates the dissolution of pozzolans and speeds up the pozzolanic reaction between $Ca(OH)_2$ and the pozzolan. The kinetic analyses in previous sections have indicated that the acceleration occurs mainly when the reaction is controlled by the dissolution of the natural pozzolan. At the same time, the introduction of Na_2SO_4 increases the concentration of SO_4^{2-} and results in the formation of more AFt than in the control pastes.

The calculations based on the properties of hydration products (Shi and Day 2000b) indicate that the generation of AFt increases the solid volume by 164%, but the formation of C–S–H increases the solid volume by 17.5%. Thus, the generation of AFt reduces voids in the structure and increases the early strength of hardened lime-pozzolan pastes very significantly (Electrical Power Research Institute 1991). Thus, the high early strength of the Na_2SO_4-activated pastes is attributed to two aspects: acceleration of early pozzolanic reaction and the formation of more AFt.

Monosulphoaluminate (AFm) was identified at 50 °C and 65 °C in the Na_2SO_4-activated pastes. Thermodynamic calculation indicates that AFt is the stable phase below 70 °C and AFm above 70 °C (Babushkin *et al.* 1985). Due to the variation in the concentration of reactants and environmental conditions, it is possible for AFm to form directly or from the conversion of AFt below 70 °C in the Na_2SO_4-activated pastes:

$$2(Al(OH)_4)^- + SO_4^{2-} + 4Ca^{2+} + 6H_2O$$
$$+4OH^- \rightarrow C_3A \cdot CaSO_4 \cdot 12H_2O \tag{10.12}$$
$$6Ca(OH)_2 + 2Al_2O_3 + 3CaO \cdot Al_2O_3 \cdot 3CaSO_4 \cdot 32H_2O$$
$$\rightarrow 3(3CaO \cdot Al_2O_3 \cdot CaSO_4 \cdot 12H_2O) + 2H_2O \tag{10.13}$$

The direct formation of AFm increases the solid volume by 79.2%, but the conversion of AFt to AFm decreases the solid volume by 7.0%. This means the Na_2SO_4-activated pastes will exhibit lower strength at higher temperatures.

10.3   Lime-fly ash cement

10.3.1   Introduction

As discussed in Chapter 3, ASTM Class F coal fly ashes are very similar to volcanic ashes in chemical composition and vitreous structure. However, they have different particle shapes: fly ashes are spherical and ground volcanic ashes are very irregular. Thus, lime-fly ash cements can be activated like lime-natural pozzolan cement. This section discusses Na_2SO_4-activated lime-fly ash cements.

10.3.2   Setting time

The effects of Na_2SO_4 on setting times of lime-fly ash cements are shown in Table 10.5. The addition of Na_2SO_4 to lime-fly ash paste does not show an obvious effect on the initial setting time, but shortens the final setting time from approximately 60 to 26 hours.

10.3.3   Compressive strength

Figure 10.7 shows the effect of Na_2SO_4 on strength development of lime-fly ash pastes at 23 °C. The control lime-fly ash paste exhibited no measurable strength up to three days, a strength of approximately 1 MPa at 7 and 28 days, and 13 MPa at 90 days. Na_2SO_4-activated pastes also exhibit no measurable strength at one day, but give 3 MPa at three days, 10 MPa at seven days and 19 MPa at 28 days.

Table 10.5 Effect of activators on the setting
times of lime-fly ash pastes at room
temperatures (Shi 1996a)

Paste	Setting times (h:m)	
	Initial	Final
Control	7:45	60:05
Control + 4% Na$_2$SO$_4$	7:20	26:05

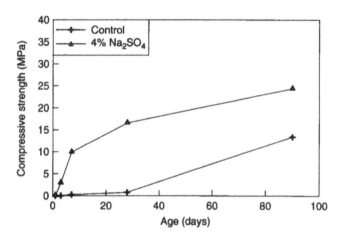

Figure 10.7 Effect of Na$_2$SO$_4$ on strength development of lime-fly ash pastes at 23 °C
(Shi 1996a).

The effect of the activators on the strength development of lime-low calcium fly ash pastes at 50 °C is shown in Figure 10.8a. At one day, the strength of Na$_2$SO$_4$-activated pastes was almost three times as high as that of the control pastes. With time, the influence of Na$_2$SO$_4$ on strength diminishes, and the strength of the Na$_2$SO$_4$-activated pastes is about 50% higher than that of the control pastes at 90 days and thereafter.

The effect of Na$_2$SO$_4$ on strength of lime-high calcium fly ash pastes at one day was the same as that on lime-low calcium fly ash pastes, as shown in Figure 10.8b: Na$_2$SO$_4$ improved the strength significantly. The effect is opposite to that on lime-low calcium fly ash pastes after one day: Na$_2$SO$_4$ continued showing a significant activation effect. There was a strength reduction of the Na$_2$SO$_4$ activated lime-high calcium fly ash pastes at 180 days and this can be attributed to the effect of long-term elevated temperature curing of pastes which coarsens the pore structure of the pastes.

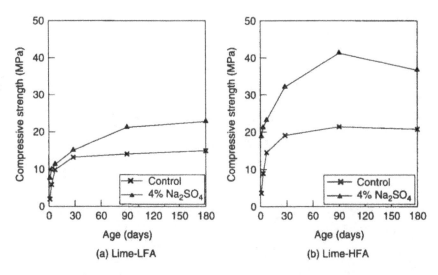

Figure 10.8 Effect of activators on strength development of lime-fly ash pastes (Shi 1998).

10.3.4 Effect of activators on pozzolanic reaction rate and microstructure

The presence of 4% Na_2SO_4 accelerates the depletion of $Ca(OH)_2$ signifi-cantly at one day, thereafter the rate of reaction diminishes. From 3 to 90 days, the $Ca(OH)_2$ content in the Na_2SO_4-activated pastes is only slightly lower than that in the control pastes. The effect of Na_2SO_4 on the depletion of $Ca(OH)_2$ in lime-high calcium fly ash pastes was different from that in lime-low calcium fly ash pastes. The addition of Na_2SO_4 slightly accelerates the consumption of lime in lime-high calcium fly ash pastes at one day but does not show significant effects thereafter. Thus, the strength increase in the presence of activators can be due to difference in the nature of hydration products.

The characteristics of fly ashes definitely have a great effect on the hydra-tion products and microstructure of hardened pastes. XRD analyses and SEM observations indicate that, for lime-low calcium fly ash pastes, C–S–H gel is still the main hydration product after the addition of Na_2SO_4, but changes other hydration products. After 3, 7 and 28 days of curing, ettrin-gite (AFt) and AFm were detected, in addition to C–S–H, in the control pastes. Weak diffraction peaks indicated that the amount of AFt was minor and no observable change happened from 3 to 28 days; While AFm diffrac-tion peaks were intensified from 3 to 28 days. Unreacted hydrated lime was also detected and its diffraction peaks seemed unchanged from 3 to 28 days.

When 4% Na₂SO₄ is added, strong AFt diffraction peaks are observed at three days and diminish slightly with time. Very weak AFm diffraction peaks are identified at three days, and diminish with time and disappear at 28 days. Diffraction peaks of unreacted hydrated lime are weaker than those in the control pastes and diminish slightly with time.

Figures 10.9 and 10.10 show the microstructure of control and Na₂SO₄-activated lime-fly ash pastes. It is obvious that the presence of Na₂SO₄ accelerates the reaction of fly ashes. All fly ash particles in the control pastes are barely corroded at three days (Figure 10.9a). At 28 days, they are more etched, but still far separated needles and bigger hexagonal plates precipitated on the surface of fly ash particles (Figure 10.10a). When Na₂SO₄ is

(a) Control pastes   (b) Na₂So₄ activated pastes

Figure 10.9 SEM observation of lime-fly ash pastes cured for three days at 23 °C (Shi 1996a).

(a) Control pastes   (b) Na₂SO₄ activated pastes

Figure 10.10 SEM observation of lime-low calcium fly ash pastes cured for 28 days at 23 °C (Shi 1996a).

added, C–S–H gel and overlapped needles fill up the voids between fly ash particles even at three days (Figure 10.9b), and become more crowded at 28 days (Figure 10.10b).

For lime-high calcium fly ash cement pastes, C–S–H, C_3AH_6 and a solid solution of C_4AH_{13} and $C_3A \cdot CaSO_4 \cdot 12H_2O$ are detected on XRD patterns. When Na_2SO_4 is added, strong AFt peaks, instead of the solid solution, appear, compared with the control pastes. At the same time, C_3AH_6 peaks are also intensified. No obvious differences can be noticed on the XRD patterns of the Na_2SO_4-activated pastes cured for 1, 7 and 28 days.

Observations using SEM, as shown in Figure 10.11, indicate that, although the morphologies of hydration products of the control and Na_2SO_4-activated lime-high calcium fly ash plates are similar, Na_2SO_4-activated lime-high calcium fly ash plates are less porous and have larger crystals than the control pastes.

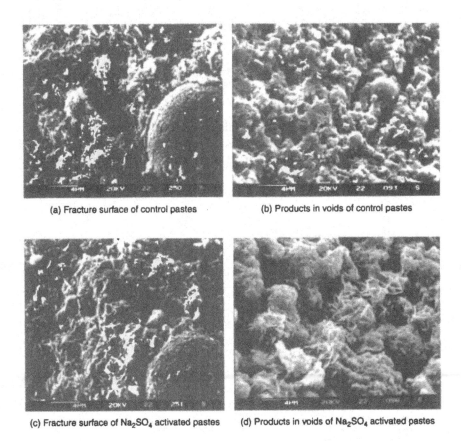

(a) Fracture surface of control pastes

(b) Products in voids of control pastes

(c) Fracture surface of Na_2SO_4 activated pastes

(d) Products in voids of Na_2SO_4 activated pastes

Figure 10.11 Microstructure of hardened lime-HFA pastes cured three days at 50 °C (Shi 1998).

10.3.5 Activation mechanism of lime-fly ash pastes

The activation mechanism of lime-fly ash cement pastes is similar to that of lime-natural pozzolan cements. The addition of Na_2SO_4 significantly accelerates the consumption of $Ca(OH)_2$ in the lime-LFA pastes at early ages. The main hydration product in the Na_2SO_4-activated pastes is AFt. The rapid consumption of $Ca(OH)_2$ and the formation of AFt contribute to the strength increase of the lime-LFA pastes.

For the lime-HFA pastes, the addition of Na_2SO_4 accelerates the consumption of $Ca(OH)_2$ at one day, and shows no effect thereafter. C_3AH_6 is the main hydration product in the control pastes. In addition to C_3AH_6, AFt forms in the Na_2SO_4-activated lime-fly cement pastes. The amount of lime consumed is almost the same in the three pastes, while AFt has the highest molar weight and lowest density, and C_3AH_6 the lowest molar weight and the highest density among C_3AH_6, AFt, and C_4AH_{13}. That may explain why Na_2SO_4-activated lime-fly ash cement pastes exhibit the highest strength and the control pastes the lowest strength.

10.4 Lime-metakaolin cement

10.4.1 Introduction

Metakaolin has been used as a pozzolanic material in concrete since 1962, when it was incorporated in concrete for Jupia Dam in Brazil (Pera 2001). The replacement of a small portion of portland cement with high-reactivity metakaolin in concrete can greatly enhance the strength development and durability of concrete (Sabir et al. 2001)

10.4.2 Heat of hydration

De Silva and Glasser (1992b) studied the heat evolution of lime-metakaolin system under the action of NaOH, Na_2SO_4, anhydrite, and combination of NaOH and anhydrate. The compositions of the formulation are listed in Table 10.6 and heat evolution curves are shown in Figure 10.12.

During the hydration of lime-metakaolin, two main exothermic peaks were measured in addition to an initial peak appearing immediately after the addition of water. When NaOH was added with a concentration of 0.5 M, the two main peaks were significantly increased. The use of Na_2SO_4 at a dosage of $SO_3/Al_2O_3 = 0.1$, instead of NaOH, showed a more significant enhanced effect on the two main heat evolution peaks than 0.5 M NaOH (De Silva and Glasser 1992b). Also, one more heat evolution peak was measured. The use of a combination of NaOH and anhydrite displayed a similar, but delayed, heat evolution curve as Na_2SO_4-activated lime-metakaolin cement. A detailed study of these exotherms in relation

Table 10.6 Composition of formulations (De Silva and Glasser 1992b)

No.	Metakaolin/Ca(OH)$_2$ (mass ratio)	Sulphate source	SO$_3$/Al$_2$O$_3$ (molar ratio)	W/S (mass ratio)	Aqueous
a	1	N/A	N/A	0.8	H$_2$O
b	1	N/A	N/A	0.8	0.5 M NaOH
c	1	B-anhydrite	0.1	0.8	H$_2$O
d	1	B-anhydrite	0.1	0.8	0.5 M NaOH
e	1	Na$_2$SO$_4$	0.1	0.8	H$_2$O

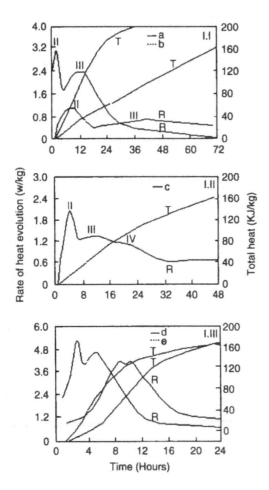

Figure 10.12 Effect of activator on heat evolution rate and total heat of hydration of lime-metkaolin cements (De Silva and Glasser 1992b).

to microstructural development and their effects on thermochemistry by De Silva and Glasser (1990) indicated that the two major heat evolution peaks of NaOH-activated lime-metakaolin cements were mainly contributed by the formation of C_4AH_{13} and C–S–H. The three major heat evolution peaks of Na_2SO_4-activated lime-metakaolin cements were due to the following three reasons: (1) formation of ettringite, (2) formation of C–S–H and C_4AH_{14} and (3) transformation of ettringite to monosulphate. Transformation of ettringite to monosulphate is an endothermic process. Thus, the appearance of peaks III and VI were due to the occurrence of an endotherm superimposed on the broad exotherm from the formation of C–S–H.

Alonso and Palomo (2001a,b) investigated the effect of different NaOH concentrations including 5, 10, 12, 15 and 18 M, hydration temperature (35, 45 and 60 °C) and lime/metakaolin ratio (50:50 and 30:70) on the heat evolution of lime-metakaolin mixtures. They found that the heat evolution peaks were greatly intensified when the NaOH concentration was increased from 5 to 10 M. However, no obvious differences could be observed from those activated materials with NaOH concentrations greater than 10 M. The increase in hydration temperature significantly accelerates and increases the heat evolution peaks. As the metakaolin-to-calcium hydroxide ratio changes, no significant differences could be observed for the time of appearance of the heat evolution peaks, but the difference is notable in the intensity of the rate of heat evolution. The heat evolution peak increases with the increase of the metakaolin-to-calcium hydroxide ratio.

10.4.3   Setting times

Table 10.7 shows the effect of three different activators – NaOH, Na_2SO_4 and Na_2CO_3 – on the setting times of lime-metakaolin cements. There is no difference in setting times between NaOH and Na_2CO_3-activated lime-metakaolin cements. However, Na_2SO_4-activated cement showed much longer initial and final setting times than NaOH- and Na_2CO_3-activated lime-metakaolin cements. This may be related to the formation of AFt during the initial hydration of Na_2SO_4-activated cement, which retards the setting of the cement.

Table 10.7 Setting times of lime-metakaolin cement with different activators (W/S = 0.35) (Jiang 1997)

Composition	Initial setting (h:m)	Final setting (h:m)
50% Metakaolin + 50% Ca(OH)$_2$ + 5% Gypsum	2:45	4:30
50% Metakaolin + 50% Ca(OH)$_2$ + 5% NaOH	1:32	2:15
50% Metakaolin + 50% Ca(OH)$_2$ + 5% Na$_2$SO$_4$	3:50	8:20
50% Metakaolin + 50% Ca(OH)$_2$ + 5% Na$_2$CO$_3$	1:20	2:10

10.4.4 Strength development

Jiang (1997) studied the effect of several activators on strength of lime-metakaolin cements at 28 days. The use of Na_2CO_3 did not show an obvious effect on the strength of lime-metakaolin cements, while the use of NaOH and Na_2SO_4 significantly increased the strength of lime-metakaolin cements, especially NaOH.

The strength development of the five formulations as listed in Table 10.10 is shown in Figure 10.13. All activators tend to improve the strength of lime-metakaolin cements at early ages, but are less or insignificant at later ages. However, strength of all the five formulations continues to increase for at least 180 days.

10.4.5 Hydration products and microstructure

De Silva and Glasser (1992b) investigated the hydration products of alkali-activated lime-metakaolin cement using XRD and DTA analyses (Table 10.8). In the lime-metakaolin cement, the hydrates present at one day are C_4AH_{13}, C–S–H and C_2ASH_8. Hydrogarnet appears in both systems after three days of hydration. As the hydration progresses, gehlenite hydrate and C_4AH_{13} disappear. The addition of NaOH significantly accelerates the process.

XRD, FTIR and NMR analyses on the samples after 24 hours of hydration indicated that when 5 M NaOH solution was used, C–S–H was the major hydration product. However, the NaOH concentration was greater than 10 M, sodium aluminosilicate gel was the main hydration product and C–S–H was the secondary hydration product (Alonso and Palomo 2001a).

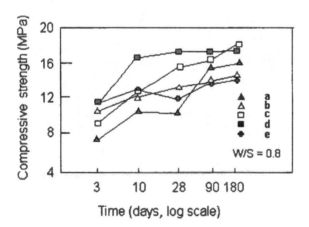

Figure 10.13 Effect of activators on strength development of lime-metkaolin cements (De Silva and Glasser 1992b).

Table 10.8 Summary of hydration products identified during the hydration of lime-metakaolin cement with different activators ($W/S = 0.8$) (De Silva and Glasser 1992b)

Composition	Detected hydration products at different ages			
	one day	28 days	90 days	180 days
50% Metakaolin + 50% Ca(OH)$_2$	CH, C$_4$AH$_{13}$, C–S–H	C$_2$ASH$_8$, C$_3$AH$_6$, C$_4$AH$_{13}$, C–S–H	C$_2$ASH$_8$, C$_4$AH$_{13}$, C$_3$AH$_6$, C–S–H	C$_2$ASH$_8$, C$_4$AH$_{13}$, C–S–H
50% Metakaolin + 50% Ca(OH)$_2$ + NaOH	CH, C$_2$ASH$_8$, C$_4$AH$_{13}$, C–S–H	C$_2$ASH$_8$, C$_3$AH$_6$, C$_4$AH$_{13}$, C–S–H	C$_3$AH$_6$, C$_4$AH$_{13}$, C–S–H	C$_3$AH$_6$, C–S–H
50% Metakaolin + 50% Ca(OH)$_2$ + Na$_2$SO$_4$	CH, AFm, C$_4$AH$_{13}$, C–S–H	C$_2$ASH$_8$, C$_3$AH$_6$, C$_4$AH$_{13}$, C–S–H	C$_2$ASH$_8$, C$_3$AH$_6$, C–S–H	C$_2$ASH$_8$, C$_3$AH$_6$, C–S–H
50% Metakaolin + 50% Ca(OH)$_2$ + NaOH + Anhydrite	CH, AFm, C$_4$AH$_{13}$, C–S–H	C$_2$ASH$_8$, C$_3$AH$_6$, C$_4$AH$_{13}$, C–S–H	C$_2$ASH$_8$, C$_3$AH$_6$, C–S–H	C$_2$ASH$_8$, C$_3$AH$_6$, C–S–H

Palomo and Glasser (1992) called NaOH-activated lime-metakaolin a chemically bonded material.

In systems with sulphates, after 24 hours of hydration, the solids present are monosulphate, C–S–H and C$_4$AH$_{13}$. Monosulphate also disappears after some time with the appearance of hydrogarnet. Sulphate is probably partly sorbed in C–S–H and partly retained in solid solution in C$_4$AH$_{13}$. Gehlenite hydrate also appears. The C$_4$AH$_{13}$ phase is either absent or present only in traces in the sulphate-activated systems, especially after longer cure durations. Some sulphate was present in the pore solution. The initial development of ettringite and its subsequent phase transformations are clearer in cement C than in alkali/sulphate-activated cement D. In all systems, most of the Ca(OH)$_2$ had reacted by 10 days.

The results also showed that alkaline activators increased the amount of calcium hydroxide reacted in the early stage (first two weeks). However, calcium hydroxide was still present at 28 days (De Silva and Glasser 1992a). Figure 10.14 shows the XRD patterns of metakaolin, Ca(OH)2, and mixture; metakaolin:Ca(OH)$_2$: NaOH = 50:50:5, at $W/S = 0.4$ cured at 25 °C and 95% RH for seven days. The solid reaction products of the mixture are identified as C–S–H (II), C$_2$ASH$_8$, C$_4$AH$_{13}$, and C$_3$AH$_6$.

Observations using SEM confirm that the nature of activators has a great effect on the morphology of hydration products of lime-metakaolin cements. It was noticed that NaOH, in the presence of sulphates, creates a featureless, massive structure at early and later stages (De Silva and Glasser 1992b).

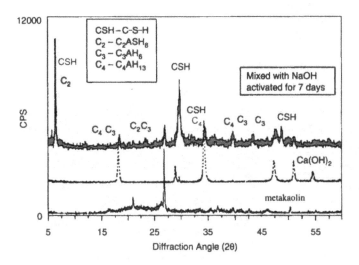

Figure 10.14 XRD patterns of metakaolin, lime and NaOH-activated lime-metakaolin cement (Jiang 1997).

10.4.6 Activation mechanism

De Silva and Glasser (1992a,b) studied mixtures of 50:50 ratio metakaolin and calcium hydroxide, with different alkali activators. Metakaolin reacts with calcium hydroxide yielding mainly products similar to the gel-like calcium silicate hydrate C–S–H obtained from portland cement. The presence of alkali undoubtedly facilitates the reaction by initial dissolution, mainly of Al species, with the subsequent formation of solid hydrates occurring by precipitation. The differences could be observed even in the early hydration stage, for the hydration products are heavily dependent on $Ca(OH)_2$ content and the second activators.

De Silva and Glasser (1992b) also noticed a high concentration of K^+ in the pore solution of the cement pastes. The K^+ came from muscovite present in the kaolin as an impurity, which is apparently activated by heating. The dissolution of K^+ into the solution enhances the solubility of aluminium and silicon species. At later stages, C–A–S–H acts as one of the hosts for released potassium. Therefore, the release of K from activated mica changes the overall chemistry, hence the hydrate balance.

10.5 Lime-blast furnace slag cement

10.5.1 Introduction

Based on the ASTM definition for pozzolans (ASTM C 618, 2003), blast furnace slag belongs to cementitious materials, not pozzolans. Lime-blast furnace slag cement is discussed here due to its similarity to lime-pozzolan cement in composition.

Lime-blast furnace slag cement was the earliest cementitious material made from slag. It was first produced in Germany, and was spread to many other countries (Moranville-Regourd 1998). It has been abandoned in most countries because of its long setting times, low early strength and fast deterioration in storage compared with modern portland cement. However, this type of cement has an excellent resistance to sulphate ground water. For example, a seawater jetty at Skinningrove (Yorkshire UK) was built with lime slag cement about 100 years ago and is still in good condition (Moranville-Regourd 1998). The use of alkaline activators can overcome some disadvantages of this type of cement.

10.5.2 Strength development

Strength is always an important criterion for most construction materials. Figure 10.15 shows the strength development of lime-slag pastes, with and without activators, from 3 to 180 days at 50 °C. In the absence of activators, lime-slag mixtures show low strength ranging from 6 MPa at three days to about 16 MPa at 90 days.

10.5.3 Reaction rate of Ca(OH)₂

The free $Ca(OH)_2$ content in hardened lime-slag mixtures at ages from 3 to 90 days is shown in Figure 10.16. Before three days, all the pastes show a rapid depletion of $Ca(OH)_2$; then, the reaction rate of $Ca(OH)_2$ slows

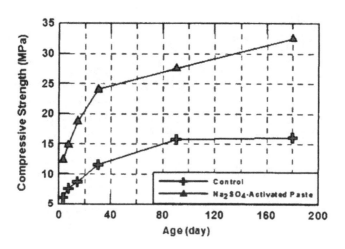

Figure 10.15 Effect of Na₂SO₄ on strength development of lime-slag cement (Shi and Day 1995c).

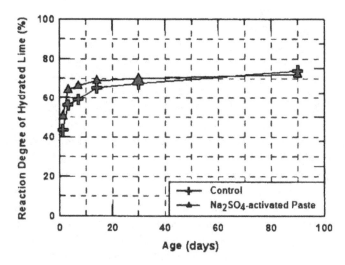

Figure 10.16 Effect of Na$_2$SO$_4$ on reaction degree of Ca(OH)$_2$ in lime-slag cement (Shi and Day 1995c).

noticeably. After 14 days, the rate of reaction of Ca(OH)$_2$ in all pastes is slow, but appears to be more rapid in the control pastes than in the Na$_2$SO$_4$-activated pastes. The major activation effect of Na$_2$SO$_4$ occurs before three days. After 14 days, control pastes show a Ca(OH)$_2$ content similar to that in the Na$_2$SO$_4$-activated pastes.

10.5.4 Hydration products and microstructure of hardened lime-slag pastes

XRD and SEM have been used to examine the microstructure of the control and Na$_2$SO$_4$-activated lime-slag cement pastes (Shi and Day 1995c). C–S–H and C$_4$AH$_{13}$ are the main hydration products in the control paste. Some C$_4$AH$_{13}$ exists, but the principal peak is small. When Na$_2$SO$_4$ is added, ettringite appears in addition to C$_4$AH$_{13}$, but the AFt and C$_4$AH$_{13}$ peak heights are also small.

Observations using SEM indicated that all slag particles in lime-slag cements are covered by foil-like C–S–H. Some crystalline products could be observed in voids after three days of hydration but are too small to be identified. When 4% Na$_2$SO$_4$ is added, many fine needles could be observed in the voids of Na$_2$SO$_4$-activated lime-slag cement pastes. These needles are likely to be AFt. Although XRD tests also indicate the presence of C$_3$A · Ca(OH)$_2$ · 12H$_2$O, no hexagonal plates were found under SEM observation.

At 30 days, control pastes show a similar but denser structure than those at three days. A major difference, however, is that many coarse needles of

AFt appear in the voids of the control pastes; no hexagonal plates were observed. Na_2SO_4-activated lime-slag pastes at 30 days are also similar to those at seven days; the slight difference is that the structure is denser and more needles appear in the voids.

10.5.5 Activation mechanism of lime-slag cement

The activation mechanism of lime-slag is roughly similar to that of lime-natural pozzolan as described above. However, some differences occur because of the high self-cementitious nature of the slag. The slag contains a high content of network modifiers (CaO and MgO); more network formers are readily dissolved into solution than in the lime-natural pozzolan or lime-fly ash systems mixtures when water is added. Thus hydrated calcium silicates or hydrated calcium aluminosilicates form quickly in the lime-slag system, and some C_4AH_{13} can also be identified. These hydration products cover the surface of the slag particles. Unlike natural pozzolan and low calcium fly ash, slag can hydrate by itself in the presence of water. The self-cementitious or self-hardening properties of slag densify and strengthen the pastes.

The addition of Na_2SO_4 accelerates the dissolution of slag, but retards the dissolution of $Ca(OH)_2$. The presence of Na_2SO_4 accelerates early hydration, which is controlled by the dissolution of slag and leads to more C–S–H forming around slag particles. At the same time, the introduction of Na_2SO_4 leads to the formation of AFt. Both factors increase the early strength of lime-slag pastes. However, the formation of more C–S–H around the slag particles in Na_2SO_4 pastes retards later hydration of lime-slag mixtures, which is controlled by the diffusion of water through the hydrated layer on the slag grain surface. This explains the differences in $Ca(OH)_2$ depletion and the pore structure changes between control and Na_2SO_4 pastes. The strength difference between the control and Na_2SO_4 pastes may be attributed to the reinforcement effect of AFt.

10.6 Summary

This chapter has discussed the activation of lime-natural pozzolan, lime-fly ash, lime-metakaolin and lime-slag cements. Generally speaking, Na_2SO_4 can be a very effective activator for all these cements. It functions in two aspects: (1) it increases the pH of the system and accelerates the reactions and (2) it results in the formation of AFt and densifies the structure. Due to the differences in chemical and mineral compositions of natural pozzolan, fly ashes, metakaolin and slag, the other hydration products and microstructure of all these cements can be very different.

Chapter 11

Other alkali-activated cementitious systems

11.1 Introduction

Previous chapters have discussed the most common alkali-activated cement and concrete materials. This chapter will describe some other alkali-activated cementing systems, which include alkali-activated metakaolin, alkali-activated fly ash, alkali-activated blast furnace slag–fly ash, alkali-activated steel slag–fly ash, alkali-activated steel slag–blast furnace slag, alkali-activated blast furnace–ladle slag, alkali-activated blast furnace slag–MgO and alkali-activated multiple component systems.

11.2 Alkali-activated metakaolin or fly ash cements

11.2.1 Introduction

In the later 1950s and the early 1960s, V.D. Glukhovsky discovered that hydraulic binders could be produced by alkaline activation of various aluminosilicate materials, in particular clay minerals, which include kaolinite and metakaolin. Metakaolin was preferably used. He called this type of hydraulic binders "soil cements" (Glukhovsky 1959, 1965, 1967). Davidovits (1988, 1991) called this type of binders "geopolymers" since they have polymeric structure. Krivenko (1997) named these binders "geocements" because the formation and characteristics of their hydration products are analogues to those of some natural minerals. The production and properties of metakaolin have been discussed in detail in Chapter 3.

Synthetic alumino-silicate minerals, such as coal fly ashes, can also be used as raw materials for polymerization (Glukhovsky 1967, Krivenko 1997, Xu and Van Deventer 2000). Recently, extensive research and development activities have been focused on fly ashes due to its low cost and availability (Silverstrim and Rostami 1997, Katz 1998, Palomo *et al.* 1999). As discussed in Chapter 3, there are two types of fly ash: Class F and Class C. Under conventional conditions, Class F fly ashes are pozzolanic materials that do not show cementitious properties. Class C fly ashes exhibit cementitious behaviour when they contact water under normal conditions. Thus, due to the differences in chemical and mineralogical compositions in Classes C and F fly ashes, it can be expected the corresponding alkali-activated fly ash cement will also behave very differently. Actually, even the same type of fly ashes from different sources or the same source but different coals can be very different as discussed in Chapter 3. This section will discuss some properties and microstructure of alkali-activated metakaolin or fly ash cements.

11.2.2 Hydration of alkali-activated metakaolin or fly ash cements

The reaction of metakaolin or fly ash with alkalis is very slow at room temperature. However, it will be accelerated very significantly as temperature increases. The reactions of metakaolin and fly ashes with different activator solutions can be monitored using a calorimeter since they are exothermic processes. The reaction process comprises three steps (Granizo and Blanco 1998, Palomo *et al.* 1999): an initial fast dissolution, which is strongly exothermic, followed by an induction period in which the heat evolution rate decreases, and finally an exothermic reaction in which cementitious materials precipitate and after which the heat evolution rate decreases.

The heat evolution is dependent on the concentration of alkaline activator, water-to-solid ratio and reaction temperature. The induction period is lengthened but the total heat evolution increases as the alkali concentration and water-to-solid ratio increases. As expected, the induction period is shortened as reaction temperature increases. The addition of chemical admixtures may accelerate or retard the heat evolution (Puertas *et al.* 2003).

Under the action of alkalis, the covalent Si–O–Si and Al–O–Al in amorphous aluminosilicate are depolymerized into monosilicate and aluminate species as follows (Shi and Day 2000b):

$$\equiv \text{Si-O-Si} \equiv + 3\text{OH}^- \rightarrow (\text{SiO(OH)}_3)^- \tag{11.1}$$

$$\equiv \text{Si-O-Al} \equiv + 7\text{OH}^- \rightarrow (\text{SiO(OH)}_3)^- + (\text{Al(OH)}_4)^- \tag{11.2}$$

The dissolved monosilicate and aluminate species condense and form amorphous to semi-crystalline three dimensional silico-aluminate structures of the poly(sialate) type (–Si–O–Al–O–), the poly(sialate-siloxo) type (–Si–O–Al–O–Si–O–), and the poly(sialate-disiloxo) type (–Si–O–Al–O–Si–O–Si–O–). Sialate is an abbreviation for silicon-oxo-aluminate. The sialate network consists of SiO_4 and AlO_4 tetrahedra linked alternately by sharing all the oxygens. Positive ions ($\text{Na}^+, \text{K}^+, \text{Li}^+, \text{Ca}^{++}, \text{Ba}^{++}, \text{NH}_4^+$ and H_3O^+) are present in the framework cavities to balance the negative charge of Al^{3+} in IV-fold coordination. Poly-sialate has this empirical formula (Davidovits 1994):

$$M_n(-(Si-O_2)_z-Al-O)_n \cdot wH_2O$$

Where M is a cation such as sodium, potassium or calcium, and n is the degree of polycondensation; z is 1, 2 or 3. Poly-sialates are chain and ring polymers with Si^{4+} and Al^{3+} in IV-fold coordination with oxygen and range from amorphous to semi-crystalline. Apart from poly-sialate (–Si–O–Al–O–), poly-sialate-siloxo (–Si–O–Al–O–Si–O–) and poly-sialate-disiloxo (–Si–O–Al–O–Si–O–Si–O–) are also possible structural units of the products when the amount of silicate reactants increases in the reaction system. Lee and Van Deventer (2002a) suggested that the mechanism of hydration of sodium silicate-activated fly ashes is similar to the dissolution and reprecipitation of fly ash in aqueous solutions of high soluble silicate concentrations. Soluble silicate speciation and distribution are likely to control the dissolution characteristics of fly ash and the formation of gel. Brouwers and Van Eijk (2002, 2003) proposed a shrinking core model to describe the dissolution of fly ash particles in alkaline environments, and the formation of products using thermodynamics. They concluded that the dissolution and formation of reaction products are dependent on pH values of the solution and chemical composition of fly ashes. Actually, the reaction steps overlap each other and the solid evolves in such a way that the dissolution of chemical species, accumulation of reaction products and polycondensation of the structures occur almost simultaneously (Davidovits 1994).

11.2.3 Microstructure development and hydration products

Sun *et al.* (2004) used environmental scanning electron microscope (ESEM) to quantitatively study the hydration process of potassium silicate-activated metakaolin cement under an 80% RH environment. An energy dispersion X-ray analysis (EDXA) was also employed to distinguish the chemical composition of hydration product. The ESEM micrographs showed that metakaolin particles pack loosely at 10 minutes after mixing, resulting in the existence of many large voids. As hydration proceeded, much gel was seen and gradually precipitated on the surfaces of these particles. At a later stage, these particles were covered by a thick layer of gel and the voids between particles were almost completely filled. The corresponding EDXA results illustrated that the molar ratios of K/Al increase while Si/Al decrease with the development of hydration. As a result, the molar ratios of K/Al and Si/Al of hydration products were 0.99 and 1.49 after four hours of hydration. The characteristics of the aluminosilicate used, the nature and concentration of activators, and the curing temperature have the most significant effects on the SiO_2/Al_2O_3 ratio of the hydration products (Kovalchuk 2002, Krivenko and Kovalchuk 2002).

Several researchers have examined the microstructure of hydrated alkali-activated metakaolin or fly ash cements using XRD, SEM, IR, NMR and other modern instruments. It is agreed that the main reaction product formed in alkali-activated metakaolin or fly ash is an amorphous aluminosilicate gel when the hydration is carried out at room temperatures. This gel is an X-ray amorphous material and difficult to be characterized. It has a 3-dimensional structure but is short-range ordered. When the curing temperature is increased but still under the atmospheric pressure, amorphous aluminosilicate gel is the main hydration product. Palomo *et al.* (1999, 2004) also call it as "zeolitic precursor" because the zeolite is the hypothetical final state of this type of material evolution. In addition, a trace amount of crystallized compounds, such as hydroxysodalite, herschelite and gismondine, could be identified in NaOH-activated fly ashes (Bakharev 2004, Fernandez-Jimenez and Palomo 2004). A metastable intermediate Al-rich phase was identified at early ages, then it evolves to a more stable Si-rich phase ("zeolite precursor") with curing time. Si occurs in a variety of environments with a predominance of $Q^4(3Al)$ and $Q^4(2Al)$ (Fernández-Jiménez and Palomo 2004, Palomo *et al.* 2004). The zeolite crystallization, the hypothetical final stage of the alkali-activated fly ashes, may happen at long term even at elevated temperatures under atmospheric pressure.

Palomo *et al.* (1999) noticed that alkali-activated fly ash cements have different microstrutures when different activators are used. When the fly ash is activated with NaOH solution, the resulting material is a very porous one and the microspheres appear to be surrounded by a crust of reaction products. The average molar ratios for the hydration products are $Si/Al = 1.5$ and $Na/Al = 0.48$. When a combination of NaOH and sodium silicate is used, it generates a microporous material, which fills the voids between fly ash particles. Also, some crystals of mullite and unreacted spheres (having high proportions of Al and/or Fe) have been detected. The average molar ratios found in the product of this reaction are $Si/Al = 2.8$ and $Na/Al = 0.46$. When the activator is a solution of potassium, the K/Al molar ratio is 1.55 and the Si/Al ratio 2.6. This different Me/Al ratio, depending on the alkali cation, means that some OH^- or Al^{3+} ions should compensate for the charges. It is interesting to remark that independent of the activator used (NaOH or NaOH + sodium silicate), the Na/Al ratio of the alkaline cement is constant. However, the Si/Al ratio of the material increases (almost doubles) when the activator is a combination of NaOH and sodium silicate.

Under hydrothermal conditions, crystallized products can form. Many researchers (Shigemoto *et al.* 1993, Singer and Berkgatit 1995, Amrhein *et al.* 1996, Querol *et al.* 1997, Grutzeck *et al.* 2004) have mixed coal fly ashes or metakaolin with alkaline solutions to synthesize zeolites under hydrothermal conditions. The nature of the raw materials, hydration time and curing temperature of alkali-activated fly ash or metakaolin dictate the nature of the zeolites formed (Grutzeck *et al.* 2004).

Figure 11.1 summarizes the hydration products of alkali-activated metakaolin or fly ashes under different curing conditions (Krivenko and Kovalchuk 2002). Under hydrothermal conditions, tobermorite gel can also be identified when the raw materials contain calcium. When cured under dry conditions, an unidentified product "phaze Z" is the main product, and a small quantity of hydrosodalite can be identified as well. Under autoclave curing conditions, hydration products similar to natural zeolites such as analcime, zeolites A, P and R are clearly observed. An increase in SiO_2/Al_2O_3 of the raw materials and the reduction of alkaline activator concentration decrease the tendency for the formation of crystalline products.

Contrary to the composition of the hydration products, the products after heating hardened alkali-activated aluminosilicate cements at $T = 800\,°C$ are dependent only on the composition of the raw materials used. Curing temperature has no effect. It was found that when the SiO_2/Al_2O_3 ratio of the raw materials varied from 2 to 8, the composition of the dehydration products changed in the following order: nepheline ($Na_2O\ Al_2O_3\ 2SiO_2$) → albite($Na_2O\ Al_2O_3\ 6SiO_2$) → α-cristobalite (SiO_2).

Lee and Van Deventer (2004) investigated the interface between natural siliceous aggregates and alkali-activated fly ashes. No apparent interfacial transition zone could be identified between the aggregates and the binder. When the activator solution contained little or no soluble silicates, the compressive strengths of the binders, mortars and concretes were

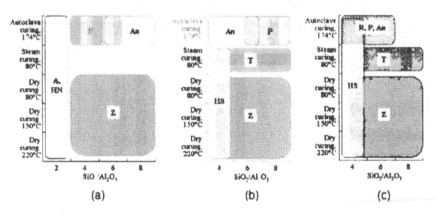

An – analcime, A – zeolite Na-A, P – zeolite P, R – zeolite R, HN – nepheline hydrate, HS – hydroxysodalite, Z – unidentified 'Phase Z', T – tobermorite gel

Figure 11.1 Effect of geocement compositions (aluminosilicate constituent: a – metakaolin, b – dry fly ash, c – fly ash of hydraulic disposal) and curing conditions on hydration products.

significantly weaker than those activated with high dosages of soluble silicates. The presence of soluble silicates in the initial activator solution was also effective in improving the interfacial bonding strengths between rock aggregates and the binder.

11.2.4   Strength of alkali-activated metakaolin or fly ash cements

Many factors affect the strength of alkali-activated metakaolin or fly ash cements. Some important factors include the nature and dosage of the activator, characteristics of metakaolin/fly ash and curing temperature/regime. They are discussed in detail in the following sections.

11.2.4.1   Nature and dosage of activators

Alkali hydroxides and waterglass or a combination of them have been studied for alkali-activated metakaolin or fly ash cements. Waterglass-activated cements often give much higher strength than alkali hydroxide-activated cements. However, high curing temperature and high concentration of alkalis are also required to achieve high strength from the activation of metakaolin or fly ash (Krivenko 1992a, Palomo *et al.* 1992, 1999, Silverstrim and Rostami 1997, Silverstrim *et al.* 1997, Popel 1999, Bakharev 2004). Figure 11.2 shows the compressive strength of alkali-activated fly ash cements at seven days after being cured at 80 °C. It can be seen that alkali-activated fly ashes can give strengths over 80 MPa very easily, which is even significantly higher than conventional portland cements. It seems that the activation reactions do not proceed in a sufficient rate to produce

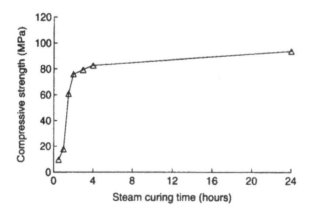

Figure 11.2 Strength development of alkali-activated fly ash cement pastes cured at 80 °C (data from Silverstrim and Rostami 1997).

high strength paste under ambient temperature, and elevated temperatures are necessary to accelerate the reactions (Xie and Xi 2001).

The modulus of waterglass is also critical for the strength development of alkali-activated metakaolin and fly ash cements. It was found (Xie and Xi 2001) that when the modulus of waterglass was reduced from 1.64 to about 1.0 by the addition of sodium hydroxide, the excessive or unreacted sodium silicate would crystallize. The tabular crystals attribute additional strength to the paste. Moreover, the unreacted spherical particles of fly ash in this structure had strong bonds with the matrix, which also helps to gain high strength. This explains why the strength of the paste increases as modulus of waterglass decreases.

The alkali metal of activator also affect the setting, hardening and strength of alkali-activated metakaolin or fly ash cement. For given conditions, it was noticed that K-based activators gave higher strength than Na-based activators (Van Jaarsveld and Van Deventer 1989). The addition of chloride salts could have a significant negative effect on strength of alkali-activated fly ash cements (Lee and Van Deventer 2002b).

The strength of alkali-activated fly ash cement increases with the increase of alkali concentration to certain value under given curing conditions and material composition (Katz 1998, Buchwald *et al.* 2003, Bakharev 2004, Sun *et al.* 2004, Fletcher *et al.* 2005). Katz (1998) obtained only 0.2 MPa when 1 M NaOH solution was used and 6.9 MPa when 4 M NaOH solution was used.

11.2.4.2 Characteristics of metakaolin and fly ashes

Metakaolin from commercial production usually has very high purity and consistent properties. However, coal fly ashes are very variable from source to source, as discussed in Chapter 2. Fernandez-Jimenez and colleagues (Fernandez-Jimenez and Palomo 2003, Fernandez-Jimenez and Puertas 2003b) characterized a group of Spanish fly ashes through chemical analysis, laser granulometry, Blaine, BET, particle size distribution, XRD and ^{29}Si MAS NMR analyses. These fly ashes were activated using an eight-molar NaOH solution and tested for strength. Results indicated that the key factors for their potential reactivity include reactive silica content, vitreous phase content and the particle size distribution. It was suggested that the fly ash should meet the following requirements for the production of quality cementing materials: a reactive silica between 40–50%; 80 to 90% of particles small than 45 μm, a high content of vitreous phase, less than 5% of loss on ignition and less than 10% of Fe_2O_3. The carbon content in fly ash, which is responsible for the loss of ignition, has a great effect on strength of activated fly ash cement. The results in Figure 11.3 indicate that the strength of alkali-activated fly ash cements decreases almost linearly with the increase of carbon content in the fly ash.

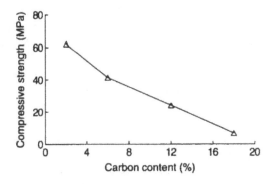

Figure 11.3 Effect of carbon content of fly ash on compressive strength of alkali-activated fly ash cement pastes (data from Silverstrim and Rostami 1997).

An increase in lime content increases strength development and ultimate strength of alkali-activated fly ash cement. When high-calcium lime or ASTM Class C fly ash is used, high strength can even be obtained at room temperature. Several patents have disclosed high strength alkali-activated fly ash cements based on Class C fly ash. For example, US Patent 4,997,484 (Gravitt *et al.* 1991) disclosed a hydraulic cement consisting mainly of Class C fly ash. In one example, the cements used contained 96.29% Class C fly ash, 1.41% potassium hydroxide, 1.28% citric acid and 1.02% borax. A concrete was prepared employing the cement and other necessary materials with the following proportions: 825 parts of cement, 1213 parts sand, 1820 parts gravel and 142 parts water. The concrete was cast in moulds and cured at room temperature, and exhibited compressive strengths as high at 70 MPa at 28 days.

US patent 5,565,028 (Roy *et al.* 1996) reported a cement made by reacting Class C fly ash with an aqueous solution having a high pH – greater than 14.3 (2.0 normal for a strong base), preferably greater than 14.69 (5.0 normal for a strong base). The cement sets rapidly at ambient temperature, which has advantages in a number of applications. Where rapid setting is not desired, retarding agents can be used to slow the setting time.

Actually, the structure formation and strength development of alkali-activated metakaolin or fly ash systems may be accelerated by adding high-calcium additives such as portland cement clinker, blast furnace granulated slag, anhydrite, etc. (Mokhort 2000).

11.2.5 Curing temperature and regime

When metakaolin or ASTM Class F fly ash is used, the strength of the cements is usually very low at room temperature. Katz (1998) investigated the activation of Class F fly ash with NaOH solution. For example, for a given water-to-fly ash ratio of 0.8, the activated fly ash cement does not show a measurable strength at 20°C, 1.75 MPa at 50°C and 6.2 at 90°C (Figure 11.4). Thus it is important to use elevated-temperature curing to obtain high strength of alkali-activated metakaolin or fly ash cement (Silverstrim and Rostami 1997, Swanepoel and Strydom 2002, Rostami and Brendley 2003, Bakharev 2004).

11.2.6   Durability

Several researchers have investigated the corrosion resistance of alkali-activated metakaolin or fly ash cements. A comparison study (Davidovits 1994) on the acid corrosion resistance of several different cements in 5% H_2SO_4 and HCl indicated that alkali-activated metakaolin cement had the best acid resistance. Portland cement and portland slag cement were destroyed easily in acidic environments. Calcium aluminate lost 30 to 60% of the mass, while alkali-activated metakaolin lost only 5 to 8% mass. In another comparison study (Silverstrim et al. 1997), laboratory results indicated that alkali-activated fly ash cements lost less than 5% of the total mass even after one year of immersion in sulphuric acid, acetic acid, hydrochloric acid and nitric acid solutions, while portland silica fume cement (containing 10% silica fume) specimens were destroyed within one to two months after immersion in these acid solutions.

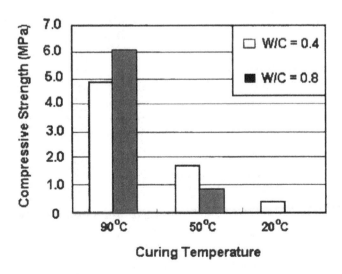

Figure 11.4 Effect of curing temperature and water-to-fly ash ratio on NaOH-activated fly ash cements (4 M) (Katz 1998).

Palomo *et al.* (1999) investigated the stability of alkali-activated metakaolin cement in 0.001 M sulphuric acid, 4.4% sodium sulphate solution and seawater. The measurement of flexural strength up to 270 days indicated that alkali-activated metakaolin cement showed very good stability in these aggressive solutions. For given conditions, although alkali-activated fly ash cement using K-based activators gave higher strength than that using Na-based activators, the latter showed better corrosion resistance than the former in 1 M HCl solution (Van Jaarsveld and Van Deventer, 1989). The authors suggested that the corrosion resistance of the material is related to the degree of polymerization. Krivenko (1999) stated that the good durability of the alkali-activated cements was due to the formation of hydration products analogous to natural minerals.

The nature of activator also affects the corrosion resistance of alkali-activated metakaolin or fly ash cements. NaOH-activated fly ash cements behaved better than sodium silicate-activated fly ash cements in acidic and sulphate solutions (Bakharev 2004). In another study, it was noticed that the use of mixed sodium-potassium alkaline activators surpassed the sodium activators in corrosion resistance (Popel 1999). However, Bakharev (2004) found that the introduction of KOH into sodium silicate had an adverse effect on durability of alkali-activated fly ash cements in sulphuric acid and sodium sulphate solutions. An essential increase in corrosion resistance is possible by the introduction of up to 10% of calcium-containing additives as well as by increasing the molar ratio SiO_2/Al_2O_3 within the cement composition.

11.3 Alkali-activated blast furnace slag-fly ash cement

11.3.1 Introduction

Alkali-activated slag cements have been discussed in previous chapters in detail. Attempts have been made to replace blast furnace slag with fly ash in alkali-activated slag cement to increase the utilization of fly ashes. This section describes the hydration and some characteristics of alkali-activated blast furnace slag-fly ash cements.

11.3.2 Heat of hydration

Measurement of heat evolution of hydration has also been used to investigate the hydration procedures of alkali-activated slag/fly ash cements. Figure 11.5 shows the heat evolution rates of NaOH-activated slag/fly ash blends, which consist of 50% blast furnace slag and 50% fly ash. LFA is an ASTM Class F fly ash and HFA is an ASTM Class C fly ash (Shi and Day 1999). The definition of the hydration peaks and the hydration periods

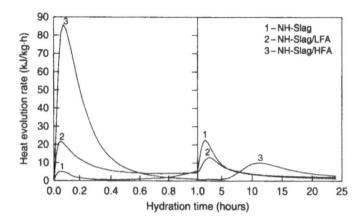

used here are the same as in Chapter 3. The initial heat evolution peak of NH-slag/LFA was four times greater than that of NH-slag, this may be explained by the rapid dissolution of fly ash in such a high alkaline environment. Partial replacement with LFA did not affect the initiation time of the accelerated hydration peak, but reduced the height of this accelerated hydration peak by approximately 50%.

With HFA, the initial peak increased from 5 to 85 kJ/kg.h, but also resulted in a retarded, lower and more diffuse accelerated hydration peak. The process of hydration of the activated slag was significantly modified due to the presence of HFA.

Figure 11.6 shows the heat evolution rates of Na_2SiO_3-activated slag/fly ash blends (Shi and Day 1999). The notations here are the same as in Figure 11.5, except that NS is used to represent Na_2SiO_3. For the NS-slag there was a small precursory peak and an initial peak during the pre-induction period and was followed by a large and broad accelerated hydration peak after the induction period. Partial replacement with LFA significantly enhanced the precursory peak, reduced the initial peak, prolonged the induction period and weakened the accelerated hydration peak. With NS-slag, the accelerated hydration peak ended at approximately 20 hours, while with NS-slag/LFA this continued beyond the end of the test (24 hours).

NS-activated slag/HFA exhibits only one strong initial peak during the pre-induction period due to the hydration of C_3A in the HFA. With HFA, no significant heat-evolution occurs even during the period from 1 to 24 hours.

The addition of 4% hydrated high calcium lime to NS-slag/LFA strengthens the precursory peak, and slightly weakens the initial peak. It shortens

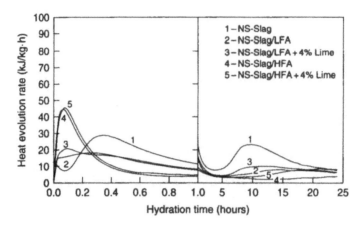

Figure 11.6 Heat evolution rates of Na₂SiO₃-activated slag–fly ash blends at 25 °C (Shi and Day 1999).

the induction period and increases the accelerated hydration peak. The addition of lime to NS-Slag/HFA does not result in a significant effect on the initial hydration peak, but shortens the induction period and appears to enhance hydration in the period after 11 hours.

11.3.3 Setting time

Like alkali-activated slag cement, the nature and dosage of activators, the nature of slag and fly ash, and slag-to-fly ash ratio can have a great effect on the setting behaviour of alkali-activated slag-fly ash cements. Table 11.1

Table 11.1 Effect of activators on setting times of activated slag–fly ash cements (Bijen and Waltje 1989)

Activator	Activator dosage (% m/m on slag and fly ash)	Water requirement for normal consistency (% m/m)	Setting time (h:m)	
			Initial	final
	6.0/4.0	45.4	3:45	5:00
Soda/Lime	7.8/5.2	46.6	1:35	3:45
	4.5/3.0	45.4	2:40	6:55
	2.0	48.0	0:23	0:50
Na₂O · SiO₂	5.0	44.0	0:20	0:25
	5.0	46.0	0:19	0:23
	7.0	45.0	0:23	0:27
Na₂O · 3.33SiO₂	5.0	46.0	1:0	24:0
Na₂O · 3.33SiO₂	5.0	51.0	1:0	24:0

lists the setting times of alkali-activated slag-fly ash cements with different activators. When soda/lime is used as an activator, the setting times of the cements are very similar to those of conventional cements. The variation of activator dosage does not show an obvious effect on the setting times. The use of $Na_2O \cdot SiO_2$ results in fast setting, showing an initial setting time of around 20 minutes and a final setting time from 23 to 50 minutes. It is surprising that the activator dosage does not show any obvious effect on the setting time of the cement materials. The other interesting phenomena are that the use of $Na_2O \cdot 3 \cdot 33SiO_2$ instead of $Na_2O \cdot SiO_2$ caused a longer setting time, especially final setting time. This is contradictory to most published results.

11.3.4 Strength development

11.3.4.1   Effect of activators

Smith and Osborne (1977) first studied alkali-activated slag–fly ash systems. A combination of 60% slag and 40% fly ash with 7% NaOH (relative to the combined mass of slag and fly ash) shows good early strength properties and little strength gain after 28 days. Dai and Cheng (1988) investigated the NaOH–BFS–fly ash and waterglass–BFS–fly ash system systematically using a factorial design method (activator – 5–11%, BFS – 45–60% and fly ash – 35–50%). No detail was given about the waterglass used. The iso-strength curves of 1:2.5 mortars at 7 and 28 days are shown in Figure 11.7. It can be observed that the waterglass-activated system showed much higher strengths than the NaOH-activated system. Some other studies obtained similar conclusions (Ionescu and Ispas 1986, Lu 1989, Shi and Day 1999).

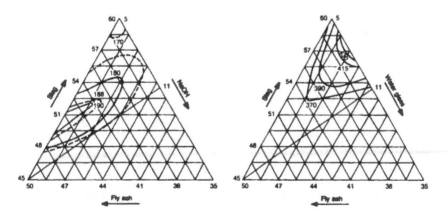

Figure 11.7 Iso-strength curves of alkali-activated fly ash–blast furnace slag cement (data from Dai and Cheng 1988).

Tang (1994) investigated the effect of slag/fly ash ratio, water-to-binder ratio and activator (waterglass and NaOH) dosage on strength development of alkali-activated slag–fly ash cement using factoral design method. He found that these factors affected the strength of the cement in the following order: slag/fly ash ratio > water/binder ratio > waterglass dosage > NaOH dosage.

Krivenko and Ryabova (1990) and Krivenko (1992a) reported results of the alkali-activated cements using 55–95% (by mass) fly ash and 5–40% slags. The slags included blast furnace slag, converter slag and cupola slag. A small quantity of lime or portland cement clinker was also added if the basicity of the mixture $((CaO+MgO)/(Al_2O_3+Fe_2O_3+SiO_2))$ was less than 0.7. Several alkaline activators, including sodium metasilicate, sodium hydroxide, soda ash and soda-alkali melt, were used. The cements exhibited strength 18–90 MPa at 28 days and up to 120 MPa at one year. When fly ash, acidic slag and blast furnace slag were used, the main hydration products of the cement were sodalite and faujasite like gels. When fly ash and basic slags were used, low-basic calcium hydrosilicates, calcite and hydrogarnets were the main hydration products.

Bijen and Waltje (1989) obtained contrary results. They found that sodium silicate-activated slag–fly ash did not show a significant strength while sodium hydroxide-activated slag-fly ash showed high early strength development. However, Pan and Zhang (1999) noticed that, for a given Na_2O dosage, waterglass was only slightly better than NaOH for the strength development of alkali-activated slag–fly ash cement. Puertas and Fernandez-Jimenez (2003) obtained a strength of 50 MPa at 28 days by using 10 M NaOH solution as an activator for a mixture 50% slag and 50% low calcium coal fly ash.

Actually, different blast furnace slags and fly ashes have different chemical and mineral compositions, and different activators have different effects on the materials with different compositions – this is the selectivity of the activators (Shi and Day 1999). So the optimum activators should be selected through testing for different systems.

11.3.4.2   Effect of fly ash

As mentioned above, the characteristics of fly ash can also have a significant effect on the strength development of alkali-activated blast furnace slag–fly ash cements. Figure 11.8 shows the strength development of NaOH-activated slag–fly ash mortars (NH + slag/LFA and NH + slag/HFA). NH + slag/LFA is composed of 50% blast furnace slag and 50% ASTM Class F fly ash, while NH + slag/HFA is composed of 50% blast furnace slag and 50% ASTM Class C fly ash. For the purpose of comparison, the strength developments of ASTM Type I portland cement (PC I), Type III portland cement (PC III), NaOH-activated blast furnace slag (NH-slag), sodium

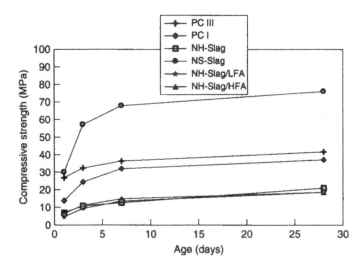

Figure 11.8 Strength development of NaOH-activated Slag/LFA mortars (Shi and Day 1999).

silicate-activated slag (NS-slag) mortars are also presented. No noticeable difference in strength development could be observed among the three series activated with sodium hydroxide. NS-slag mortars give a higher strength than Type III portland cement mortars from 1 to 28 days; this improvement is very pronounced after one day. NH-slag mortars gave strengths much lower than Type I portland cement mortars.

Figure 11.9 shows the strength development of Na_2SiO_3-activated slag–fly ash mortars. The notation here is similar to Figure 11.8, except that NS is used to represent sodium silicate. NS-slag–LFA mortars exhibit strengths lower than NS-slag mortars. The much lower strengths at one day are particularly significant when compared to the PC III and the NS-Slag results. The relative strength reduction diminishes with time. The strength of NS-slag–LFA mortars is 33% at one day and 90% at 28 days of the NS-slag mortars. NS-slag–LFA had a strength lower than PC I at one day, but higher than PC III at three days and thereafter. The strength of NS-slag–HFA mortars is even lower than NS-slag–LFA mortars, and was 25% at one day and 60% at 28 days of the strength of NS-slag mortars. However, in an earlier study (Narang and Chopra 1983), it was found that the replacement of slag with fly ash did not show any obvious effect on strength although it retarded the setting.

The fineness of fly ashes also has a significant effect on strength development of alkali-activated slag–fly ash cement and concrete. A concrete with strengths of 47.1 MPa at seven days and 76 MPa at 28 days is

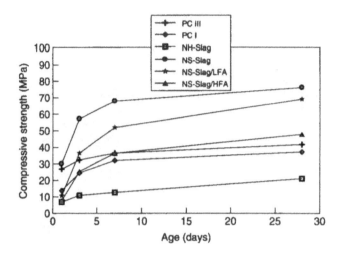

Figure 11.9 Strength development of Na$_2$SiO$_3$-activated slag/HFA mortars (Shi and Day 1999).

produced when cured in a fog room within the temperature range of 11–20 °C. The mixing proportions used are: ground fly ash (Blaine fineness about 4000 cm^2/g) : BFS : Na$_2$O · SiO$_2$: sand : coarse aggregate : water = 1.0 : 0.66 : 0.18 : 3.33 : 2.63 : 0.45. If non-ground fly ashes are used, the strength at 28 days is only about 12.9–16.2 MPa. It shows that fly ashes play a very important role in the strength development of this type of concrete (Lu 1989). Pan and Zhang (1999) confirmed this.

11.3.4.3   Effect of other additives on strength development

To improve the early strength development of alkali-activated slag–fly ash cement, use of additives has been investigated. Results in Figures 11.8 and 11.9 indicated that the replacement of slag with fly ash in Na$_2$SiO$_3$-activated cementing materials significantly decreased the early strength. It was found that the addition of 4% hydrated high calcium lime almost doubled the strength of NS-slag/LFA mortars at one day, but the effect decreased drastically thereafter (Shi and Day 1999). The relative strength was only 95% at three days and 88% at 28 days compared to the mortars without lime.

The addition of lime to NS-slag/HFA mortars resulted in a relative strength of 140% at one day, but the relative strength decreased almost linearly to 98% at seven days and then to 92% at 28 days. Thus, the addition of lime was very effective in increasing the early strength of Na$_2$SiO$_3$-activated slag/fly ash blends. When NaOH is used as an activator, the addition of additional hydrated lime can still slightly increase the strength at 7 and 28 days, but hydrated lime cannot replace NaOH (Kostic and Skendervoic 1992). Pan and Zhang (1999) found that the addition of a small percentage of zeolite can also slightly improve the strength of alkali-activated slag–fly ash cement.

11.3.4.4 Microstructure and corrosion resistance

It seems that the main hydration products of alkali-activated slag–fly ash cements are C–S–H and alkali aluminosilicate hydrate gel (Lu 1992, Puertas and Fernandez-Jimenez 2003). In NaOH-activated slag cement, the Ca/Si ratio of C–S–H is similar to that of the slag, while the Ca/Si ratio of C–S–H in alkali-activated slag–fly ash cements is much lower than that of the slag used. Most fly ash particles in alkali-activated slag–fly ash cements are seriously dissolved or corroded even after one day of hydration (Shi and Day 1999, Puertas and Fernandez-Jimenez 2003).

It is reported that alkali-activated slag–fly ash cements have very good resistance to acidic, sulphate and seawater attacks (Lu 1992). Some other studies have noticed that the strength of the cement decreases, but its corrosion resistance increases with the increase of fly ash replacement (Krivenko 1992b, Blaakmeer 1994), which can be explained by the low Ca/Si ratio of C–S–H in the alkali-activated slag–fly ash cements.

11.4 Alkali-activated blast furnace slag-steel slag system

Ground blast furnace slag has been widely used as a cement replacement in portland cement concrete. One problem for the direct replacement of portland cement with steel slag is that the free CaO and MgO in steel slag can cause soundness expansion of the concrete. However, the soundness expansion will not be a concern for the replacement of blast furnace slag with steel slag in alkali-activated slag cement since blast furnace slag can consume the free CaO or MgO. Also, the replacement can improve some properties of the alkali-activated slag cements due to the high contents of lime and crystalline compounds in steel slag.

It is well known that blast furnace slag can give much higher strength than conventional cements in the presence of alkaline activators. As mentioned above, steel furnace slag exhibits cementing property when it contacts water. It can be expected that steel slag will also show much better cementitious property in the presence of alkaline activators. Figure 11.10 shows the effect of replacement of different amounts of blast furnace slag with steel slag on strength development of alkali-activated blast furnace slag–steel slag mortars activated by 5% $Na_2O \cdot SiO_2$ (in weight of Na_2O) that has a modulus of

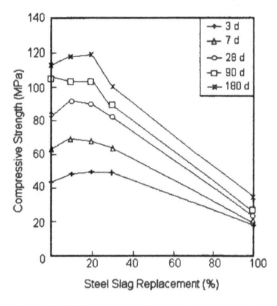

Figure 11.10 Effect of steel slag content on strength development of alkali-activated steel slag–blast furnace slag cements (data from Bin 1988).

1.7 (Bin *et al.* 1992). It can be seen that 20% replacement of blast furnace slag in alkali-activated BFS cement with steel slag gives higher strengths than are obtained with pure blast furnace slag. With the increase in amount of replacement, the strength of the cement decreases. Although alkali-activated steel slag cement develops low strength, which does not meet most practical requirements, according to the results, it can be anticipated that even if 50% blast furnace slag is replaced with steel slag, the cement may still meet the strength requirements. Lu (1989) examined the microstructure of the alkali-activated blast furnace slag–steel slag pastes and found that the alkali-activated blast furnace slag–steel slag pastes made up with proper amount of blast furnace slag and steel slag showed a denser structure than pure alkali-activated blast furnace slag pastes. He suggested that the appropriate replacement of blast furnace slag with steel slag could improve the intrinsic properties of the hardened pastes. At the same time, the proper replacement of blast furnace slag with steel slag also reduces the shrinkage and increases the abrasion resistances of the cement pastes and concretes (Bin *et al.* 1989). A recent study indicated that alkali-activated steel slag–blast furnace slag cement could show excellent corrosion resistance (Shi 1999).

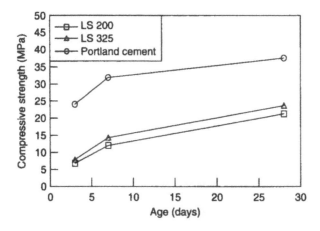

Figure 11.11 Effect of fineness of ladle slag on strength development of ladle slag-based cement mortars (Shi 2002).

11.5   Alkali-activated blast furnace slag-ladle slag cement

Ladle slag consists mainly of γ-C_2S and does not display obvious cementitious property under normal hydration condition. However, it shows significant cementing properties under the activation of alkalis. Its cementitious property increases with the fineness of the slag. Figure 11.11 shows the mortar strength of ladle slag-based cementing materials with ground blast furnace slag and a chemical activator cured at room temperature. One obvious feature of these results is that the strength increases with the fineness of ladle slag. This means that ladle slag does contribute to the strength of those materials. The other obvious feature is that these ladle slag-based cementing materials showed very low strength at three days, but they had faster strength gain rate than the reference cement – portland cement.

It is well known that β-C_2S mineral hydrates and gains strength slowly, and γ-C_2S is an inert mineral under normal hydration conditions. It is reported (Krivenko 1994b) that the potential cementitious property of both β-C_2S and γ-C_2S can be significantly increased by chemical activators under room temperature curing conditions.

11.6   Alkali-activated-blast furnace slag–MgO system

If MgO powder is mixed with a solution of $MgCl_2$, it will harden and give a very high strength (up to 200 MPa). This mixture is called magnesium oxychloride cement or Sorel cement. Because this hardened paste has very

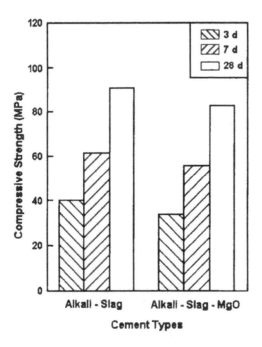

Figure 11.12 Compressive strength of alkali-activated slag and alkali-activated slag–MgO cements (Shi *et al.* 1993).

poor water resistance (Kurdowski and Sorrentino 1983), its applications are greatly restricted. Chinese researchers are experimenting with this type of binder, mixing it with about 50% (by volume) wood filler to produce bamboo reinforced wood-like plates and beams for packing boxes and other applications in dry environments.

MgO has been used to replace ground blast furnace slag in alkali-activated slag cements (Gu 1991, Shi *et al.* 1993). Preliminary results indicated that even when as much as 40% BFS was replaced by MgO under the composite activation of K_2CO_3 and waterglass (Gu 1991, Shi *et al.* 1993), the strength of mortars was only slightly lower than that of control samples (Figure 11.12). No details were given about the dosage of activators. This research is still in progress. It can be expected that this binder will show much better water resistance than Sorel cement.

Summary

This chapter has discussed several alkali-activated metakaolin, alkali-activated fly ash, alkali-activated blast furnace slag–fly ash, alkali-activated steel slag–fly ash, alkali-activated steel slag–blast furnace slag, alkali-activated blast furnace–ladle slag, alkali-activated blast furnace slag–MgO and alkali-activated multiple component systems. Pozzolanic materials can also exhibit cementitious properties in the presence of alkaline activators. Since the characteristics of fly ash, blast furnace slag and steel slag vary significantly from source to source, or even from time to time for a given source, the cements made with these materials will also vary significantly. Of course, the nature and dosage of activators have a great effect on the hydration, microstructure and performance of these cements.

Chapter 12

Applications and case studies

12.1 Introduction

Since the discovery of alkali-activated cements and concretes in 1958, they have been commercially produced and used in a variety of construction projects in the former Soviet Union, China and some other countries. A lot of experience has been gained from design, production and applications during the past 40 years, which is valuable for the further development and applications of alkali-activated cements and concretes.

During 1999–2000 years, a group of scientists in Ukraine inspected several concrete structures built with alkali-activated slag cement and concrete, which include drainage anti-soil-slipping collector/tubing built in 1966 in Odessa, Ukraine; silage trenches, slopes of railway embankment and cast-in-situ concrete sites built in 1982, one two-storey and another 15-storey residential buildings built in 1960, concretes and constructions fabricated in 1999–2000, special concrete pavements for heavy loaded trucks (50–60 tons) built in 1984, 24-storey residential building built in 1994 and pre-stressed reinforced concrete slippers for railway built 1988. In all these cases, alkali-activated slag cement concrete is still performing well and exceeds the performance of portland cement concrete used in the same area. Performance testing and microstructural examination of the samples from these structures indicate that the properties of the concrete depend upon the raw materials used, service conditions and age. This chapter discusses the production and application experience of alkali-activated slag cement in China and the field inspection results of several alkali-activated slag cement concrete structures in the former Soviet Union.

12.2 Alkali-activated portland-blast furnace slag-steel slag cement and concrete in China

As discussed in Section 8.5, use of Na_2SO_4 instead of gypsum decreases the setting time and increases the early strength of the cement significantly. Na_2SO_4-activated portland-blast furnace slag-steel slag cement has been commercially manufactured and applied in China since 1988. Anyan Steel Slag Cement Plant in Anyan City, Henan Province, was the first plant to manufacture and market it. Several months of monitoring of the cement quality indicated that Na_2SO_4-activated portland-blast furnace slag-steel slag cement showed more consistent performance than conventional portland-blast furnace slag-steel slag cement. Since Na_2SO_4 was interground with the other components, the cement can be used in concrete mixtures in the same manner as conventional portland cement from both the materials handling and mixture design aspects. Results from concrete testing indicated that the concrete made with Na_2SO_4-activated portland-blast furnace slag-steel slag cement exhibited excellent workability, shorter setting time, higher early strength and smoother surface. The cement was used for all kinds of construction purposes. The following sections describe the experiences from several field projects.

12.2.1 Office and retail building

A 6-storey office and retail building with measurement of 8.6 m × 31.5 m was built in Yinshan County, Hubei Province in 1988, using Na_2SO_4-activated portland-blast furnace slag-steel slag cement concrete produced by Anyan Steel Slag Cement Plant (As shown in Figure 12.1). The design compressive strength was 20 MPa. The concrete mixing proportion was: cement: sand:crushed limestone:water = 1:1.8:4.23:0.44. The concrete mixture was mixed on site using a small concrete mixer and had a slump of 30–50 mm. The building was an in-situ cast monolithic concrete structure.

Figure 12.1 A 6-storey office and retail building built with Na_2SO_4-activated portland-blast furnace slag-steel slag cement concrete.

The side forms were removed one day after casting, and bottom forms were removed seven days after the casting. The concrete surface was very smooth and no indication of cracking could be observed. The average actual strength was 24.1 MPa at 28 days, 20% higher than the design strength. As discussed in Section 9.5, the strength of portland-blast furnace slag-steel slag cement concrete continues to develop with time very significantly even after 28 days. It can be expected that the strength at later ages would be much higher than 24 MPa.

12.2.2 Precast concrete beams and columns for workshop

A workshop with an area of 3500 m² was built in Yinshan County, Hubei Province in 1988, using Grade 425 Na_2SO_4-activated portland-blast furnace slag-steel slag cement produced by Anyan Steel Slag Cement Plant (as shown in Figure 12.2). The cross section of the columns was 400 × 400 mm. The beams had a cross sectional area of 350 × 450 mm and a span of 12.6 m. The concrete design strength was 30 MPa. The concrete mixing proportion is cement:sand:crushed limestone:water = 1:2.0:3.9:0.50. All the concrete beams were precast on site and then assembled together. Concrete forms were removed one day after casting. The average actual strength was 35.9 MPa at 28 days, almost 30% higher than the design strength.

12.2.3 Concrete irrigation ditch

An irrigation ditch was built using Grade 325 Na_2SO_4-activated portland-blast furnace slag-steel slag cement in Lushan County, Henan Province, as

Figure 12.2 A workshop built with Na_2SO_4-activated portland-blast furnace slag-steel slag cement concrete.

Figure 12.3 An irrigation ditch built with Na₂SO₄-activated portland-blast furnace slag-steel slag cement concrete.

shown in Figure 12.3. It is 2 m wide at the bottom and 500 m long. The concrete design strength was 20 MPa. The concrete mixing proportion is cement:sand:crushed limestone:water = 1:1.9:4.3:0.56. The concrete was mixed on site and exhibited an average strength of 26.5 MPa at 28 days, almost 30% higher than the design strength.

12.3 Structural alkali-activated slag concrete

12.3.1 High-storey residential buildings in the city of Lipetsk, Russia

Several high story residential buildings were built using alkali-activated slag cement concrete by the industrial enterprise "Tsentrmetallurgremont" between 1986 and 1994. Three high-storey buildings in the city of Lipetsk, Russia, were inspected in 2000 – one 24-storey residential building on Berezina Street, 12 (Figure 12.4), a 20-storey building on Yesenina Avenue and a sixteen-storey building on Levoberezhnaya Street.

The exterior walls of the three buildings were cast-in-situ in a whole piece using alkali-activated slag cement concrete. The floor slabs, stairways and other structures were pre-cast using alkali-activated slag cement concrete. The cast-in-situ concrete was transported from mixing station to construction site using concrete truckers and cured by electroheating. These precast elements were cast and steam-cured in a precast plant. The design strength of the concrete after steam curing was 25 MPa. The characteristics of the raw materials and their mixing proportions are summarized in Table 12.1.

Figure 12.4 A 24-storey building built with alkali-activated slag cement concrete on Berezinsa street 2, city of Lipetsk, Russia, 1994.

The buildings were inspected in 2000. As it was difficult to take cores from the structures, the achieved 10 cm × 10 cm × 10 cm concrete cubes made during the construction and stored in laboratory, were tested for compressive strength. These cubes exhibited compressive strength from 32 to 37 MPa, compared with the design strength of 25 MPa. Visual observations indicated that the alkali-activated slag cement concrete structures were in good condition with no cracking, deterioration or defects on the surface (Ilyin 1994). The mixture proportions of the concrete and inspection results are also summarized in Table 12.1.

12.3.2 A storehouse in Krakow, Poland

In 1974 a storehouse was built in Kraków using reinforced alkali-activated slag concrete floor slabs and wall panels (Małolepszy 1989, Małolepszy

Table 12.1 Characteristics of the raw materials, concrete mix proportions and properties of the fresh and hardened concrete used in the residential buildings in the city of Lipetsk, Russia

Raw materials	Material	Proportions	Characteristics
	Blast furnace slag	450 kg/m^3	$Mb = 1$; $S = 400$–420 m^2/kg
	Activator solution (or calculated dry mass)	160 l/m^3 (19.2–24 kg/m^3)	Soda solution from coke gas cleaning of hydrogen sulfide, $\rho = 1120$–1150 kg/m^3
	Fine aggregate	325 kg/m^3	Pit sand with $F = 2$–2.5
		325 kg/m^3	River sand $F = 1.75$
	Coarse aggregate	1160 kg/m^3	Dolomite limestone with size 5–20 mm
	Activator solution to slag ratio	0.35	
Design strength	25 MPa		
Fresh concrete	Slump 8–9 cm		
Characteristics of field inspected concrete	Strength		32–37 MPa (achieved 10 cm × 10 cm × 10 cm cubic samples made during construction, but cured indoor)
	pH of field concrete		10.77
	Water absorption		8.12%
	Depth of carbonation		3–5 mm
	Visual observation		no visual cracking, scaling and other defects.

et al. 1994, Deja 2002a). The concrete compositions for these products included ground granulated blast furnace slag – 300 kg/m^3, sand–gravel mix – 1841 kg/m^3, Na_2CO_3 – 18 kg/m^3 and water – 140 kg/m^3. These products were cast and cured in hot air at 70 °C for six hours in a heating tunnel. Two floor and wall panels made with conventional portland cement concrete were also installed together with the alkali-activated slag concrete products to compare their properties and to control the joint of these two kinds of concrete under loading in terms of normal climatic factors. The wall elements were not covered with any plaster.

This building has been monitored for many years. Cylinders of 100 mm diameter were cut off from the outside wall panels. These 100 mm diameter cylinders were tested for compressive strength and carbonated depth microstructure. The compressive strength and carbonated depths of the concrete cores are summarized in Table 12.2.

The microstructure of these concrete cores was examined using an environmental scanning electron microscope. Generally, dense C–S–H phase

Table 12.2 Compressive strength and carbonated depth of concrete cores (Deja 2002a)

Samples	Compressive strength (MPa)		Carbonated depth (mm)
	28 days	27 years	27 years
1	22.1	43.4	10.1
2	21.7	41.9	10.3
3	21.2	46.2	9.4
4	22.3	41.1	11.8
5	23.1	45.1	12.9
6	22.8	41.6	11.7
Average	22.6	43.2	11.0

was identified as the main hydration product, some $CaCO_3$ and thomsonite were also detected. No microcracks were observed. The interface between cement paste and gravel aggregate looked very dense. Traces of interaction between cement paste and aggregate could also be identified.

The pore structure measurements indicated that there was little difference between the outer surface layer, exposed to water and CO_2 and the interior. No indication of the corrosion of steel reinforcement was observed.

12.3.3 Prestressed reinforced concrete railway sleepers, Tchudovo railway station, Russia

In 1988, prestressed reinforced alkali-activated slag cement concrete sleepers were constructed not too far from the Tchudovo Railway Station, which is located on the railway between St Petersburg and Moscow. The alkali-activated slag cement concrete sleepers were about 20-m long.

In year 2000, a field inspection of the alkali-activated slag cement concrete sleepers indicated that they were in good working condition after 13 years of service. One sleeper was dismantled from the railway bed and delivered to the building materials department of the St. Petersburg State Communications University of Russia for further investigation. The mixing proportions of the concrete and the test results from the alkali-activated slag concrete sleeper are summarized in Table 12.3.

Table 12.3 Characteristics of the raw materials, concrete mix proportions and properties of the fresh and hardened concrete in prestressed reinforced concrete railway sleepers, Tchudovo railway station, Russia

Raw materials	Material	Proportions	Characteristics
	Blast furnace slag	$500\,kg/m^3$	$Mb = 1; S = 400\,m^2/kg$
	Alkaline activator solution (or calculated on dry mass)	$160\,l/m^3$ (or $40\,kg/m^3$)	Sodium metasilicate, $\rho = 1250\,kg/m^3$
	Fine aggregate	$600\,kg/m^3$	Sand $F = 1.75$
	Coarse aggregate	$1200\,kg/m^3$	Crushed stone with size 5–20 mm
	Activator solution to slag ratio	0.35	
Design strength	45 MPa		
Fresh concrete	Slump = 8–9 cm		
Characteristics of field inspected concrete	Strength	82 MPa	
	pH of Field concrete	11.3	
	Water absorption	7.8%	
	Depth of carbonation	8–12 mm	
	Visual observation	no visual cracking, scaling or other defects	

12.4 Masonry blocks, city of Mariupol, Ukraine

The commercial production of alkali-activated slag cement concretes and products such as blocks and hemiblocks was started in the Mill 'Stroydetal' of the association "Azovshelezobeton" in 1960. These blocks and hemi-blocks have been used for the construction of residential houses, garages, fences, etc. One 2-storey and another 15-storey apartment buildings were built completely with the blocks on the left bank of Kalmius River in Liporskiy, Pashkovskiy streets, Kievskiy Lane, Victory Avenue, etc. Exterior walls of the two-storey apartment building were plastered. The exterior walls of the 15-storey apartment building were covered by ceramic tiles. In addition to residential buildings, blocks and hemiblocks were also used for construction of a long fence on the bank of the Kalmius River along the buildings of the Association "Azovshelezobeton" and in other districts of the City of Mariupol.

Several residential buildings, built between 1960–1966 in the City of Mariupol, were inspected in 2000. The characteristics of the raw materials and properties of the concrete are summarized in Table 12.4.

The block manufacture plant was closed down in about 1980, but "Metallurgical Mill Ilycha" re-started the production of ready mixed alkali-activated slag cement concrete and concrete products in 1999. Its ready mixed alkali-activated slag cement concrete was mainly used for its own

Table 12.4 Characteristics of the raw materials, mixing proportions and properties of the fresh and hardened concrete of masonry Blocks, City of Mariupol, Ukraine

Raw materials	Material	Proportions	Characteristics
	Blast furnace slag	460 kg/m³	Mb = 1.12 S = 310–330 m²/kg
	Alkaline activator	165 l/m³	Soda-alkali solution,
	(or calculated on	(or	ρ = 1150 kg/m³
	dry matter)	24.7 kg/m³)	
	Fine and coarse	1420 kg/m³	Granulated blast furnace
	aggregate		slag
	Activator solution to slag ratio	0.36	
Design strength	7.5–10 MPa		
Fresh concrete	Slump = 9–10 cm		
Characteristics of field inspected concrete	Field concrete strength	26 MPa	
	Density	1960 kg/m³	
	pH of field concrete	9.08	
	Water absorption	6.6%	
	Visual observation	good, no cracks, scaling and other defects	

needs of cast-in-situ or precast slab driveway for heavy vehicles near the lime production facility, and the machine shop of the lime production plant. It also manufactures several precast products such as foundation blocks, slabs, cap plates, etc. Some of the products from Metallurgical Mill Ilycha were also examined in year 2000.

12.5 Concrete pavements

12.5.1 Heavy duty road to the Magnitnaya mountain quarry, city of Magnitogorsk, Russia

Two roads built in the City of Magnitogorsk, Russia, in 1984 were inspected in 1999. The first one was about 6 km long and a heavy-duty road to the Magnitnaya Mountain Quarry, on which vehicle loads typically weighed 60–80 tons (as shown in Figure 12.5). The half of the road for loaded vehicles had concrete bed depth of around 45–50 cm, the other half of the road for unloaded vehicles had a concrete bed depth of 25–30 cm. The second one was about 5 km long, near the vehicle refuelling station "Shuravi" and had a concrete bed depth of around 25–30 cm.

The fresh concrete mixtures for the road construction had a slump of 9–12 cm. The concrete mixtures were delivered to the site by concrete

Figure 12.5 Cast-in-situ heavy-duty alkali-activated slag concrete road to Magnitnaya mountain quarry, city of Magnitogorsk, Russia.

truckers and compacted by vibration strip. The concrete set and hardened under natural weather conditions.

During inspection, an attempt was made to drill out cores from the roads. However, it failed. A large piece of concrete was taken by pick-hammer and sawed into 7 cm × 7 cm × 7 cm cubes in the Kiev Mill "Granit" for further investigation. The characteristics of the raw materials and properties of the concrete are summarized in Table 12.5.

12.5.2   Concrete road and fountain basin built by industrial enterprise of trust "Ternopolpromstroy", City of Ternopol

A 330 m long and 3–4 m wide cast-in-situ alkali-activated slag concrete road on the industrial street and a cast-in-situ alkali-activated slag concrete fountain basin were built by Industrial Enterprise of Trust "Ternopolprom-stroy", City of Ternopol, between 1984 and 1990. In 1999, these road and fountain basin were inspected. The same structures built with port-land cement concrete were also inspected at the same time. It was found that the road and basin built with alkali-activated slag cement concrete exhibited very good working conditions, while those built with portland cement concrete deteriorated seriously (Figures 12.6 and 12.7) (Brodko 1999).

Table 12.5 Characteristics of the raw materials, mixing proportions and properties of the fresh and hardened concrete for the heavy duty roads to the Magnitnaya mountain quarry, city of Magnitogorsk, Russia

Raw materials	Material	Proportions	Characteristics
	Blast furnace slag	500 kg/m³	Mb = 0.91; S = 400 m²/kg
	Portland cement	25 kg/m³	
	Alkaline activator	175 l/m3 (or	Soda-alkali
	(or calculated as	28 kg/m³)	solution,
	dry matter)		$\rho = 1160\,kg/m^3$
	Fine aggregate	700 kg/m³	Sand
	Coarse aggregate	1060 kg/m³	Crushed stone
	Activator solution to slag ratio	0.35	
Design strength	30 MPa		
Fresh concrete reinforcement	Slump = 9–12 cm		
	Near surface layer: grid with pace of 10 cm and $d = 5$ mm		
	Lower layer	All steel, $d = 12$ mm	
Characteristics of field concrete (2000)	Strength	86.1 MPa	
	Density	2405 kg/m³	
	PH of field concrete	11.12	
	Water absorption	7.96%	
	Depth of carbonation	10–15 mm	
	Visual observation	good, no reinforcement corrosion, no cracking, scaling and other defects	

(a) General view

(b) Close view

Figure 12.6 Comparison of cast-in-situ alkali-activated slag concrete (left side) and ordinary portland cement concrete road (right side) near the industrial enterprise of trust "Ternopolpromstroy", city of Ternopol, 1984.

12.6 Drainage collector No. 5, 1966, city of Odessa, Ukraine

In order to prevent the erosion of soil from the steep slope into the sea because of a high ground water level, a 33 km long drainage collector (No. 5) was built along the bank of the sea in the City of Odessa in 1965. Drainage was led to the sea to reduce the level of ground water. The design and construction of the collector were similar to those for uderground railway tunnels. Thus, the construction of the collector was complicated and expensive, and the technical requirements for materials were very high. Around 40 alkali-activated slag cement concrete pipes were fabricated by the "Kievmetrostroy" Integrated Plant and were used in collector No. 5.

After construction, the drainage collector was monitored by hydrogeologists of anti-soil slipping administration of Odessa and research scientists of Kyiv National University of Construction and Architecture (Glukhovsky 1967). The results of the examination in year 2000 (over 34 years of service), as shown in Table 12.6, confirmed that alkali-activated slag cement concrete pipes exhibited good appearance; no visual corrosion indication could be observed even though the concrete protective layer was only 3 mm.

12.7 Silage trenches in the village of Orlyanka, Zaporozhye, Oblast, Ukraine, 1982

Zaporozhoblagrodorstroy in Zaporozhye, Oblast, Ukraine, started commercial production of alkali-activated slag concretes and concrete products in 1972, and supplied them to the whole Zaporozhye district for 20 years.

Silage trenches were built for the milk farmers in the Village of Orlyanka of Vasilyevka region in 1982. Access roads, bottoms and fences of the silage

(a)   (b)

Figure 12.7 Comparison of cast-in-situ alkali-activated slag concrete (left side) and ordinary portland cement concrete fountain basin (right side) on the S. Bandera Avenue, City of Ternopol, 1990.

Table 12.6 Characteristics of the raw materials, concrete mix proportions and properties of the fresh and hardened concrete for concrete pipes in the drainage collector No. 5, 1966, City of Odessa, Ukraine

Raw materials	Material	Proportions	Characteristics
	Blast furnace slag	$500\,kg/m^3$	$Mb = 1.1$; $S = 350\,m^2/kg$
	Alkaline activator (calculated on dry matter)	$185\,l/m^3$ ($30\,kg/m^3$)	Soda-potash solution, $\rho = 1150\,kg/m^3$
	Fine aggregate	$1500\,kg/m^3$	River sand
	Coarse aggregate	–	
	Activator solution to slag ratio	0.37	
Design strength	40 MPa		
Fresh concrete	Slump = 8–9 cm		
Characteristics of field inspected concrete (2000)	Strength	62 MPa	
	Density	$2120\,kg/m^3$	
	pH of Field concrete	11.57	
	Water absorption	4.76%	
	Depth of carbonation	2–3 mm	
	Visual observation	good, no cracking, scaling and other defects	

trenches where built with $2\,m \times 3\,m \times 0.2\,m$ reinforced alkali-activated slag concrete panels. The total area was about $20\,m \times 60\,m$ and used about 2000 tons of alkali-activated slag concrete. Characteristics of the raw materials, concrete mix proportions and properties of the fresh and hardened concrete for the silage trenches are summarized in Table 12.7.

The fermentation of silage would result in the formation of aggressive organic acids, which corrode concrete. Also, the traffic of vehicles, tractors, bulldozers, etc. would cause dynamic loads on the concrete panels. However, the field inspection indicated that the part of the trench built with alkali-activated slag concrete panels did not show any indication of deterioration, while deteriorated concrete and exposed rusty reinforcements could be observed on the trench built with portland cement concrete panels, as shown in Figure 12.8.

12.8 Autoclaved aerated concrete

In 1978, Constructions Plant in Berezovo, Russia, set up a pilot-scale production of the building products using autoclaved foamed alkali-activated slag concrete was launched. It used a waste from soda production, which contains 65% NaOH and 25% KOH, as an activator. Several types of slag

Table 12.7 Characteristics of the raw materials, concrete mix proportions and properties of the fresh and hardened concrete for silage renches in the village of Orlyanka, Zaporozhye, Oblast, Ukraine

Raw materials	Material	Proportions	Characteristics
	Blast furnace slag	$550\,kg/m^3$	$Mb = 1.13$; $S = 400–420\,m^2/kg$
	Alkaline activator (calculated on dry matter)	$163\,l/m^3$ $(24.4\,kg/m^3)$	Soda-potash solution $(\rho = 1150\,kg/m^3)$, chemical industry waste from nitrogen production.
	Fine aggregate	$517\,kg/m^3$	Crushed granite
	Coarse aggregate	$1150\,kg/m^3$	Crushed granite
	Activator to slag ratio	0.3	
Design strength	30 MPa (after steam curing)		
Fresh concrete	Slump = 9–10 cm		
Characteristics of field inspected concrete (2000)	Field concrete Strength	39 MPa	
	Density	$2340\,kg/m^3$	
	PH of field concrete	10,9	
	Water absorption	8.04%	
	Depth of carbonation	3–8 mm	
	Visual observation	good, no cracking, scaling and other defects	

(a)   (b)

Figure 12.8 Silage trenches made with reinforced alkali-activated slag cement and ordinary portland cement concrete panels, village of Orlyanka, Zaporozhye Oblast, 1982.

Table 12.8 Properties of autoclaved aerated alkali-activated slag concrete

Slag type	Density (kg/m³)	Strength (MPa)			Initial modulus of elasticity (GPa)
		Compressive (Cube)	Tensile (Cube)	Compressive (Prism)	
Basic blast	300	1.1	0.087	–	–
Furnace slag	600	7.1	0.49	5.0	2.53
	1200	32.4	1.5	25.5	6.85
Acidic blast	300	1.6	0.11	–	–
Furnace slag	600	7.9	0.55	5.1	2.60
	1200	35.8	2.15	29.3	7.19
Acidic nickel	800	0.83	0.06	–	–
Slag	600	3.7	0.18	2.4	1.96
	1200	27.8	1.34	19.8	6.33

were used. Their fineness was approximately $520 \, m^2/kg$. The characteristics of the concretes are given in Table 12.8.

12.9   Refractory concrete

During the period between 1981 and 1983, both pilot and commercial scale testing were conducted on the refractory properties of alkali-activated slag cements and kaolin fibres. They were used in the pumps for feeding aluminium melts at the Institute for Foundry Problems of the Academy of Sciences of Ukraine, Kiev, at the Aluminium Alloys Plant in Mtsensk, Russia. It was also used in the production of a crucible of a magnetodynamic metering pump for feeding aluminium alloys in the City of Kazan, Russia. The refractory concrete was composed 3% blast furnace slag, 10% ground serpentinite rock, 50% kaolin fibres crushed in a vibration mill and 10% sodium metasilicate solution with density of $1250 \, kg/m^3$. The slag and rock were ground together to a specific surface of $450 \, m^2/kg$. The temperature of the aluminium melt was 867–950 °C. The experience indicated that the alkali-activated slag cement based refractory materials had a service life of nine months compared with one month when a high-alumina cement and kaolin fibres are used as refractory materials.

12.10   Oil-well cement

Alkali-activated slag cements were used as oil-well cements for grouting of the casing strings/columns (245 m in height) of the oil wells in the Republics of the Soviet Union (Kirgizia and Tajikistan). The grout was used at a depth of 3505–2052 m, at a temperature from 50 to 80 °C and pressure between 28 and 61.5 MPa. Alkali-activated cement mortars consisting of ground granulated slag, soda ash and sand were used. The dosage of soda ash was between 0.5 and 3.5% by mass of the slag. The properties of the grouting mortar were:

Water content	40–50%
Density	1720–2010 kg/m^3
Flow value	21–24 cm
Initial setting time	2 hrs. 30 min. to 3 hrs. 30 min.
Final setting time	3 hrs. 30 min. to 4 hrs. 40 min.
Flexural strength (one day)	2.8–5.2 MPa
Compressive strength (one day)	11.0–18.5 MPa

The use of alkali-activated slag cements for oil-well cementing applications has also been reported in several other publications (Brylicki *et al.* 1992, 1994, Malolepszy and Deja 1994). Actually, drilling fluid and mud can be mixed with alkali-activated slag to form a cementitious slurry for cementing operations (Javanmardi *et al.* 1993, Nahm *et al.* 1994a,b, Wu *et al.* 1996). This slag-mix technology was initially developed by Shell Oil Co. in 1991 and has been successfully used for more than 163 cementing jobs including primary, temporary abandonment and sidetrack plug cements (Wu *et al.* 1996).

Recently, Sugama and Brothers (2004) attempted to use sodium silicate-activated slag cement for completing geothermal wells containing highly concentrated H_2SO_4 and some CO_2. They used a 20% (by mass) sodium silicate solution (SiO_2/Na_2O mol ratio of 3.22) as the alkali activator, the cements autoclaved at temperatures up to 200 °C displayed an outstanding compressive strength of more than 80 MPa, and a minimum water permeability of less than 3.0×10^{-5} darcy. The combination of C–S–H and tobermorite phases was responsible for strengthening and densifying the autoclaved cement. At 300 °C, an excessive growth of well-formed tobermorite and xonotlite crystals generated an undesirable porous microstructure, causing the retrogression of strength and enhancing water permeability. Therefore, alkali-activated slag cements have good potential as an acid-resistant geothermal well cement at temperatures up to 200 °C. Alkali-activated ultrafine slag has been used as a grout for dam and other applications (Clarke and Helal 1989).

12.11 Stabilization/Solidification of hazardous, radioactive and mixed wastes

12.11.1 Introduction

Solidification/stabilization (S/S) is typically a process that involves the mixing of a waste with a binder to reduce the contaminant leachability by both physical and chemical means and to convert the hazardous waste into an environmentally acceptable waste form for land disposal or construction use. The S/S of contaminants by cements includes the following three aspects: (a) chemical fixation of contaminants – chemical interactions between the hydration products of the cement and the contaminants, (b) physical adsorption of the contaminants on the surface of hydration products of the cements, and (c) physical encapsulation of contaminated waste or soil (low permeability of the hardened pastes) (Shi *et al.* 1992b, 1994). The first two aspects depend on the nature of the hydration products and the contaminants, and the third aspect relates to both the nature of the hydration products and the density and physical structure of the paste. S/S has been widely used to dispose of low-level radioactive, hazardous and mixed wastes, as well as remediation of contaminated sites. According to the US Environment Protection Agency (US EPA), S/S is the best-demonstrated available technology (BDAT) for 57 hazardous wastes (USEPA 1993). About 30% of the superfund remediation sites used S/S technologies according to a USEPA report in 1996 (USEPA 1996). Of all the binders, cementitious materials are the most widely used for S/S of wastes.

As described in Chapters 4 and 8, the main hydration product of alkali-activated slag cement is C–S–H with a low Ca/Si ratio, and no $Ca(OH)_2$ exists. Also, alkali-activated slag cements exhibit much better resistance in aggressive environments than portland cement. The former also has much less porous structure and higher stability at high temperatures than the latter. For example, under hydrothermal conditions, the main hydration products of portland cement are $C_2SH(A)$ and $Ca(OH)_2$, which give the cement paste a very porous and unstable structure. However, the main hydration products of alkali-activated slag cement are C–S–H (B) and tobermorite or xonotlite, which give the hardened pastes a much less porous and more stable structure (Shi *et al.* 1991b). Also, C–S–H (B), tobermorite or xonotlite has an obvious cation ion exchange capacity and enhances the chemical fixation of contaminants (Komarneni and Roy 1985, Shrivastava and Glasser 1985, Komarneni *et al.* 1986, 1988). This means that alkali-activated slag cement is better than portland cement for S/S of wastes.

12.11.2 Stabilization/Solidification of hazardous wastes with alkali-activated slag cements

Heavy metals, such as Zn^{2+}, Pb^{2+}, Cd^{2+} and Cr^{6+}, can be well stabilized in NaOH-, Na_2CO_3- and sodium silicate-activated slag cements (Malolepszy and Deja 1995, Deja 2002b). These alkali-activated slag cements could immobilize these heavy metals very well regardless of activators. Cho *et al.* (1999) investigated the leachability of Pb^{2+} and Cr^{6+} immobilized in NaOH- and sodium silicate-activated slag cement pastes. They noticed that the leachability of Pb^{2+} and Cr^{6+} in alkali-activated slag cement pastes varied with curing conditions, but was very small. There is a very good relationship between the diffusion coefficient of Cr^{6+} and the pore volume with a radius less than 5 nm.

Qian et al. (2003a) found that low concentrations of Hg^{2+} ions had little effect on the compressive strength, pore structure and degree of hydration of alkali-activated slag cements; however, the addition of 2% Hg^{2+} ions into the alkali-activated slag cement showed an evident retardation on early hydration and reduction of early compressive strength, but no negative effects were noticed after hydration for 28 days. The results also show that up to 2% of Hg^{2+} ions can be effectively immobilized in the alkali-activated slag cement matrix, with the leaching meeting the Toxicity Characteristic Leaching Procedure (TCLP) mercury limit. Two mechanisms, physical encapsulation and chemical fixation, are assumed to be responsible for the immobilization of mercury in the alkali-activated slag cement matrix.

The effects of Zn^{2+} on the alkali-activated slag cement matrix depend on Zn^{2+} ion concentrations (Qian et al. 2003b). At low Zn^{2+} ion concentrations, little negative influences on the compressive strength, setting time and distribution of pore structure were observed. Moreover, low concentrations of Zn^{2+} ion could be effectively immobilized in the alkali-activated slag cement matrix. For 2% Zn-doped AAS matrix, the hydration of alkali-activated slag cement paste was greatly retarded and leaching from this matrix was higher than TCLP zinc limit even at 28 days. Based on the analyses of hydration products, the chemical fixation mechanisms are considered responsible for the immobilization of Zn^{2+} ions in the alkali-activated slag cement matrix.

Many substances can significantly interfere with the hydration of cement, as discussed in Roger and Shi (2004). This is the basic principle for the use of different cement chemical admixtures such as retarders, accelerators, superplasticizers, etc. to obtain some special properties of cements and concrete. It has been reported that heavy metals show much less interference with the hydration of alkali-activated slag cements than with portland cement. Shi et al. (1997a,b) investigated S/S of an electrical arc furnace dust with portland cement and an alkali-activated slag cement using an adiabatic calorimeter. The electrical arc furnace dust is from the production of a speciality steel, which contains a high concentration of variety of heavy metals. When, 30% electrical arc furnace dust is added, it retarded the hydration of the cement, but did not show an obvious effect on the hydration of cement at later ages. As the electrical arc furnace dust content is increased from 30 to 60%, it retarded the hydration of portland cement very significantly and the solidified waste forms did not show measurable strength after six months of hydration. When electrical arc furnace dust is mixed with alkali-activated slag-based cement, the early hydration of the cement was retarded more obviously as the electrical arc furnace dust content increased. However, the cement continued to hydrate with time and released more heat as the electrical arc furnace dust content increased. It seems that the presence of arc furnace dust retarded the early hydration but was beneficial to the later hydration of the slag (Caldwell *et al.* 1995).

A full-scale demonstration indicated that the alkali-activated slag worked well during the field operation (Shi *et al.* 1997b, Stegemann *et al.* 1997). Although the solidified waste form cured in laboratory behaved better than that in the field, the latter still demonstrated very less porous structure and low leachability (Stegemann 1995, Caldwell *et al.* 1999a,b). If the waste material is acidic, it may need to be neutralized before alkali-activated cements are used (Ipatti 1992).

12.11.3 Stabilization/Solidification of hazardous wastes with alkali-activated metakaolin or fly ash cements

Several studies have reported the use of alkali-activated calcined metakaoline or fly ash cements as effective binders for S/S of wastes (Comrie and Davidovits 1988, Khalil and Merz 1994, Van Jaarsveld *et al.* 1996, Palomo and Palacios 2003, Palacios and Palomo 2004, Siemer *et al.* 2001, Siemer 2004). Lead inhibits the hydration of portland cement. However, it was found that alkali-activated fly ash cements could effectively immobilize lead because it is precipitated as a highly insoluble silicate (Palomo and Palacios 2003, Palacios and Palomo 2004). The nature of the initial fly ash does have a significant impact on the effectiveness of lead fixation in these systems. Bankowskia *et al.* (2004) used alkali-activated metakaolin to stabilize brown coal ash containing high concentrations of heavy metals.

Fernandez-Jimenez *et al.* (2004) examined the alkali-activated fly ash matrices containing arsenic using a combination of scanning and transmission electron microscopy along with energy-dispersive X-ray analysis of argon-milled sections. It was found that little arsenic was incorporated in the matrix of hydration products but it was apparently associated with iron derived from the hydration of the fly ash. There was no association of As with added Fe_2O_3.

Soluble boron salts retard the setting and hardening of portland cement, and negatively affect its durability characteristics as well. It was found that a high concentration of boron salts did not affect the properties of alkali-activated fly ash cements (Palomo and De la Fuente 2003). Experimental results indicated that leaching indexes and diffusion coefficients of boron in activated fly ash-based matrix was 100 times lower that those in portland cement-based matrix. The authors felt that the boron present in the system would precipitate as compounds such as $NaB(OH)_4$ or other kinds of sodium borates, which would fill the voids in the matrix. The main factor affecting the leaching degree was the solubility of these compounds in the usually rather high alkaline matrix.

In portland cement-based systems, chromium can be stabilized in the form of Ca_2CrO_4, which also enhances the strength of the cement. However, it was found (Palomo and Palacios 2003) that all Cr^{6+} in alkali-activated fly ash matrix could be leached out due to the formation of $Na_2CrO_4 \cdot 4H_2O$, which is highly soluble. The formation of $Na_2CrO_4 \cdot 4H_2O$ also had a negative effect on the activation of fly ashes.

12.11.4 Stabilization/Solidification of radioactive wastes

Caesium (Cs) is the most difficult radionuclide to be stabilized in radioactive wastes. Several laboratory studies have confirmed that the caesium leachability from alkali-activated slag pastes is much lower than that from portland cement pastes (Wu *et al.* 1991, Krivenko *et al.* 1993, Shen *et al.* 1994, Shi *et al.* 1992b, Shi and Day 1996a). To evaluate the leachability of Cs^+ in alkali-activated slag cements, monolithic specimens are suspended in deionized water in teflon containers at testing temperature, are transferred to other containers with fresh deionized water at specified intervals, and the concentration of Cs^+ in original solutions was measured. (Shi *et al.* 1992b, Shi and Day 1996a). The caesium leachability L_t (cm^{-1}) at time t was calculated by the following equation:

$$L_t = (a_t/A)(F/V) \tag{12.1}$$

where
a_t = mass of leached Cs^+ at time t (g);
A = total original mass of Cs^+ in the specimens (g);
F = total surface area of the monolithic specimens (cm^2);
V = volume of the monolithic specimens (cm^3).

Figure 12.9 shows the leached fraction of Cs^+ in portland cement and alkali-activated slag cement pastes containing 0.5% $CsNO_3$ after 28 days of moist curing at 25 °C. The results indicate that the Cs^+ in portland cement pastes shows much higher leached fraction than that in alkali-activated slag cement pastes at the same temperature. As the temperature increases from 25 to 70 °C, the leached fraction of Cs^+ in both pastes escalates. The leached fraction of Cs^+ in portland cement pastes at 25 °C is even higher than that from alkali-activated slag cement pastes at 70 °C. The calculation using Arrhenius' Equation indicated that the Cs^+ leaching activation energy of portland cement pastes is 19 kJ/mol compared with 25 KJ/mol for alkali-activated slag cement pastes. The lower leached fraction and higher leaching activation energy of Cs^+ in alkali-activated slag cement pastes than in portland cement pastes can be attributed to the less porous structure and lower Ca/Si ratio in C–S–H.

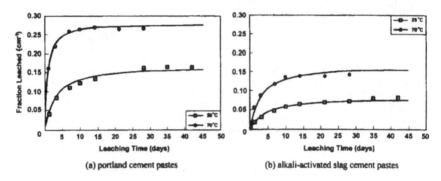

Figure 12.9 Leached fraction of Cs$^+$ in hardened portland and alkali-activated slag cement pastes (Shi *et al.* 1994).

A partial replacement of slag with zeolites or metakaolin increases the porosity of the hardened cement pastes, but decreases the leached fraction of Cs$^+$ and Sr^{2+} in the hardened pastes (Shen *et al.* 1994, Qian *et al.* 2002). The authors attributed the decrease in leached fraction to the formation and adsorption properties of (Al + Na) substituted C–S–H and self-generated zeolite precursor.

12.11.5 Stabilization/Solidification of wastes under hydrothermal conditions

Under hydrothermal conditions, zeolites form when mixing coal fly ashes or metakaolin with alkaline solutions. Zeolites are a family of complex aluminosilicates having a three-dimensional network structure containing channels and cavities. They are thermodynamically stable and their channels and cavities can immobilize a variety of contaminants.

In one study, cementitious materials with high fly ash contents were used to solidify highly alkaline low-level radioactive waste solutions (Brough *et al.* 2001). The temperature of the solidified materials during adiabatic curing reached up to 90 °C. Zeolites were identified in the solidified waste materials.

Hydroceramics are hydrous materials consisting mainly of zeolites and have been designed for the treatment of sodium bearing wastes (Siemer *et al.* 2001, Siemer 2004). The raw materials include waste (typically 30 wt% dry-basis), metakaolin or Class F fly ash, ~5% powdered vermiculite (to enhance ^{137}Cs fixation), ~0.5% sodium sulphide (redox buffer and heavy metal precipitant), plus ~10% sodium hydroxide dissolved in enough water to produce a stiff paste.

The hydroceramic waste forms are then autoclaved at 90 °C or 190 °C, then a dense matrix having sufficient strength to withstand the rigours of stacking in a repository environment is formed. But best of all, the matrix is extremely insoluble. The primary load-limiting characteristic of the hydroceramic waste forms is that cancrinite (or sodalite) represents maximal waste loading; i.e., no more than 25% of the formulation's sodium can be in forms other than hydroxide, silicate or aluminate.

The hydroceramic waste forms were specifically developed to deal with the reprocessing waste at the Idaho National Engineering and Environmental Laboratory (INEEL). INEEL waste is uniquely suited for cementitious solidification because of its overwhelming amounts of sodium. A portion of the reprocessing waste generated at other DOE fuel reprocessing facilities could be directly processed into hydroceramic waste forms, but the majority (the "supernates" and "salt cakes") are unfit unless they are pretreated in a way that re-speciates the sodium.

12.12   Summary

This chapter has described some applications of alkali-activated cements and concrete for different purposes, and field inspection results of some structures after many years of service. Some types of alkali-activated cements can be used as conventional portland cements. The special properties of alkali-activated cements make them suitable for some special applications. The field inspection results indicate that alkali-activated cements and concretes can behave much better than conventional portland cements and concretes. Alkali-activated cements are much better than portland cement for stabilization solidification of hazardous and radioactive wastes due to their less porous structure and better resistance to corrosive environments.

Chapter 13

Standards and specifications for alkali-activated cements, concretes and products

13.1 Introduction

Alkali-activated slag cements and concretes have been used in construction since 1958 in the former Soviet Union. As discussed in previous chapters, alkali-activated slag cement and concrete possess many unique features compared with conventional portland cements and concretes. The manufacture, applications and specifications for alkali-activated slag cement and concrete are different from those for ordinary portland cement and concrete. Over 60 standards related to alkali-activated slag cements and concretes were developed in the Soviet Union. This chapter summarizes some specifications and standards related to the raw materials, cements, concretes, structures and manufactures for alkali-activated slag cements and concretes in the former Soviet Union.

13.2 Specifications on the ingredients for alkali-activated slag cements

13.2.1 Slags

Ground granulated blast furnace slags or phosphorus slags are the most suitable slags for making alkali-activated slag cements. They should have a glass content of at least 70% and a modulus of basicity (Mb) between 0.7 and 1.2 as defined as follows:

$$Mb = \frac{CaO + MgO}{SiO_2 + Al_2O_3} \tag{13.1}$$

Non-ferrous slags from the production of lead, nickel and copper can also be used. According to TU 67-648-84 (1984), the hydraulic properties of non-ferrous slag are evaluated based on the Eqn (13.2) and can be classified into three categories as shown in Table 13.1:

$$Cq = \frac{CaO + MgO + Al_2O_3 + Fe_2O_3 + 1/2FeO}{SiO_2 + 1/2FeO} \tag{13.2}$$

Table 13.1 Classification of granulated non-ferrous slags

Slag characteristics	Specification (%)		
	Category I	Category II	Category III
Cq	>1	0.7–1	<0.7
SiO$_2$	26–32	33–52	33–52
CaO + MgO	>20	>17	>8
FeO	>30	>30	>35

Based on the type of furnace used and the type of steel made, steel slags include open-hearth, converter, electrical furnace, cupola and silico-manganese slags (Petropavlovsky 1990). Their typical compositions are shown in Table 13.2. Since the chemical and mineralogical compositions of steel slag vary in a wide range (Shi 2004b), it is very difficult to use them in conventional cement and concrete. However, a wide range of steel slags can be used in the production of alkali-activated slag cements since alkalis can activate the potential cementitious properties of the steel slags effectively (Shi 2002). Based on the chemical and physical composition of steel slags, they can be classified into three catalogues: glassy akermanite-melilite slag with a basicity Mb $\leq$ 0.8 (cupola and silicomanganese slags); crystallized monticellite-diopside slag with 0.8 < Mb < 1.5 (open-hearth and electric furnace steel slags); and crystallized orthosilicate slag with Mb $\geq$ 1.5 (open-hearth and converter slags). TU 14-113 UzSSR 11-91 (1991) specifies steel slags for use in alkali-activated slag cements.

A slag can be ground in a cement plant or in a concrete precast plant. The grinding mill can be a ball mill, jet mill or other mills. The moisture content of slag before being fed into a grinding mill should not exceed 1%. The temperature of flue gases at the inlet of the drum should not exceed 500°C. Temperature of the feeding materials should not exceed 80°C, and

Table 13.2 Chemical composition of steel slags (Petropavlovskii 1990)

Slag	Oxide content, % by mass						
	SiO$_2$	CaO	Al$_2$O$_3$	MgO	FeO+Fe$_2$O$_3$	MnO	Mb*
Open-hearth	13.9–26.2	24.7–43.6	2.8–5.2	6.9–11	11.4–26.3	2.9–10	0.8–1.5 > 1.5
Converter	16.3–16.7	44.9–46.7	1.9–4.8	3.5–5.5	18.4–25.9	4–4.6	> 1.6
Electric furnace	18.8–19.8	39–43.9	3.1–7.6	15.7–16.7	10.8–20.8	1.2–2.8	0.8–1.5 > 2
Cupola	38.3–48.2	26.1–31.5	10.6–11.3	4.3–5	3–4.9	2.7–11.5	0.6–0.8
Silico-manganese	42–45	20–22	12–15	2–3	–	16–18	0.35–0.45

* M$_b$ – see Eqn (13.1).

should not exceed 100°C after grinding. The fineness of the ground slag should not be less than 400 m²/kg (by Blaine apparatus). Additives, which meet specifications RST UkrSSR 5024-83 and OST 67-11-84 (1984) can be used during the grinding of steel slag.

13.2.2 Alkaline activators

A general factor for acceptability of an alkaline compound for use in alkali-activated slag cements is its ability to produce an alkaline solution immediately after being dissolved, like alkali hydroxides, non-silicate salts of weak acids and alkali silicates; or cation-exchange processes during hardening such as alkali metal sulphates and zeolites.

In addition to these commercial products as described in Chapter 3, alkali-containing industrial wastes and by-products can also be used in the production of alkali-activated slag cements (Glukhovsky et al. 1988). The waste and by-product alkalis from different sources can be divided into 4 groups as listed in Table 13.3.

Table 13.3 Classification of alkali-containing by-products and wastes from different sources (Glukhovsky et al. 1988)

Manufacturing process	Group			
	I	II	III	IV
Metal working, machine-manufacturing industry, foundry				
Pickling of castings from ceramics, burnt-on sand, scale in alkali metal melts	+	+		
Oxidation of metals in alkali solution	+			
Degreasing of metals in NaOH-solution	+			
Chemical cutting in alkali solution	+			
Alkaline etching of aluminium	+			
Chemical, organic chemical, biochemical industry				
manufacture of				
kaprolactam (kapron)	+			
nitrogen fertilizers	+	+		+
Soda	+		+	
isopropyl alcohol $(CH_3)_2CHOH$	+			
Ammonia NH_3	+			
anhydrous sodium sulphate	+			
acrylic plastic	+			
Oxygen	+			
metallic sodium	+			
cation-active surfactant	+			
Polyethylenepolyamines	+			

tetraethyl lead Pb(C$_2$H$_5$)$_4$	+		
methyl-cellulose	+		
Phenol	+		
potassium permanganate KmgO$_4$	+		
Ethylene	+		
Dyes	+		
sodium sulphite Na$_2$SO$_3$	+		
chemical fibres		+	
chlor-methane		+	
sodium bicarbonate	+		
Naphthol	+	+	
benzol C$_6$H$_6$		+	
cake and by-product process		+	
Superphosphate		+	
folic acid/pteroylglutamic acid (PGA) C$_{19}$H$_{19}$N$_7$O$_6$		+	
viscose high-modulus staple fibres			+
resorcinol C$_6$H$_4$(OH)$_2$			+
barium salts	+		
Paper and cellulose making industry			
manufacture of cellulose	+		+
Cement industry			
cement kiln dust (from electrofilters)	+		
Metallurgical industry			
Manufacture of			
alumina from bauxites	+		+
titanium dioxide	+		
alumina from nephelines and sienites	+		
Petroleum refining industry			
refining of straight rum petrol fractions from sulphur-containing compounds		+	
manufacture of aluminisilicate catalysts			+
Food industry			
washing of containers/ equipment, etc.		+	
washing of working clothes		+	
Regeneration of ion-exchange resins		+	

The by-products in Group I can be used directly without any pretreatments. Examples include soda-alkali melts from the manufacture of kaprolaktam, used alkaline solutions and solids from pickling of castings in alkali metal solutions and melts, alkaline solutions from oxidation of metals, alkaline solutions from cleaning of the air from CO$_2$, etc.

The by-products in Group II usually have a low alkali concentration and require concentration and removal of some undesirable organic matters. For example, sodium salt of adipic acid needs to be heated to remove organic matters; wastes from vacuum-carbonate purification of coking gas requires concentration; wastes from petroleum refining industry should be treated by carbonization to increase concentration of alkaline compounds and to remove organic matters.

The materials in Group III are in the form of sludge or slurry and contain toxic substances. These wastes can be used only after detoxication of these toxic substances. Typical examples are sludges and slurries from the manufacture of sodium sulphide, isopropyl alcohol, phenol, etc.

Group IV includes the wastes from manufacture of aluminium. Among them, the most widely available waste is sodium sulphate.

Several standards and specifications regulate the alkaline wastes to be used in the production of alkali-activated cements, including TU 6-18-35-85 (1985), TU 14-14-145-85 (1985) TU 6-16-29-45-86 (1986), TU 37.002.0442-88 (1988) and TU 6-46-91 (1991).

Activator(s) is usually dissolved or mixed with mixing water prior to use. RST 5024-83 (1983) and OST 67-11-84 (1984) provide the guidelines for dissolution of activators. The activator dosage is controlled through the measurement of density of the activator solution. Since the activator dosage or the solution density determines the strength of designed cements and concretes, the density of the activator solution should be constantly controlled and adjusted by taking account of moisture contained in the aggregates, as specified in RSN 336-84 (1984).

The preparation of alkaline activator solution may require equipment for crushing/grinding, measuring and dissolving; tanks for storage of solutions that are concentrated and diluted to required densities; as well as storage rooms and conveyers for feeding solid/dry raw materials and solutions. A mixing tank should be equipped with a unit for mechanical or pneumatical agitation with heating using a vapour. In case constituents are mixed in two steps, two tanks for alkaline activator solutions of different densities are needed. To prevent crystallization of dissolved activator in the mixing tanks and pipelines, the temperature should not be lower than 60 °C; for this, heat registers should be installed in the tanks and the pipelines should be placed in heat jackets. The pipelines should be arranged with connection units to blow them through using a vapour in case of crystallization of the solution in the pipes. The design of pipelines should provide that the solution would not remain in the pipes after completing the work. Temperature of the ready-to-use solution should be higher than 20 °C and, above all, should not exceed the temperature of the surrounding environment.

13.3   Standards for alkali-activated slag cements

All standards for alkali-activated slag cements are developed based on aluminosilicate components. They include TU 67-Ukr-181-74 (1974);

TU 204 UzbSSR 1-83 (1983); RST UkrSSR 5024-83 with amendments of 30.05.89 (1983); OST 67-10-84 (1984); TU 14-11-228-87 (1987); TU 67-1020-89 (1989); TU 7 BelSSR 5-90 (1990); TU 559-10.20-001-90 (1990); TU 14-11-228-90 (1990); TU 10.20 UkrSSR 169-91 (1991); TU 10.15 UzSSR 04-91 (1991) and DSTU BV 2.7-24-95 (1995). In these existing standards, alkali-activated slag cements are classified based on their strength, as summarized in Table 13.4. Ordinary portland cement or other substance with high basicity is added to accelerate strength development and to reduce shrinkage of the cement and concrete. The dosage of these additives increases with the decrease in density of the alkaline activator solutions or the basicity of slag, but should not exceed the limits as set in OST 67-11-84 (1984) and RST 5024-83 (1983).

Non-granulated cupola and converter steel-making slags, belite sludge and high-calcium ashes containing portland clinker minerals are allowed to be used in alkali-activated slag cements after they are tested. The proportions of these additives should not exceed 25% by the mass of the slags as described in Table 13.4.

Depending upon the nature of the alkaline activator used, the cements made with blast furnace and phosphorus slags are classified into different grades, based on their standard mortar strengths, as shown in Table 13.5. Different grades of cement made with non-ferrous slags are summarized in Table 13.6.

Table 13.4 Designation and specifications of alkali-activated slag cements

Cement designation	Content of OPC (% by mass)	Slag requirements
SAAC 0	0	Blast furnace slag with $M_b \geq 0.6^*$
SAAC 2	2 ± 1	Blast furnace slag with $M_b > 1.05^*$
SAAC 4	4 ± 1	Blast furnace slag with $M_b \leq 0.95^*$
SAAC 6	6 ± 2	Blast furnace slag with $0.95 \leq M_b \leq 1.05^*$
SAAC T5	5 ± 2	Phosphorus slag
SAAC CO	0	Lead toxic matters
SAAC H0	0	Nickel
SAAC H7	7 ± 2	Nickel
SAAC M0	0	Copper
SAAC M5	5 ± 2	Copper

$^*M_b$ – see Eqn (13.1).

Table 13.5 Grades of alkali-activated slag cements made with blast furnace and phosphorus slags

Cement grade

SAAC 0	SAAC 2	SAAC 4	SAAC 6	SAAC T5
Soda ash				
300	300	300	300	300
400	400	400	–	400
–	500	500	500	
Soda ash melt				
300	300	300	300	300
400	400	400	–	400
500	500	500	500	500
–	–	–	600	–
Sodium soluble silicate ($M_s = 1$)				
300	–	300	300	300
400	400	–	–	–
500	500	500	500	500
600	–	–	–	–
–	700	–	700	700
–	–	800	–	–
900	–	–	–	–
–	1000	–	1000	1000
–	–	–	1100	–
1200	–	1200	–	–
Sodium soluble silicate ($M_s = 2$)				
300	300	–	300	300
400	–	400	400	–
500	500	500	500	500
600	–	–	–	–
700	700	700	700	700
800	–	800	800	–
900	900	900	–	900
–	1000	–	–	1000
Sodium soluble silicate ($2 \leq M_s \leq 3$)				
300	–	300	300	–
400	400	–	400	–
500	–	500	–	–
600	600	–	600	–
700	700	700	700	–
800	–	800	–	–
900	900	–	–	–

Table 13.6 Grades of alkali-activated slag cements made with non-ferrous slags

Cement type				
SAAC C0	SAAC H0	SAAC H7	SAAC M0	SAAC M5
Soda ash				
300	300	300	–	300
–	–	400	–	–
Soda ash melt				
300	300	300	–	300
400	–	–	–	400
Sodium soluble silicate ($M_s = 1$)				
300	300	300	300	300
400	400	400	400	400
500	600	600	–	–
Sodium soluble silicate ($M_s = 2$)				
300	300	300	300	300
400	400	400	400	400
500	500	500	500	500
600	600	600	–	600
700	700	700	–	–
800	800	800	–	–
–	–	900	–	–
–	–	1000	–	–
Sodium soluble silicate ($2 \leq M_s \leq 3$)				
300	300	300	300	300
400	400	400	–	400
–	–	500	–	–

13.4   Alkali-activated slag cement concretes

Any type of concrete can be produced using alkali-activated slag cements like portland cement concretes. DSTU BV 2.7-24-95 (1995), OST 67-10-84 (1984), TU 67-1019-89 (1989) and TU 10.20.186-92 (1992) provide guidelines and requirements for the production of normal weight concretes using alkali-activated slag cements. Based on the strength of concrete, concretes are classified into many grades as summarized in Table 13.7.

When frost resistance is required, the concretes can also be classified based on their frost resistance, which include F200, F300, F500, F600, F700, F800, F900 and F1000. They are also classified based on water permeability requirement such as W4, W6, W8, W10, W12, W14, W16, W18, W25 and W30.

The deformation, which includes shrinkage and creep, of the normal-weight alkali-activated slag cement concretes should not exceed the limits as listed in Tables 13.8 and 13.9.

Table 13.7 Concrete grades and strength requirements

Grade of concrete	Compressive strength (MPa)	Grade of concrete	Compressive strength (MPa)
B10	20	B55	80
B15	25	B60	90
B20	30	B70	100
B25	40	B80	110
B30	50	B90	120
B40	60	B100	130
B50	70	B110	140

Table 13.8 Shrinkage ($\varepsilon_{sh(t)}$) and creep (C_n) limits for normal-weight alkali-activated slag cement concretes of class B50 and lower

Consistency, sec., by USSR standard 10181.1–81	$C_n \cdot 10^5$, MPa^{-1}						$\varepsilon_{sh(t)} \cdot 10^5$	
	Class of concrete						Class of concrete	
	B10	B15	B25	B30	B40	B50	B10–B15	B25–B50
85...30	14.0	10.8	7.7	6.2	5.2	4.5	28	27
15...10	16.2	12.4	8.9	7.2	6.0	5.3	29	33
8...4	18.2	14.0	10.1	8.1	6.8	5.9	35	40
4...2	19.2	14.8	10.7	8.5	7.2	6.2	38	43
∠2	24.0	18.0	12.0	9.5	7.5	6.3	41	46

Table 13.9 Shrinkage ($\varepsilon_{sh(t)}$) and creep (C_n) of normal-weight alkali-activated slag cement concretes of class B55 and greater

Consistency, sec., by USSR standard 10181.1–81	$C_n \cdot 10^5$, MPa^{-1}							$\varepsilon_{sh(t)} \cdot 10^5$	
	Class of concrete							Class of concrete	
	B55	B60	B70	B80	B90	B100	B110	B10–B15	B25–B50
50...40	3.60	3.10	2.70	2.50	2.30	1.90	1.85	31	35
35...25	3.80	3.35	2.85	2.65	2.39	2.09	1.94	32	36
20...15	3.95	3.50	3.05	2.90	2.52	2.15	2.00	33	37
15...10	4.20	3.60	3.07	2.88	2.61	2.28	2.11	34	38
8...4	4.70	4.00	3.50	3.28	2.97	2.52	2.34	41	44
4...2	4.80	4.20	3.57	3.36	3.04	2.60	2.40	44	47
∠2	5.40	4.70	4.10	3.70	3.40	2.90	2.70	46	49

Alkali-activated slag cement of grades from 300 to 900 can be used for the production of concretes of grades from B10 to B90. It is recommended to use cements SAAC B0 and SAAC B4 of grades from 900 to 1200, SAAC B2 and SAAC T5 of grades from 900 to 1000, and SAAC B6 of grades from 900 to 1100 to produce concrete of grades B50 to B110.

Either crushed stone or gravel can be used as coarse aggregates, and sand as fine aggregate in normal-weight alkali-activated slag cement concrete. Ultra fine sands, industrial wastes and other materials containing less than 25%, by mass in total, dusts and clays can also be used as fine aggregates. The proportion of clayey particles should not exceed 5% by mass. The use of aggregates containing gypsum and anhydrate grains is not allowed. The properties of the hardened normal alkali-activated slag cement concretes are tested in the same way as for portland cement concretes.

The strength of alkali-activated slag cement concretes is controlled by the nature and dosage of the alkaline activator or the density of the alkaline activator solution. Soluble sodium silicates can be used to produce very high strength concrete. When they are used, the basicity of the slag and slag content has little effect on strength, especially when concrete is steam-cured. The compressive strengths of the sodium silicate-activated slag cement concrete after steam-curing do not change obviously with time. When alkali carbonates are used as activators, the nature of slag has an obvious effect on strength. Also, the concretes continue to gain strength with time after steam curing.

The ratio of strength form prism specimens to that from cube specimens of alkali-activated slag concrete is higher compared to that of portland cement concretes. It is 0.783 for alkali silicate-activated slag cement concrete and 0.753 for alkali carbonate-activated slag cement concrete. The ratio usually increases with time and can be adjusted by introducing portland cement clinker.

Concretes with densities from 200 to 1800 kg/m^3 can be produced using alkali-activated slag cements and expanded clay, blast furnace granulated slag, shell limestone, slag pumice, expanded perlite and vermiculites, wood, etc. as aggregates. The specifications for lightweight alkali-activated slag cement concrete can be found in TU 65-484-84 (1984).

13.5 Structures and structural elements made with alkali-activated slag cement concretes

Very extensive experiments had been conducted on alkali-activated slag cement concrete elements in the former Soviet Union to establish testing and design criteria for alkali-activated slag cement concrete products. The standard design processes were developed in two steps. First, many standards and specifications were developed for the design and testing of specific products, which include technical specification for pipes (1970),

TU-33 UzSSR-03-82 (1982), TU 65 UkrSSR 196-86 (1987), TU 203.1-1-88 (1988), TU 234 UkrSSR 212-89 (1989) and TU 12.0175921.003-89 (1989). After extensive experiences were obtained, some general design standards and specifications were developed for structures and structural elements made from alkali-activated slag cement concrete, they include TU 67-1010-88 (1989), OST 67-12-84 (1984), TU 33 UzSSR 01-84 (1984) and RST UkrSSR 5026-84 (1984).

Several structural design codes for alkali-activated slag cement concrete have also been developed: Recommended Design Codes/Norms For Structures Made From Slag Alkaline Cement Concretes (1983), VSN 65.12-83 (1983), TU 67-1021-89 (1989) and TSNIIOMTP (1985a).

13.6   Recommendations on production and use of alkali-activated slag cement concretes and structures

During the past decades, a series of standards and specifications have been developed for the production of alkali-activated slag cements, concretes, plain structure and reinforced concrete structures, which include RTU 153-65 (1966); VTU 183-74 (1974); VSN 2-97-77 (1978); Donetsky Promstroiniiproekt (1980); RSN 25-84 (1984); RSN 336-84 (1984); RSN 344-87 (1987); RDIBS (1988); RSN 336-90 (1984); RSN 354-90, (1991). The following sections summarize some recommendations and specifications for the production of alkali-activated slag cements, concretes and structures.

13.6.1   Production of concrete mixtures

In order to produce alkali-activated slag cement concrete mixtures, the concrete ready mix plants or precast plants need to have additional facilities for handling alkaline activator solutions. The ingredients, such as slag, aggregates, activators and water, should be measured to an accuracy of 1%, and mixed in concrete mixers with forced action, and the use of gravity concrete mixers is allowed for making coarse aggregate concrete mixes.

The required mixing time is determined experimentally in a pilot production trial. In order to keep the workability or consistency of the concrete mixtures for a longer period of time, the alkaline activator solution can be introduced in two steps at an interval of 2 to 3 minutes, as recommended in RST UkrSSR 5024-83 (1983). The temperature of the concrete mixture at the discharge from a concrete mixer in summer should not exceed the temperature of the surrounding air and not be lower than $20 \pm 5\,°C$. The transportation and handling of alkali-activated slag cement concrete mixtures is similar to that for portland cement concrete mixtures.

Alkali-activated slag cement concrete mixtures should not be allowed to mix with any other types of concrete mixtures.

13.6.2   Placement

The mould preparation (cleaning, greasing), embedding of reinforcement and fittings, placement and finishing are similar to those for conventional prefabricated concrete product plants. The following materials can be used as greases: mixture of used oils or straw oil and petrolatum, motor oil (cup grease) and other greasing aids after being tested.

The concrete mixtures can be compacted like conventional portland cement concrete mixtures. The coefficient of compaction (a ratio of its real density to a calculated theoretical one) for a normal weight concrete should not be less than 0.98, and not less than 0.97 for fine aggregate concrete. A static pressure on the concrete mixtures by means of loading, vibrostamps, vibropresses and other forms should not exceed 0.025 MPa, as recommended by specifications. It should not be allowed to mix with any other concrete mixtures during placement and finishing.

13.6.3   Curing

Curing has effects on the strength and deformation characteristics of alkali-activated slag cement concretes. Steam curing of the alkali-activated slag cement concrete made with acidic blast furnace slag and soda ash increase the strength (by 70%) and the modulus of elasticity of the concrete. For the same concrete mixture, when 4% portland cement clinker is added, steam curing is not so important. Actually, both the strength and modulus of elasticity of the concrete cured at room temperature are even better.

When sodium disilicate is used as an alkaline activator, with or without a portland cement clinker additive, the modulus of elasticity of the alkali-activated slag cement concrete cured at room temperature is better compared to that cured at elevated temperatures.

Steam curing regime has a significant effect on the properties of the concrete. However, under any steam curing regime, there should be a pre-set time period and the curing temperature should be increased gradually. The pre-set time before steam curing is dependent on the type of alkali-activated slag cement used. The curing temperature starts to rise when the concrete has the highest heat evolution rate. When using the alkali-activated slag cements made from basic blast furnace slags, the following curing regimes are recommended:

- a curing regime of $4 + 3 + 6 + 3$ hours (four hours of presetting time, three hours of temperature rise time, six hours of constant temperature time and three hours of temperature decrease time) with the highest temperature of 80–95 °C for the products with a thickness of more than 20 cm, and soda-alkali melt, soda ash or soluble sodium silicate $(2 \leq Mb \leq 3)$ as an activator(s).

- A curing regime of $2+3+0+3$ hours with the highest temperature of $75-85\,°C$ for the products with a thickness of less than 20 cm and soluble sodium silicates $(1 \leq Mb \leq 2)$ as an activator(s).

When neutral or acid blast furnace slag or phosphorus slag is used, the curing time should be determined by experiments.

Figure 13.1 shows the four types of steam curing regimes for alkali-activated slag cement concretes. The curing regimes are determined by considering the type of aggregate used. Concretes made with regular aggregates are less sensitive to changes in the steam curing regimes. Two hours of constant temperature curing at $45-50\,°C$ during the temperature increase stage, as shown in Figures 13.1b and 13.1d, results in inconsiderable strength increase in concrete strength. However, increase of that curing temperature decreases the modulus of elasticity as compared to that of the concrete steam-cured with a regime of $2+3+6+3$ hours. The decrease in the highest curing temperature to below $80\,°C$ can decrease the shrinkage of the concrete.

According to RST UkrSSR 5024-83, when a monolithic concrete made with soda ash and soda melt is cured at temperatures above $15\,°C$, the following slag alkali-activated cements are recommended to be used: SAAC2, SAAC4, SAAC6 and SAACT5.

The alkali-activated slag cement concrete mixtures can be used at below $0\,°C$. In one case, the foundations and walls of a storehouse were placed at $-8\,°C$. The strength of the concrete, after 20 days at temperatures varying

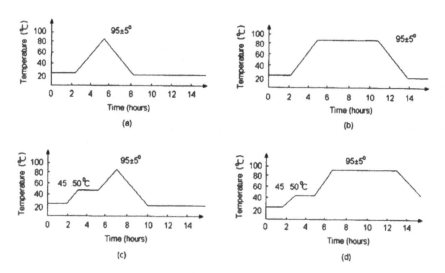

Figure 13.1 Recommended steam curing regimes for alkali-activated slag cement concretes.

between −8 °C and −12 °C, was 30% of the designed strength. The surveillance on the structure over 7 years of time period showed that the concrete strength had increased by 2 times (20–25 MPa) compared with those cured under controlled curing conditions.

For the concretes hardened at 0 °C to −15 °C, a quick-hardening alkali-activated slag cement concrete made with caustic alkalis or sodium metasilicate should be used. The temperature difference between the concrete and the environment during demoulding or transportation of precast concrete elements to a storehouse should meet the requirements for regular portland cement concrete products.

13.7 Use of alkali-activated slag cements and concretes for special applications

Because of the unique properties of alkali-activated slag cement and concrete, as described in previous chapters, they can be used for many applications where conventional cement and concrete are not applicable. Based on extensive lab research and field application experiences, several specifications have been developed for the use of alkali-activated slag cement and concrete in special applications, which include:

1   Technical Specification for Construction of Road Basements Made with Slag Alkaline Binder Stabilized Soils (The Ministry of Industrial Construction of the USSR 1979);
2   Guide for manufacturing and use of high-consistency slag alkaline cement concretes for construction of anti-filtration screens (Donetsky Promstroiniiproekt 1980, VSN 67-247-83 1984, TU 67-629-84 1984);
3   Guide for the use of a slag alkaline binder in strengthening stone materials for road construction (MUSSREOAR 1986, TU U V.2.7-16403272.006-98 1998).

13.8 Summary

This chapter summarizes standards, specifications and guidelines developed in the former Soviet Union and Ukraine. These standards, specifications and guidelines cover raw materials, production of cement and concrete, mix design and structural design of concrete products. They are very useful recourses for other regions or countries for the development and utilization of alkali-activated cements and concretes.

References

AASHTO T 277, 1990, Standard Specifications for Test Method for Rapid Determination of the Chloride Permeability of Concrete, Part II Tests, American Association of States Highway and Transportation Officials (AASHTO), Washington, D.C., USA.

Aborin, A. V., Brykov, A. S., Danilov, V. V. and Korneev, V. I., 2001, The influence of hydrated sodium silicate on hardening of cement compositions. *Tsement (Cement) (in Russian)*, St. Petersburg, Russia, 5–6, 40–43.

ACI 209R-92, 1993, Prediction of Creep, Shrinkage and Temperature Effects in Concrete Structures, American Concrete Institute, Detroit, MI, USA.

Akimov, V. A., Kryzhanovsky, I. I. and Morozova, L. V., 1984, Optimization of concrete mixtures based on the slag-alkaline cements. *II National Scientific and Practical Conference on Slag-Alkaline Cements, Concretes and Structures* (edited by Glukhovsky), Kiev, USSR, 207–208.

Alexander, K. M., 1960, Reactivity of ultrafine powders produced from siliceous rocks. *Journal of American Concrete Institute*, 57, 557–569.

Alonso, S. and Palomo, A., 2001a, Calorimetric study of alkaline activation of calcium hydroxide-metakaolin solid mixture. *Cement and Concrete Research*, 31(1), 25–30.

Alonso, S. and Palomo, A., 2001b, Alkaline activation of metakaolin and calcium hydroxide mixtures: influence of temperature, activator concentration and solids ratio. *Materials Letters*, 47(1), 55–62.

Amrhein, C., Haghnia, G. H., Kim, T. S., Mosher, P. A., Gagaiena, R. C., Amanios, T. and De La Torre, L., 1996, Synthesis and properties of zeolites from coal fly ash. *Environmental Science and Technology*, 30, 735–742.

Andersson, R. and Gram, H.-E., 1987, Properties of alkali-activated slag concrete. *Nordic Concrete Research*, 6, 7–18.

Apers, J. and Pletinck, M., 1985, A lime-pozzolan cement industry in Rwanda. *J. Appropriate Technology*, 11(4), 22–23.

Ashbridge, A. H., Jones, T. R. and Osborne, G. J., 1996, High performance metakaolin concrete: results of large scale trials in aggressive environments. International Congress on Concrete in the Service of Mankind. June 24–28. Dundee, UK, 13–24.

Astapov, N. I., 1976, Investigation of Density and Strength of the Slag-Alkaline Cement Concretes of High Brands, Ph.D. Thesis, Kiev Civil Engineering Institute, Kiev, USSR.

ASTM C 109, 2003, Test Method for Compressive Strength of Hydraulic Cement Mortars (Using 2-in. or 50 mm Cube Specimen), Annual Book of ASTM Standards, Vol. 04.01. Cement; Lime; Gypsum, American Society for Testing & Materials (ASTM), Philadelphia, USA.

ASTM C 311-03, 2003, Standard Test Methods for Sampling and Testing Fly Ash or Natural Pozzolans for Use as a Mineral Admixture in Portland-Cement Concrete, Annual Book of ASTM Standards, Vol. 04.01. Cement; Lime; Gypsum, American Society for Testing & Materials (ASTM), Philadelphia, USA.

ASTM C 469, 2003, Test Method for Static Modulus of Elasticity and Poisson's Ratio of Concrete in Compression, Annual Book of ASTM Standards, Vol. 04.02. Concrete and Aggregate, American Society for Testing & Materials (ASTM), Philadelphia, USA.

ASTM C 595, 2003, Specification for Blended Hydraulic Cements, Annual Book of ASTM Standards, Vol. 04.01. Cement; Lime; Gypsum, American Society for Testing & Materials, Philadelphia (ASTM), USA.

ASTM C 618, 2003, Standard Specification for Coal Fly Ash and Raw or Calcined Natural Pozzolan for Use as a Mineral Admixture in Concrete, Annual Book of ASTM Standards, Vol. 04.01. Cement; Lime; Gypsum, American Society for Testing & Materials, Philadelphia.

ASTM C 666, 2003, Test Method for Resistance of Concrete to Rapid Freezing and Thawing, Annual Book of ASTM Standards, Vol. 04.02. Concrete and Aggregate, American Society for Testing & Materials (ASTM), Philadelphia, USA.

ASTM C 989, 2003, Specification for Ground Blast-Furnace Slag for Use in Concrete and Mortars, Annual Book of ASTM Standards, Vol. 04.02. Concrete and Aggregate, American Society for Testing & Materials (ASTM), Philadelphia, USA.

ASTM C 1073, 2003, Test Method for Hydraulic Activity of Ground Slag by Reaction with Alkali, Annual Book of ASTM Standards, Vol. 04.02. Concrete and Aggregate, American Society for Testing & Materials (ASTM), Philadelphia, USA.

ASTM C 1202, 2003, Test Method for Electrical Indication of Concrete's Ability to Resist Chloride Ion Penetration, Vol. 04.02. Concrete and Aggregate, Annual Book of American Society for Testing Materials Standards (ASTM), Philadelphia, USA.

Atkins, M., Bennett, D., Dawes, A., Glasser, F., Kindness, A. and Read D., 1991, A thermodynamic model for blended cements, Research Report for Department of Environment, DoE/HMIP/RR/92/005, Aberdeen University, UK.

Babushkin, V. I., Matveyev, G. M. and Mchedlov-Petrossyan, O. P., 1985, Thermodynamics of Silicates (NY: Springer-Verlag).

Bakharev, T., 2004, Effect of Curing Regime and Type of Activator on Properties of Alkali-activated Fly Ash. In Nanotechnology in Construction, Edited by P. J. M. Bartos, J. J. Hughes, P. Trtik and W. Zhu, Royal Society of Chemistry, Cambridge, UK, pp. 249–262.

Bakharev, T., Sanjayan, J. G. and Cheng, Y. B., 1999a, Alkaline activation of Australian slag cements. Cement and Concrete Research, 29(1), 113–120.

Bakharev, T., Sanjayan, J. G. and Cheng, Y. B., 1999b, Effect of elevated temperature curing on properties of alkali-activated concrete. Cement and Concrete Research, 29(10), 1619–1626.

Bakharev, T., Sanjayan, J. G. and Cheng, Y. B., 2000, Effect of admixtures on properties of alkali-activated slag concrete. Cement and Concrete Research, 30(9), 1367–1374.

Bakharev, T., Sanjayan, J. G. and Cheng, Y. B., 2001a, Resistance of alkali-activated slag concrete to alkali-aggregate reaction. *Cement and Concrete Research*, 31(2), 331–334.

Bakharev, T., Sanjayan, J. G. and Cheng, Y. B., 2001b, Resistance of alkali-activated slag concrete to carbonation. *Cement and Concrete Research*, 31(9), 1277–1283.

Bakharev, T., Sanjayan, J. G. and Cheng, Y. B., 2002, Sulphate attack on alkali-activated slag concrete. *Cement and Concrete Research*, 32(2), 211–216.

Bakharev, T., Sanjayan, J. G. and Cheng, Y. B., 2003, Resistance of alkali-activated slag concrete to acid attack. *Cement and Concrete Research*, 33(1), 1607–1612.

Bankowskia, P., Zoua, L. and Hodges, R., 2004, Reduction of metal leaching in brown coal fly ash using geopolymers. *Journal of Hazardous Materials*, B114(1), 59–67.

Baragano, J. R. and Rey, P., 1980, The study of a non-traditional pozzolan – copper slags. *7th International Congress on the Chemistry of Cement*, Paris, France, Vol. II, III-37-III-42.

Barber, J. C., 1975, Solid Wastes From Phosphorus Production. In Mantell (ed.) *Solid Wastes: Origin, Collection, Processing, and Disposal* (New York: A Wiley-Interscience Publication, John Wiley & Sons), 927–947.

Batalin, B. S., Rzhanitsyn, Yu. P. and Mokrushin, A. N., 1979, Criterion equation of strength of the slag-alkaline concrete. *1st National Scientific Conference on Slag-Alkaline Cements, Concretes and Structures* (edited by Glukhovsky) Kiev, Ukraine, 82–83.

Bazhenov, Yu. M., 1975, *Methods of Concrete Mix Design of Different Types* (Moscow: Stroyizdat), 85–86.

Beaudoin, J. J. and Brown, P. W., 1992, The structure of hardened cement paste. *9th International Congress on the Chemistry of Cement*, New Delhi, India, I, 485–525.

Belie, D., Verselder, H. J., Blaere, B. D., Nieuwenburg, D. V., and Verschoore, R., 1996, Influence of the cement type on the resistance of concrete to feed acids. *Cement and Concrete Research*, 26(11), 1717–1725.

Belitsky, I. V., 1994, Design of slag alkaline concrete mixes. *1st International Conference on Slag-Alkaline Cements and Concretes* (edited by Krivenko), Kiev, Ukraine, 861–869.

Belitsky, I. V., Sakata, A. and Goto, S., 1993a, Kinetics of the hydration of slag in the slag-alkaline cements. *3rd Beijing International Symposium on Cement and Concrete*, Beijing, P.R. China, 2, 1028–1031.

Belitsky, I. V., Sakata, A. and Goto, S., 1993b, The role and behaviour of alkaline activators in the slag-alkaline cement based on the soluble silicate glass. *3rd Beijing International Symposium on Cement and Concrete*, International Academic Publishers, Beijing, P.R. China, I.2, 1038–1042.

Bensted, J., 1983, Early Hydration of Portland Cement – Effects of Water/Cement Ratio, Cement and Concrete Research, 13(4), 493–498.

Bensted, J., 1987, Some applications of conduction calorimetry to cement hydration, *Advances in Cement Research*, 1(1), 35–44.

Bijen, J. and Waltje, H., 1989, Alkali activated slag-fly ash cements. *3rd International Conference on the Use of Fly Ash, Silica Fume, Slag & Natural Pozzolans in Concrete*, ACI SP-114, Trondheim, Norway, 565–1578.

Bin, Q., 1988, Investigation of Alkali-Steel and BFS Slag Cements. M.Sc. Thesis, Nanjing Institute of Chemical Technology, Nanjing, P.R. China.

Bin, Q., Wu, X. and Tang, M., 1989, An investigation on alkali-BFS-steel slag cements. *2nd Beijing International Symposium on Cements and Concretes*, Beijing, P.R. China, 2, 288–294.

Bin, Q., Wu, X. and Tang, M., 1992, High strength alkali steel-iron slag binder. *9th International Congress on the Chemistry of Cement*, New Delhi, India, III, 291–297.

Blaakmeer, J., 1994, Diabind: An alkali activated slag fly ash binder for acid resistant concrete. *1st International Conference on Alkaline Cements and Concretes* (edited by Krivenko), 1, Kiev, Ukraine, 347–360.

Bogue, R. H., 1955, *The Chemistry of Portland Cement* (New York: Reinhold Publication Corp.).

Bolen, W. P., 2000, Pumice and pumicite. *U.S. Geological Survey Minerals Yearbook*, 61.1–61.4.

Botvinkin, O. K., 1955, *About Multiplicity in Glasses Structure* (Moscow: Gosizdat Stroitelnyh Materialov).

Breck, D. W., 1974, *Zeolite Molecular Sieves. Structure, Chemistry, and Use* (New York, London, Sydney, Toronto: A Wiley-Interscience Publication, John Wiley & Sons).

Bregg, U. and Klaringboul, G., 1967, *Crystalline Structure of Minerals* (Moscow: Mir Publisher).

Breton, D., Carles-Gibergues, A., Ballivy, G. and Grandet, J., 1993, Contribution to the formation mechanism of transition zone between rock-cement paste. *Cement and Concrete Research*, 23(2), 335–346.

Brodko, O. A., 1992, Slag Alkaline Cements and Concretes with Increased Acid Resistance, Ph.D. Thesis, Kiev Civil Engineering Institute, Kiev, Ukraine.

Brodko, O. A., 1999, Experience of exploitation of the alkaline cement concretes. *2nd International Conference on Alkaline Cements and Concretes* (edited by Krivenko), Kyiv, Ukraine, 657–684.

Brough, A. R. and Atkinson, A., 2000, Automated identification of the aggregate-paste interfacial transition zone in mortars of silica sand with Portland or alkali-activated slag cement paste. *Cement and Concrete Research*, 30(6), 849–854.

Brough, A. R. and Atkinson, A., 2002, Sodium silicate-based, alkali-activated slag mortars, Part I. Strength, hydration and microstructure. *Cement and Concrete Research*, 32(6), 865–879.

Brough, A. R., Holloway, M., Sykes, J. and Atkinson, A., 2000, Sodium silicate based slag mortars: Part II: The retarding effect of sodium chloride or malic acid. *Cement and Concrete Research*, 30(9), 1375–1380.

Brough, A. R., Katz, A., Sun, G. K., Struble, L. J., Kirkpatrick, R. J. and Young, J. F., 2001, Adiabatically cured alkali-activated cement-based wasteforms containing high levels of fly ash – Formation of zeolites and Al-substituted C-S-H. *Cement and Concrete Research*, 31(10), 1437–1447.

Brouwers, H. J. H. and Van Eijk, R. J., 2002, Reactivity of fly ash: extension and application of a shrinking core model. *Concrete Science and Engineering*, 14(1), 106–113.

Brouwers, H. J. H. and Van Eijk, R. J., 2003, Chemical Activation of fly ash. *11th International Congress on the Chemistry of Cement*, Durban, South Africa, Vol. 2, 791–800.

Brouxel, M., 1993, The Alkali-aggregate reaction rim: Na$_2$O, SiO$_2$, K$_2$O and CaO chemical distribution. *Cement and Concrete Research*, 23(2), 309–320.

Brylicki, W., Małolepszy, J. and Stryczek, S., 1992, Alkali activated cementitious material for drilling operation. *9th International Congress on the Chemistry of Cement*, New Delhi, India, III, 312–318.

Brylicki, W., Malolepszy, J. and Stryczek, S., 1994, Industrial scale application of the alkali activated slag cementitious materials in the injection sealing works. In Goumans, van der Sloot and Aalbers (eds) *Environmental Aspects of Construction with Waste Materials* (Amsterdam: Elsevier Science), 841–849.

Buchwald, A., Kaps, Ch. and Hohmann, M., 2003, Alkali-activated binders and pozzolanic cement binders – compete binder reaction or two sides of the same reaction? *Proceedings of the 11th International Congress on the Chemistry of Cement*, May, Durban, South African, 1238–1246.

Budnikov, P. P. and Gorshkov, V. S., 1965, To a Question on the Use of Alumothermic Slags. *J. Stroitel'nye Materialy (in Russian) (Building Materials)*, Moscow, 18–20.

Byfors, K. Klingstedt, Lehtonene, V., Pyy, H. and Romben, L., 1989, Durability of concrete made with alkali-activated slag. *Third International Conference on the Use of Natural Pozzolans, Fly Ash, Blast Furnace Slag and Silica Fume in Concrete*, ACI SP-114, 2, 1429–1466.

Caldarone, M. A., Gruber, K. A. and Burg, R. G., 1994, High reactive metakaolin: A new admixture. *Concrete International*, 16(11), 37–40.

Caldwell, R., Shi, C. and Stegemann, Julia A., 1995, Solidification formulation development for a specialty steel electric arc furnace dust. *1st International Conference on Stabilization and Solidificaion*, Nancy, France, pp. 148–158.

Caldwell, R., Stegemann, J. A. and Shi, C., 1999a, Effect of Curing on Field-Solidified Waste Properties, Part II: Chemical Properties. *Waste Management and Research*, 17(1), 37–43.

Caldwell, R., Stegemann, J. A. and Shi, C., 1999b, Effect of Curing on Field-Solidified Waste Properties, Part I: Physical Properties, *Waste Management and Research*, 17(1), 44–49.

Carino, N. J., 1984, Maturity method: Theory and application. *Cement, Concrete and Aggregate*, 6(2), 61–73.

Charchenko, I., Krivenko, P., Runova, R., Kochevykh, M. and Rudenko, I., 2001, The conditions for formation of the Friedel's salt in the binding systems containing a cement kiln dust, *Collection of Works on Building Materials and Articles and Sanitary Technics*, Kiev, 16, 26–31.

Chatterjee, M. K. and Lahiri, D., 1967, Pozzolanic activity in relation to specific surface of some artificial pozzolans. *Trans. Indian Ceramic Soc.*, 26, 65–74.

Chen, X. and Yang, N., 1989, Influence of Polymeric Structure of Granulated Blast furnace slag on Their Hydraulic Activities. *2nd Beijing International Symposium on Cement and Concrete*, Beijing, P.R. China, 1, 346–351.

Chen, Z. and Liao, X., 1992, The selection of stimulation agents for alkali-slag cement. *9th International Congress on the Chemistry of Cement*, New Delhi, India, III, 305–310.

Cheng, J. J., 2003, A study on the setting characteristics of sodium silicate – activated slag pastes. *Cement and Concrete Research*, 33(7), 1005–1011.

Chang, J. J., Yeih, W. and Hung, C. C., 2005, Effects of gypsum and phosphoric acid on the properties of sodium silicate-based alkali-activated slag pastes. *Cement & Concrete Composites*, 27(1), 85–91.

Cheng, Q.-H., Tagnit-Hamou, A. and Sarkar, S. L., 1991, Strength and microstructural properties of water glass activated slag. *Materials Research Society Symposium*, 49–54.

Cheng, Q.-H. and Sarkar, S. L., 1994, A Study of rheological and mechanical properties of mixed alkali activated slag pastes. *Advanced Cement Based Materials*, 1, 178–184.

Cheron, M. and Landinois, 1968, The Role of Magnesia and Alumina in the Hydraulic Properties of Granulated Blast-Furnace Slags, 1968, *5th International Congress on the Chemistry of Cement*, Tokyo, Japan, 3, 227–285.

Chinese Academy of Building Materials, Physical Testing Methods for Cements (3rd ed.) 1985, Chinese Academy of Building Materials (Beijing: Chinese Construction Press).

Cho, J. W., Ioku, K. and Goto, S., 1999, Effect of Pb^{II} and Cr^{VI} on the hydration of slag-alkaline cement and the immobilization of these heavy metal ions. *Advances in Cement Research*, 11, 111.

Cincotto, M. A., Melo, A. A. and Repette, W. L., 2003, Effect of Different Activators Type and Dosages and Relation with Autogenous Shrinkage of Activated Blast Furnace Slag Cement. In: G. Grieve and G. Owens (eds), *Proceedings of the 11th International Congress on the Chemistry of Cement*, Durban, South African, pp. 1878–1888.

Clarke, W. J. and Helal, M., 1989, Alkali-activated Slag and Portland/Slag ultrafine cements. In: *Specialty Cements with Advanced Properties*, Materials Research Society Symposium Proceedings, 179, 219–232.

Coad, J. R., 1974, Natural pozzolans. In: Spence (ed.) *Lime & Alternative Cements, One-day Seminar on Small-scale Manufacturing of Cementitious Materials*, Intermediate Technology Development Group, London, 46–48.

Coale, R. D., Wolhuter, C. W., Jochens, P. R. and Howat, D. D., 1973, Cementitious properties of metallurgical slags. *Cement and Concrete Research*, 3(1), 81–92.

Collins, F., 1999, *High Early Strength Concrete Using Alkali-activated Slag*, Ph.D. Thesis, Department of Civil Engineering, Monash University, Australia.

Collins, F. G. and Sanjayan, J. G., 1999a, Workability and mechanical properties of alkali-activated slag concrete. *Cement and Concrete Research*, 29(3), 455–458.

Collins, F. G. and Sanjayan, J. G., 1999b, Effect of ultra fine materials on workability and strength of concrete containing alkali-activated slag as the binder. *Cement and Concrete Research*, 29(2), 459–462.

Collins, F. G. and Sanjayan, J. G., 1999c, Strength and shrinkage properties of alkali-activated slag concrete containing porous coarse aggregate. *Cement and Concrete Research*, 29(4), 607–610.

Collins, F. G. and Sanjayan, J. G., 1999d, Strength and shrinkage properties of alkali-activated slag concrete placed into a large column. *Cement and Concrete Research*, 29(5), 659–666.

Collins, F. G. and Sanjayan, J. G., 2000a, Cracking tendency of alkali-activated slag concrete subjected to restrained shrinkage. *Cement and Concrete Research*, 30(5), 791–798.

Collins, F. G. and Sanjayan, J. G., 2000b, Effect of pore size distribution on drying shrinkage of alkali-activated slag concrete. *Cement and Concrete Research*, 30(9), 1401–1406.

Collins, F. G. and Sanjayan, J. G., 2001a, Early age strength and workability of slag pastes activated by sodium silicates. *Magazine of Concrete Research*, 52(5), 321–326.

Collins, F. G. and Sanjayan, J. G., 2001b, Microcracking and strength development of alkali-activated slag concrete. *Cement & Concrete Composites*, 23, 345–352.

Comrie, D. C. and Davidovits, J., 1988, Long term durability of hazardous toxic and nuclear waste disposals, Geopolymer '88, First European Conference on Soft Mineralurgy, Compiegne, France, 1, 125–134.

Costa, U. and Massazza, F., 1974, Factors affecting the reaction with lime of Italian pozzolans. *Il Cemento*, 74, 131–139.

Cotaworth, R. P., 1980, The Slag Pelletizer and Its Application to Metallurgical Slags, *International Symposium on Metallurgical Slags*, CIMM, Halifax, Nova Scotia, Paper 5.8, pp. 1–14.

Craf, O., 1960, Die Eigeuschaften des Betons, Lweite neubearbeitete Auflage, Berlin, Springer-Verlag Goettingen, pp. 10–16.

Dai, L. and Cheng, J., 1988, An investigation on BFS-fly ash-alkali systems. *Bulletin of Chinese Silicate Society (in Chinese)*, 16, 25–32.

Dave, N. G., 1981, Pozzolanic wastes and their activation to produce improved lime pozzolana mixtures. *2nd Australian Conference on Engineering Materials*, Sydney, Australia, 623–638.

Davidovits, J., 1981, Synthetic mineral polymer compound of the silicoaluminates family and preparation process, *US Patent*, 4, 472,

Davidovits, J., 1988, Geopolymeric reaction in archaeological cements and in modern blended cements. *GÉOPOLYMÈRE '88 – International Conference*, Saint-Quentin, France, 1, 93–105.

Davidovits, J., 1991, Geopolymers: Inorganic polymeric new materials. *J. Therm. Anal.*, 37(8), 1633–1656.

Davidovits, J., 1994, Properties of geopolymer cements. *1st International Conference on Alkaline Cements and Concretes* (edited by Krivenko), Kiev, Ukraine, 1, 131–149.

Day, R. L., 1992, Pozzolans for Use in Low-cost Housing: A State of the Art Report, Department of Civil Engineering, Research Report No. CE92-1, The University of Calgary, Canada.

Day, R. L. and Shi, C., 1994, Correlation between the strength development of lime-natural pozzolan cement pastes and the fineness of natural pozzolan. *Cement and Concrete Research*, 24(8), 1485–1491.

Deja, J., 2002a, Carbonation Aspects of Alkali Activated Slag Mortars and Concretes. *Silicates Industriels*, 67(3–4), 37–42.

Deja, J., 2002b, Immobilization of Cr^{6+}, Cd^{2+}, Zn^{2+} and Pb^{2+} in alkali-activated slag binders. *Cement and Concrete Research*, 32, 1971–1977.

Deja, J. and Malolepszy, J., 1989. Resistance of alkali-activated slag mortars to chloride solution. *3rd International Conference on the Use of Fly Ash, Silica*

Fume, Slag & Natural Pozzolans in Concrete, ACI SP-114, Trondheim, Norway, 1547–1561.

Deja, J. and Malolepszy, J., 1994, Long-term resistance of alkali-activated slag mortars to chloride solution (Supplementary Paper). *3rd CANMET/ACI International Conference on Durability of Concrete*, Nice, France, 657–671.

Deja, J., Malolepszy, J. and Jaskiewicz, G., 1991, Influence of chloride corrosion on durability of reinforcement in the concrete. *2nd International Conference on the Durability of Concrete*, Montreal, Canada, 511–521.

Demoulian, E., Gourdin, P., Hawthorn, F. and Vernet, C., 1980, Influence of slags chemical composition and texture on their hydraulicity. *7th International Congress on the Chemistry of Cement*, Paris, France, 2 , III-89-94.

Deng, M., 1989, MgO Based Expansive Cements, M.Sc. Thesis, Nanjing Institute of Chemical Technology, Nanjing, P.R. China.

Deng, Y., Wu, X. and Tang, M., 1989, High strength alkali-slag cement. *Journal of Nanjing Institute of Chemical Technology (in Chinese)*, P.R. China, 11(2), 1–7.

Dent Glasser, L. S. and Lachowski, E. E., 1980, Silicate species in solution, Part 1, Experimental observations. *Journal of the Chemical Society, Dalton Transactions*, 393–398.

De Silva, P. S. and Glasser, F. P., 1990, Hydration of cements based on metakaolin: thermochemistry. *Advances in Cement Research*, 3(12), 166–177.

De Silva, P. S. and Glasser, F. P., 1991, Pozzolanic activation of metakaolin. *Advances in Cement Research*, 4(16), 167–178.

De Silva, P. S. and Glasser. F. P., 1992a, Pozzolanic activation of metakaolin. *Advances in Cement Research*, 4(16), 167–178.

De Silva, P. S. and Glasser, F. P., 1992b, The hydration behaviour of metakaolin-$Ca(OH)_2$ – sulphate binder. *9th International Congress on the Chemistry of Cement*, IV, 671–677.

Diamond, S., 1983, On the glass present in low-calcium and high-calcium fly ashes. *Cement and Concrete Research*, 13(3), 459–463.

Din, Z., 1979, *The Physical Chemistry of Silicates* (Bejing: Chinese Construction Industry Press).

Donetsky Promstroiniiproekt, 1980, *Guide for manufacturing and use of High-consistency slag alkaline cement concretes for construction of anti-filtration screens*, Donetsk-Makeevka, USSR.

Douglas, E. and Mainwaring, P. R., 1985, Hydration and pozzolanic activity of nonferrous slags. *American Ceramic Society Bulletin*, 64(5), 700–706.

Douglas, E. and Brandstetr, J., 1990, A preliminary study on the alkali activation of granulated blast furnace slag. *Cement and Concrete Research*, 20(5), 746–756.

Douglas, E., Bilodeau, A. and Brandstetr, J., 1991, Alkali activation of granulated blast furnace slag concrete: preliminary investigation. *Cement and Concrete Research*, 21(1), 101–108.

Douglas, E., Bilodeau, A. and Malhotra, V. M., 1992, Properties and durability of alkali-activated slag concrete. *ACI Materials Journal*, 89(5), 509–516.

Dove, P. M. and Rimstidt, J. D., 1995, Silica-Water Interactions. In Heaney, Prewitt and Gibbs (eds) *Silica – Physical Behavior, Geochemistry and Materials Applications* (Washington, D.C.: Mineralogical Society of America).

Dron, R., 1974, Experimental and theoretical study of the CaO–Al$_2$O$_3$–SiO$_2$–H$_2$O system (Supplementary paper). *6th International Congress on the Chemistry of Cement held in Moscow in 1974*, Moscow, USSR, 2(1), 208–211.

DSTU BV 2.7-24-95, 1995, National Standard of Ukraine. Technical Specification for a Binder, Slag Alkaline, The State Committee of the Ukraine for Urban Planning and Architecture, Kiev, USSR.

Duchesene, J. and Berube, M. A., 1994, The effectiveness of supplementary cementing materials in suppressing expansion due to ASR: another look at the reaction mechanism, Part 2: Pore solution chemistry. *Cement and Concrete Research*, 24(2), 221–230.

EPRI, Commercialization Potential of AFBC Concrete: Part 2, Vol. 2: Mechanistic Basis for Cementing Action, 1991, EPRI GS-7122, Project 2708-4, Electrical Power Research Institute, Palo Alto, California.

Escalante-Garcia, J. J., Gorokhovsky, A. V., Mendoza, G. and Fuentes, A. F., 2003, Effect of geothermal waste on strength and microstructure of alkali-activated slag cement mortars. *Cement and Concrete Research*, 33(10), 1567–1574.

Everett, F. D., 1967, Potential pozzolanic materials. *Symposium on International Mineral Exploration and Development – Forum on Geology of International Minerals*, University of Kansas, USA, Publication, 34, 156–161.

Feng, N., 1993, Properties of Zeolite Mineral Admixture Concrete. In Sarkar, Harsh and Ghosh (eds) *Progress in Cement and Concrete, Mineral Admixture in Cement and Concrete* (New Delhi, India: ABI Books Pty. Ltd.), 396–447.

Fernandez-Jimenez, A., Lachowski, E. E., Palomo, A. and Macphee, D. E., 2004, Microstructural characterisation of alkali-activated PFA matrices for waste immobilization, *Cement & Concrete Composites*, 26(X), 1001–1006.

Fernandez-Jimenez, A. and Palomo, A., 2003, Alkali-activated fly ashes: Properties and characteristics. *11th International Congress on the Chemistry of Cement*, Durban, South Africa, Vol. 3, 1322–1340.

Fernández-Jiménez, A. and Palomo, A., 2004, Activation of Fly Ashes: A General View. *8th CANMET/ACI International Conference on Fly Ash, Silica Fume, Slag, and Natural Pozzolans*, Las Vegas, May 23–29, ed. V. M. Malhotra, American Concrete Institute, Michigan, U.S., 351–365.

Fernandez-Jimenez, A. and Puertas, F., 1997a, Influence of the activator concentration on the kinetics of the alkaline activation process of a blast furnace slag. *Materiales de Construction*, 47(246), 31–41.

Fernandez-Jimenez, A. and Puertas, F., 1997b, Alkali-activated cement: kinetics studies. *Cement and Concrete Research*, 27(3), 359–368.

Fernandez-Jimenez, A., Puertas, F. and Arteaga, A., 1998, Determination of kinetic equations of alkaline-activation of blast furnace slag by means of calorimetric data. *Journal of Thermal Analysis*, 52, 945–955.

Fernandez-Jimenez, A. and Puertas, F., 2002, The alkali-silica reaction in alkali-activated granulated slag mortars with reactive aggregate. *Cement and Concrete Research*, 32(7), 1019–1024.

Fernandez-Jimenez, A. and Puertas, F., 2003a, Effect of activator mix on the hydration and strength behaviour of alkali-activated slag cements, *Advances in Cement Research*, 15(3), 129–136.

Fernandez-Jimenez, A. and Puertas, F., 2003b, Characterisation of fly ashes – potential reactivity as alkaline cements. *Fuel*, 82(18), 2259–2265.

Fernández-Jiménez, A., Puertas, F., Sobrados, I. and Sanz, J., 2003, Structure of calcium silicate hydrates formed in alkaline-activated slag: influence of the type of alkaline activator. *Journal of American Ceramic Society*, 86(8), 1389–1394.

Firsov, N. N., 1984, The proportioning of the slag-alkaline cement concrete mixtures. *II National Scientific and Practical Conference on Slag-Alkaline Cements, Concretes and Structures* (edited by Glukhovsky), Kiev, USSR, 206–207.

Flemming, H.-C., 1995, Eating away at the infrastructure – the heavy cost of microbial corrosion. *J. Water Quality International*, 4, 16–19.

Fletcher, R. A., MacKenzie, K. J. D., Nicholson, C. L. and Shimada, S., 2005, The composition range of aluminosilicate geopolymers. *Journal of the European Ceramic Society*, 25(9), 1471–1477.

Forss, B., 1983a, F-Cement, A New Low-Porosity Slag Cement. *Silicates Industriels*, No. 3, pp. 79–82.

Forss, B., 1983b, Experiences from the use of F-cement – a binder based on alkali-activated blastfurnace slag. In G. M. Idorn, Steen Rostam (eds), Alkalis in Concrete, Danish Concrete Association, Copenhagen, Denmark, pp. 101–104.

Frearson, J. P. H. and Uren, J. M., 1986, Investigation of a ground granulated blast furnace slag containing merwinitic crystallization. *2nd International Conference on the Use of Fly Ash, Silica Fume, Slag and Natural Pozzolans in Concrete*, ACI SP-91, American Concrete Institute, Detroit, 2, 1401–1421.

Frenkel, I. M. and Shakhmuratyan, E. A., 1974, The concretes based on slag-alkaline binders. In *Collection of Works on the Industry of Precast Reinforced Concrete* (Moscow: VNIIESM), 7, pp. 10–11.

Frigione, G., 1986, Manufacture and Characteristics of Portland Blast-Furnace Slag Cements, Blended Cements (edited by Frohnsdorff). ASTM STP 897, American Society for Testing and Materials, Philadelphia, USA, 15–28.

Gao, Q., Zhang, Z. and Zhang, X., 1989, The relationship of structure and pozzolanic activity of kaolinite at different calcination temperatures. *2nd Beijing International Symposium on Cement and Concrete*, Beijing, P.R. China, 1, 377–382.

GB 13590-92, 1992, Chinese National Standard for Steel and Iron Slag Cement, P.R. China.

Garrett, D. E., 1992, *Natural Soda Ash – Occurrences, Processing and Use* (New York: Van Nostrand Reinhold).

Geiseler, J., 1996, Use of steel works slag in Europe. *Waste Management*, 16(1–3), 59–63.

General Chemical Industrial Products Inc., 2003, *Soda Ash, General Chemical Industrial Products Inc.*, Parsippany, NJ, USA.

Gerasimchuk, V. L., 1982, The Influence of Aggregates Properties on the Slag Alkaline Cement Concrete Structure and Strength, Ph.D. Thesis, Kiev Civil Engineering Institute, Kiev, USSR.

Ghosh, A. and Pratt, P. L., 1981, Studies of the hydration reaction and microstructure of cement – fly ash paste. *MRS Symposium*.

Gifford, P. and Gillott, J. E., 1996a, Freeze-thaw durability of activated blast furnace slag cement concrete. *ACI Materials Journal*, 93(3), 242–245.

Gifford, P. and Gillott, J. E., 1996b, Alkali-silica reaction (ASR) and alkali-carbonate reaction (ACR) in alkali-activated blast furnace slag cement (ABFSC) concrete. *Cement and Concrete Research*, 26(1), 21–26.

Gifford, P. M. and Gillott, J. E., 1997, Behaviour of mortar and concrete made with activated blast furnace slag cement. *Canadian Journal of Civil Engineering*, 24, 237–249.

Ginstling, A. M. and Brounshtein, B. I., 1950, Concerning the diffusion kinetics of reaction in spherical particles. *Journal of Applied Chemistry (USSR) (English Translation)*, 23(12), 1327–1328.

Gjorv, O. E., 1989, Alkali activation of a Norwegian granulated blast furnace slag. *Proceedings of the 3rd International Conference on the Use of Fly Ash, Silica Fume, Slag & Natural Pozzolans in Concrete*, SP-114, Vol. 2, pp. 1501–1518.

Glukhovsky, V. D., 1959, *Soil Silicates (Gruntosilikaty)* (Kiev, USSR: Budivelnik Publisher).

Glukhovsky, V. D., 1965, Soil Silicates, Their Properties, Technology of Manufacturing and Fields of Application, Doct. Tech. Sc. Degree Thesis, Kiev Civil Engineering Institute, Kiev, USSR.

Glukhovsky, V. D., 1967, *Soil Silicate Articles and Constructions (Gruntosilikatnye virobi i konstruktsiii)* (Kiev: Budivelnik Publisher).

Glukhovsky, V. D. (ed.), 1979, *Alkaline and Alkaline–Alkali-earth Hydraulic Binders and Concretes* (Kiev, USSR: Vysscha Shkola Publisher).

Glukhovsky, V. D., 1981, *Slag Alkaline Fine Aggregate Concretes* (Kiev, USSR: Vysscha Shkola Publisher).

Glukhovsky, V. D. and Pakhomov, V. A., 1978, *Slag-alkali Cements and Concretes* (Kiev: Budivelnik Publisher).

Glukhovsky, V. D. and Raksha, V. A., 1979, In Glukhovsky (ed.) *Alkaline and Alkaline-Alkali-Earth Hydraulic Binders and Concretes* (Kiev: Vysscha Shkola Publisher), pp. 121–124.

Glukhovsky, V. D., Rostovskaya, G. S., Raksha, V. A. and Chirkova, V. V., 1974, Activation of hardening of hydraulic binders by sodium compounds. *Abstracts of Reports of The National Meeting*, Ufa, USSR, 14–16.

Glukhovsky, V. D., Rostovkaya, G. S. and Rumyna, G. V., 1980, High strength slag-alkali cement. *7th International Congress on the Chemistry of Cements*, Paris, France, III, V-164–168.

Glukhovsky, V. D., Krivenko, P. V., Starchuk, V. N., Pashkov, I. A. and Chrkova, V. V., 1981, *Slag Alkaline Concretes Made with Fine Aggregates* (edited by Glukhovsky) (Kiev, USSR: Vysscha Shkola Publisher).

Glukhovsky, V. D., Krivenko, P. V., Rumyna, G. V. and Gerasimchuk, V. L., 1988, *The Manufacture of Concretes and Structures from Slag Alkaline Binders*, Budivel'nik Publisher, Kiev.

Gontcharov, N. N., 1984, Corrosion Resistance of Slag Alkaline Cements and Concretes in Organic Aggressive Environments, Ph.D. Thesis, Kiev Civil Engineering Institute, Kiev, USSR.

Gordon, S. S., 1969, *Structure and Properties of Heavyweight Concretes with Various Aggregates* (Moscow: Stroyizdat Publisher).

Govorov, A. A., 1976, *The Processes of Hydrothermal Hardening of Slag Dispersions* (Kiev: Naukova Dumka Publisher).

Granizo, M. L. and Blanco, M. T., 1998, Alkaline activation of metakaolin. *Journal of Thermal Analysis*, 52, 957–965.

Gravitt, B. B., Heitzmann, R. F. and Sawyer, J. L., 1991, Hydraulic cement and composition employing the same, US Patent 4,997,484, March 5.

Greenberg, S. A., 1961, Reaction between silica and calcium hydroxide solutions, Part I: Kinetics in the temperature range 30 to 85 °C. *Journal of Physical Chemistry*, 65, 12–16.

Gregg, S. J., 1961, *The Surface Chemistry of Solids* (London: Chapman and Hall Ltd.).

Grutzeck, M. W., 1997, Characteristics of C–S–H gels. *10th International Congress on Chemistry of Cement*, Gothenburg, Sweden, 2, 2ii067, 10pp.

Grutzeck, M. W., Kwan, S. and DiCola, M., 2004, Zeolite formation in alkali-activated cementitious systems. *Cement and Concrete Research*, 34(6), 949–955.

Gu, J., 1991, Hydration mechanism, properties and application of alkali-slag cement. *Cement and Concrete Product (in Chinese)*, 5, 8–11.

Guo, J. and Liang, C., 1980, A study on reaction mechanisms of zeolite during the hydration of cement, *Journal of Chinese Ceramic Society*, 8(3), 242–257.

Hakkinen, T., 1986, Properties of alkali-activated slag concrete, VTT Research. Notes. No. 540, Technical Research Centre of Finland (VTT), Finland.

Hakkinen, T., 1987, Durability of alkali-activated slag concrete. *Nordic Concrete Research*, 6, 81–94.

Hakkinen, T., 1993, The influence of slag content on the microstructure, permeability and mechanical properties of concrete: Part 2 technical properties and theoretical examinations. *Cement and Concrete Research*, 23(3), 518–530.

Hakkinen, T., Pyy, H. and Koskinen, P., 1987, Microstructural and permeability properties of alkali-activated slag concrete, VTT Research Report. No. 486, Technical Research Centre of Finland (VTT), Finland.

Hansen, T. C., 1986, Physical structure of hardened cement pastes – a classic approach. *Materials and Structure*, 19(114), 423–436.

Hansen, W. and Almudaiheen, J. A., 1987, Ultimate drying shrinkage of concrete – influence of major parameters, *ACI Materials Journal*, 84(3), 217–223.

Hazra, P. C. and Krishnaswamy, V. S., 1987, Natural Pozzolans in India, Their Utility, Distribution and Petrography, *Records of the Geological Survey of India*, 87(4), 675–706.

Hearn, N., Hooton, R. D. and Mills, R. H., 1994, Pore Structure and Permeability. In Klieger and Lamond (eds) *Significance of Tests and Properties of Concrete and Concrete-Making Materials*, STP 169C, ASTM, Philadelphia, USA, pp. 240–262.

Helmuth, R., 1987, Fly Ashes in Cement and Concrete, *Portland Cement Association*, Skokie, IL, USA.

Hemmings, R. T. and Berry, E. E., 1988, On the glass in coal fly ashes: Recent advances. *Proceedings of Materials Research Society*, Vol. 113, 3–38.

Hobbs, D. W., 1988, *Alkali-silica Reaction in Concrete*. Thomas Telford, London.

Hoebbel, D. and Ebert, R., 1988, Sodium silicate solution – structure, properties and problems. *Zeitschrift für Chemie*, 28(2), 41–51.

Hogan, F. J., 1983, Study of grinding energy required for pelletized and water granulated slag. *Silicates Industriels*, 3, 71–78.

Hogan, F. J., 1985, The effect of blast furnace slag cement on alkali-aggregate reactivity: A literature review. *Cement, Concrete, and Aggregate*, 7(2), 100–107.

Hogan, F. J. and Rose, J. H., 1986, ASTM Specification for ground blast furnace slag: Its development, use and future. *2nd International Conference on the Use of Fly Ash, Silica Fume, Slag and Natural Pozzolans in Concrete*, Madrid, Spain, SP-91, 2, 1551–1576.

Hong, S. Y. and Glasser, F. P., 1999, Alkali binding in cement pastes. *Cement and Concrete Research*, 29(12), 1893–1904.

Hong, S. Y., Kia, J. C. and Kim, J. K., 1993, Studies on the hydration of alkali-activated slag. *3rd Beijing International Symposium on Cement and Concrete*, Beijing, P.R. China, 2, 1059–1063.

Hooton, R. D., Gruber, K. and Boddy, A., 1997, The chloride penetration resistance of concrete containing high-reactivity metakaolin. *PCI/FHWA International Symposium on High Performance Concrete*, New Orleans, LA, USA, 172–183.

Hrazdira, J. and Kalousek, D., 1994, Effect of alkaline admixtures on properties of ground slag binders. *1st International Conference on Alkaline Cements and Concretes* (edited by Krivenko), Kiev, Ukraine, 1, 593–600.

Iler, R. K., 1979, *The Chemistry of Silica – Solubility, Polymerization, Colloid and Surface Properties and Biochemistry* (New York, London, Sydney, Toronto: A Wiley-Interscience Publication, John Wiley & Sons).

Ilyin, V. P., 1994, Durability of materials based on slag-alkaline binders. *1st International Conference on Alkaline Cements and Concretes* (edited by Krivenko), 2, Kiev, Ukraine, 789–835.

Ilyukhin, V. V., Kuznetsov, V. A., Lobatchov, A. N. and Bakshutov, V. S., 1979, *Hydrosilicates of Calcium. Synthesis of Monocrystals and Crystal Chemistry* (Moscow: Nauka Publisher).

Ionescu, I. and Ispas, T., 1986, Properties and durability of some concretes containing binders based on slag and activated ashes. *2nd International Conference on the Use of Fly Ash, Silica Fume, Slag & Natural Pozzolans in Concrete*, ACI SP-91, Madrid, Spain, 1475–1493.

Ipatti, A., 1992, Solidification of ion-exchange resins with alkali-activated blast-furnace slag. *Cement and Concrete Research*, 22(2–3), 282–286.

Isozaki, K., Iwamoto, S. and Nakagawa, K., 1986, Some properties of alkali-activated slag cements. *CAJ Review*, 120–123.

Jambor, J., 1963, Relation between phase composition, overall porosity and strength of hardened lime-pozzolan pastes. *Magazine of Concrete Research*, 15(45), 131–142.

Javanmardi, K., Flodberg, K. D. and Nahm, J. J., 1993, Mud to Cement technology proven in offshore drilling project. *Oil & Gas Journal* (Feb. 15), 49–57.

Jawed, I., Skalny, J. and Young, J. F., 1983, Hydration of Portland Cement. In Barnes (ed.), *Structure and Performance of Cements* (London and New York: Applied Science Publishers), pp. 237–317.

Jennings, H., 1986, Aqueous solubility relationships for two types of calcium silicate hydrate. *Journal of American Ceramic Society*, 69, 614–618.

Ji, Y.-J., 1991, The fracture properties of alkali-slag concrete. *International Symposium on Concrete Engineering*, Nanjing, P.R. China, 459–464.

Jiang, W., 1997, Alkali-activated Cementitious Materials: Mechanisms, Microstructure and Properties. Ph.D. Thesis, The Pennsylvania State University, Pennsylvania, US.

Jiang, W., Silsbee, M. R. and Roy, D. M., 1997, Alkali activation reaction mechanism and its influence on microstructure of slag cement. *10th International Congress on the Chemistry of Cement*, Gothenburg, Sweden, 3, 3ii100, 9pp.

Jolicoeur, C., Simard, M. A., Sharman, J., Zamojska, R., Dupuis, M., Spiratos, N., Douglas, E. and Malhotra, V. M., 1992, Chemical activation of blast-furnace slag, An overview and systematic experimental investigation. In Malhotra (ed.) *Advances in Concrete Technology*, Ministry of Supply and Services, Ottawa, Canada, 471–502.

Jumppaenen, U.-M., Diederichs, U. and Hinrichsmeyer, K., 1986, Materials properties of F-concrete at high temperatures, VTT Research Report 452, Technical Research Centre of Finland (VTT), Finland.

Kalifa, P., Chene, G. and Galle, C., 2001, High-temperature behaviour of PHC with polypropylene fibers from spalling to microstructure. *Cement and Concrete Research*, 31(10), 1487–1499.

Kalousek, G. L., 1944, Studies of portions of the quaternary system soda-lime-silica-water at 25 °C. *Journal of Research of the National Bureau of Standards*, 32, 285–502.

Kalousek, G. L., Davis, C. W. and Schmertz, W. E., 1949, An investigation of hydrating cements and related hydrous solids by differential thermal analysis. *Journal of ACI*, 20(10), 693–706.

Kasai, Y., Tobinai, K., Asakura, E. and Feng, N., 1992, Comparative study on natural zeolites and other inorganic admixtures in terms of characterization and properties of mortars. *4th International Congress on the Use of Fly Ash, Silica Fume, Slag and Natural Pozzolans in Concrete*, ACI SP-132, American Concrete Institute, Detroit, 1, 615–634.

Katz, A., 1998, Microscopic study of alkali-activated fly ash. *Cement and Concrete Research*, 28(2), 197–208.

Khalil, M. Y. and Merz, E., 1994, Immobilization of intermediate-level wastes in geopolymers, *J. Nucl. Mater.*, 211, 141–148.

Kirkpatrick, R. J., 1988, MAS NMR Spectroscopy of Minerals and Glasses. In Reviews in Mineralogy, Vol. 18, Spectroscopic Methods in Mineralogy and Geology, Mineralogical Society of America, Washington, D.C., pp. 99–159.

Knudsen, T., 1980, On particle size distribution in cement hydration. *7th International Congress on the Chemistry of Cement*, Paris, France, II, 170–175.

Komarneni, S. and Roy, D. M., 1985, New Tobermorite Cation Exchangers. *Journal of Materials Science*, 20, 2930.

Komarneni, S., Roy, R. and Roy, D. M., 1986, Pseudomorphism in Xonotlite and Tobermorite with Co^{2+} and Ni^{2+} Exchange for Ca^{2+} at 25 °C. *Cement and Concrete Research*, 16, 47.

Komarneni, S., Breval, E., Roy, D. M. and Roy, R., 1988, Reactions of Some Calcium Silicates With Metal Ions. *Cement and Concrete Research*, 18, 204.

Kondo, R., Lee, K. and Daimon, M., 1976, Kinetics and mechanism of hydrothermal reaction in lime-quartz-water system. *Journal of Ceramic Society (Japan)*, 84(11), 573–578.

Kondo, R. and Ohsawa, S., 1968, Studies on a Method to Determine the Amount of Granulated Blast Furnace Slag and the Rate of Hydration of Slag in Cement, *Proceedings of the 5th International Congress on the Chemistry of Cement*, Vol. IV, pp. 225–262, Tokyo.

Korneev, V. I. and Brykov, A. S., 2000, Synthesis and Characteristics of Properties of the Alkali Metal Hydrosilicates, Reports of the II Int. Meeting on the Chemistry and Technology of Cement, Moscow, 4-8 December, Vol. 2, pp. 27–31.

Korneev, V. I. and Danilov, V. V., 1996, *Water Soluble Glass*, Stroiizdat SPb Publisher, St. Petersburg.

Kostic, N. and Skenderovic, B., 1992, Investigation of the Possibility of Making Hydraulic Binder by Chemical Activation of Domestic Ash and Mixture of Slag and Domestic Ash. *9th International Congress on the Chemistry of Cement*, New Delhi, India, III, 325–330.

Kostick, D. S., 1993a, Soda Ash, Annual Report, U.S. Department of the Interior Bureau of Mines.

Kostick, D. S., 1993b, Sodium Sulphate, Annual Report, U.S. Department of the Interior Bureau of Mines.

Kostick, D. S., 1994, Soda Ash, Industrial Minerals and Rocks, Society for Mining, Metallurgy, and Exploration, Inc., Littleton, Colorado.

Kovalchuk, G. Yu., 2002, Heat Resistant Gas Concrete Based on Alkaline Aluminosilicate Binder. Ph.D. Thesis, Kyiv National University of Civil Engineering and Architecture, Kyiv, Ukraine (in Ukrainian).

Krivenko, P. V., 1986, Synthesis of Cementitious Materials in a System $R_2O–Al_2O_3–SiO_2–H_2O$ with Required Properties, DSc(Eng) Thesis, Kiev Civil Engineering Institute, Kiev, Ukraine.

Krivenko, P. V., 1992a, Alkaline cements. *9th International Congress on the Chemistry of Cement*, New Delhi, India, IV, 482–488.

Krivenko, P. V., 1992b, *Special Slag Alkaline Cements* (Kiev: Budivelnik Publisher), 19–54.

Krivenko, P. V., 1994a, Influence of physico-chemical aspects of early history of a slag alkaline cement stone on stability of its properties. *1st International Conference on Reinforced Concrete Materials in Hot Climates*, United Arab Emirates University, Dubai, United Arab Emirates.

Krivenko, P. V., 1994b, Alkaline cements. *1st International Conference on Alkaline Cements and Concretes* (edited by Krivenko), Kiev, Ukraine, 1, 11–130.

Krivenko, P. V., 1997, Alkaline cements: Terminology, classification, aspects of durability. *10th International Congress on the Chemistry of Cement*, Gothenburg, Sweden, 4, 4iv046, 6pp.

Krivenko, P. V., 1999, Alkaline cements and concretes: Problems of durability. *2nd International Conference on Alkaline Cements and Concretes* (edited by Krivenko), Kiev, Ukraine, 3–43.

Krivenko, P. V. and Kovalchuk, G. Yu., 2002, Heat resistant fly ash based geocements. International Conference on Geopolymer-2002 "Turn potential into profit", October 28–29, 2002, Melbourne, Australia, Proceedings on CD (file:Paper_015.pdf), 11pp.

Krivenko, P. V., Pushkaryeva, E. K. and Brodko, O. A., 1991b, Acid resistant slag alkaline binders of hydration hardening, *J. Tsement (Cement)(in Russian)*, Leningrad, USSR, 11–12, 16–23.

Krivenko, P. V. and Pushkaryeva, E. K., 1993, *Durability of the Slag Alkaline Cement Concrete* (Kiev: Budivelnik Publisher), 187–198.

Krivenko, P. V. and Ryabova, A. G., 1990, The fly ash alkaline binders, *Tsement (Cement)(in Russian)*, Leningrad, USSR, 11, 14–16.

Krivenko, P. V., Skurchinskaya, Zh. V. and Sultanov, A. A., 1984, Peculiarities of the processes of hydration and structure formation of the slag alkaline binders based on non-ferrous slags. *II National Conference on Slag Alkaline Cements, Concretes and Structures* (edited by Glukhovsky), Kiev, Ukraine, 25–26.

Krivenko, P. V., Skurchinskaya, J. V. and Lavrinenko, L. V., 1993, Environmentally Safe Immobilization of Alkali Metal Radioactive Waste, Concrete 2000 (edited by R. K Dhir and M. R. Jones), London: E&FN Spon, pp. 1579–1589.

Krizan, D. and Zivanovic, B., 2002, Effects of dosage and modulus of water glass on early hydration of alkali-slag cements. *Cement and Concrete Research*, 32(8), 1181–1188.

Kurdowski, W. and Sorrentino, F., 1983, Special Cements. In Barnes (ed.) *Structure and Performance of Cements*, II, 471–554.

Kurdowski, W., Duszak, S. and Trybalska, B., 1994, Corrosion of Slag Cement in Strong Chloride Solutions. *1st International Conference on Alkaline Cements and Concretes* (edited by Krivenko), 2, Kiev, Ukraine, 961–970.

Kutti, T. and Malinowski, R., 1982, Influence of the curing conditions on the flexural strength of alkali-activated blast furnace slag mortars. *Chalmers University Publication*, Gothenburg, Sweden, 10, 9–18.

Kutti, T., Malinowski, R. and Srebnik, M., 1982, *Investigation of Mechanical Properties and Structure of Alkali Activated Blast-Furnace Slag Mortars*, 47(6), 149–153.

Kutti, T., 1992, Hydration Products of Alkali-activated Slag. *Proceedings of 9th International Congress on the Chemistry of Cement*, New Delhi, India, IV, 468–474.

Lachowski, E. E., Mohan, K. and Taylor, H. F. W., 1980, Analytical electron microscopy of cement pastes, II. Pastes of Portland cement and clinkers. *Journal of American Ceramic Society*, 63(7–8), 447–452.

Latina, N. I., Puzhanov, G. T. and Lavrinenko, V. P., 1984, The use of numerical method for experimental design when choosing mix proportions of the slag-alkaline concretes. *II National Scientific and Practical Conference on Slag-Alkaline Cements, Concretes and Structures* (edited by Glukhovsky), Kiev, USSR, 209–210.

Lea, F. M., 1974, *The Chemistry of Cement and Concrete* (3rd edn) (London: Edward Arnold).

Lee, C. Y., Lee, H. K. and Lee, K. M., 2003, Strength and microstructural characteristics of chemically activated fly ash-cement systems. *Cement and Concrete Research*, 33(3), 425–431.

Lee, W. K. W. and Van Deventer, J. S. J., 2002a, Structural reorganisation of class F fly ash in alkaline silicate solutions, Colloids and Surfaces A: Physicochem. Eng. Aspects, 211(1), 49–66.

Lee, W. K. W. and Van Deventer, J. S. J., 2002b, The effects of inorganic salt contamination on the strength and durability of geopolymers. *Colloids and Surfaces A: Physicochem. Eng. Aspects*, 211, 115–126.

Lee, W. K. W. and Van Deventer, J. S. J., 2004, The interface between natural siliceous aggregates and geopolymers. *Cement and Concrete Research*, 34(2), 195–206.

Lentz, C. W., 1964, Silicate minerals as sources of trimethylsilyl silicates and silicate structure analysis of sodium silicate solutions. *Inorg. Chem.*, 3, 574–579.

Li, D. and Wu, X., 1992, Improvement of early strength of steel slag cement. *Jiangsu Building Materials (in Chinese)*, 4, 24–27.

Li, D., Fu, X., Wu, X. and Tang, M., 1997, Durability study of steel slag cement. *Cement and Concrete Research*, 27(7), 983–987.

Li, Y. and Sun, Y., 2000, Preliminary study on combined-alkali-slag paste materials. *Cement and Concrete Research*, 30(6), 963–966.

Lorenz, W., 1985, The use of volcanic rocks as construction raw materials. *Natural Resources and Development*, 22, 7–24.

Lu, C., 1989, The preliminary research of the fly ash-slag-alkali concrete. *2nd Beijing International Symposium on Cements and Concrete*, Beijing, P.R. China, 2, 232–239.

Lu, C., 1992, The research and the reactive products and mineral phase for FKJ cementitious material. *9th International Congress on the Chemistry of Cement*, New Delhi, India, III, 319–324.

Lu, P., 1989, Origin and development of microstructure of alkali-BFS-SS paste. *2nd Beijing International Symposium on Cements and Concrete*, Beijing, P.R. China, 1, 339–345.

Lu, P. and Young, J., F., 1993, Slag-portland cement based DSP paste. *Journal of American Ceramic Society*, 76(5), 1329–1334.

Luke, K. and Glasser, F. P., 1987, Selective Dissolution of Hydrated Blast Furnace Slag Cements. *Cement and Concrete Research*, 17(2), 273–282.

Malek, R. I. A, Licastro, P. M., Roy, D. M. and Langton, C. A., 1986, Slag cement-low level radioactive waste forms at Savannah River Plant, Ceramic Bulletin, 65, 1578–1583.

Malek, R. I. A. and Roy, D. M., 1997b, Synthesis and characterization of new alkali-activated cements. *10th International Congress on the Chemistry of Cement*, Vol. 1, li024, 8pp.

Malek, R. I. A. and Roy, D. M., 1997a, Durability of Alkali Activated Cementitious Materials. In Scrivener and Young (eds), *Mechanisms of Chemical Degradation of Cement-based Systems* (London: FN Spon), 83–89.

Malhotra, V. M., Ramachandran, V. S., Feldman, R. F. and Aitcin, P.-C., 1987, *Condensed Silica Fume in Concrete* (Boca Raton, Florida: CRC Press, Inc.).

Malolepszy, J., 1986, Activation of synthetic melitite slags by alkalis. *8th International Congress on the Chemistry of Cement*, Rio de Janeiro, Brazil, 4, 104–107.

Małolepszy, J., 1989, Hydration and properties of alkali-activated binders; *Zeszyty Naukowe AGH* (53), Krakow (in Polish).

Malolepszy, J., 1993, Some Aspects of Alkali Activated Cementitious Materials Setting and Hardening, *Proceedings of 3rd Beijing International Symposium on Cement and Concrete*, International Academic Publishers, Beijing, China, Vol. 2, pp. 1043–1046.

Malolepszy, J. and Deja, J., 1988, The influence of curing conditions on the mechanical properties of alkali-activated slag binders. *Silicates Industrials*, 53, 179–186.

Malolepszy, J. and Deja, J., 1994, Industrial application of slag alkaline concretes. *1st International Conference on Alkaline Cements and Concretes* (edited by Krivenko), Kiev, Ukraine, 2, 987–1001.

Malolepszy, J. and Deja, J., 1995, Effect of heavy metals immobilization on properties of alkali-activated slag mortars. *Proceedings of the 5th International Conference on the Use of Fly Ash, Silica Fume, Slag & Natural Pozzolans in Concrete*, SP-153, Farmington Hills, MI, 1087–1102.

Malolepszy, J. and Deja, J., 1999, Durability of alkali activated slag mortars and concrete. *2nd International Conference on Alkaline Cements and Concretes* (edited by Krivenko), Kiev, Ukraine, 685–697.

Malolepszy, J., Deja, J. D. and Brylicki, W., 1994, Alkali-Activated Slag Cements – a Useful Material for Environmental Protection. *2nd International Conference on Alkaline Cements and Concretes* (edited by Krivenko), Kiev, Ukraine, 2, 989–1001.

Malolepszy, J. and Nocun-Wczelik, W., 1988, Microcalorimetric studies of slag alkaline binders. *Journal of Thermal Analysis*, 33, 431–434.

Malolepszy, J. and Petri, M., 1986, High strength slag-alkaline binders. *8th International Congress on the Chemistry of Cement*, Rio de Janeiro, Brazil, 4, 108–111.

Malquori, G., 1960, Portland-pozzolan cement. *4th International Symposium on the Chemistry of Cement*, Washington, USA, II, 983–1000.

Mannion, L. E., 1983, Sodium Carbonate Deposits. In Lefond (ed.) *Industrial Minerals and Rocks* (5th edn) (New York: AIME), 1187–1206.

Mantel, D. G., 1994, Investigation into the hydraulic activity of five granulated blast furnace slags with eight different Portland cements. *ACI Materials Journal*, 91(5), 471–477.

Margesson, R. D. and Englang, W. G., 1971, Processes for pelletization of metallurgical slag. US Patent No. 3,594,142.

Maso, J. C., 1980, The bond between aggregates and hydrated cement paste. *7th International Congress on the Chemistry of Cement*, Paris, France, I, VII - 1/4 - 1/15.

Mass, H. and Peters, K.-H., 1978, Effect of granulation on the hydraulic properties of blast furnace slag. *Zement Kalk Gips*, 31(6), 300–301.

McCarthy, G. J., Swanson, K. D., Keller, L. P. and Blatter, W. C., 1984, Mineralogy of Western fly ashes. *Cement and Concrete Research*, 14(3), 471–478.

McCarthy, G. J., Swanson, K. D. and Steinwand, S. J., 1988, X-ray diffraction analysis of fly ash. *Advances in X-Ray Analysis*, 31, 331–342.

McDowell, J. F., 1986, Hydrogarnet-gelhelnite hydrate cements from $CaO–Al_2O_3–SiO_2$ glass. *8th International Congress on the Chemistry of Cement*, Rio de Janeiro, Brazil, IV, 423–428.

McIlveen, S., Jr and Cheek, R. L., Jr, Sodium Sulphate Resources, Industrial Minerals and Rocks, Society for Mining, Metallurgy, and Exploration, Inc., Littleton, Colorado, 1994.

Mehta, P. K., 1986, *Concrete – Structure, Properties and Materials* (Englewood Cliffs, NJ: Prentice-Hall Inc.).

Mesto, J., 1982, The alkali aggregate reaction of alkali-activated Finnish blast furnace slag. *Silicates Industrial*, 47(4–5), 123–127.

Meyerson, R., 2001, Compressive Creep of Prestressed Concrete Mixtures with and without Mineral Admixtures, M.Sc. Thesis, Virginia Polytechnic and State University, USA.

Millers, J. G. and Oulton, T. D., 1970, Prototropy in kaolinite during percussive grinding, *J. Clay and Clay Minerals*, 18, 313–323.

Ministry of Industrial Construction of the USSR, 1979, Manual of Practice for the Construction of Road Basements Made with Slag Alkaline Binder Stabilized Soils (in the conditions of Western Siberia), The Ministry of Industrial Construction of the USSR, Glavomskpromstroy, Trust Spetsstroy, Omsk, USSR.

Minnick, L. J., Webster, W. C. and Purdy, E. J., 1971, Predictions of the effect of fly ash in Portland cement mortar and concrete. *Journal of Materials*, 6(1), 163–187.

Mindess, S. and Young, J. F., 1981, *Concrete* (Englewood Cliffs, NJ: Prentice-Hall International Inc.).

Mokhort, N. A., 2000, The formation of structure and properties of the alkaline geocements. *Tagunsbericht 14. Internationale Baustofftagung*, Weimar, Germany, 1-0553-1-0560.

Monosi, S. and Collepardi, M., 1990. Research on identification in concretes damaged by $CaCl_2$ attack. *IL Cemento*, 87(1), 3–8.

Montgomery, D. G. and Wang, G., 1991, Instant-chilled steel slag aggregate in concrete – strength related properties. *Cement and Concrete Research*, 21(6), 1083–1091.

Moran, W. T. and Gilliland, J. L., 1950, Summary of methods ford determining pozzolanic activity. ASTM STP-99, 109–131.

Moranville-Regourd, M., 1998, Cements Made from Blast Furnace Slag. In Lea's *Chemistry of Cement and Concrete* (edited Hewlett) (London: Arnold), pp. 633–699.

Mortureux, B., Hornain, H., Gautier, E. and Regourd, M., 1980, Comparison of the reactivity of different pozzolans. *7th International Congress on the Chemistry of Cement*, Paris, France, IV/110–115.

MUSSREOAR, 1986, Guide for the use of a slag alkaline binder in strengthening stone materials while making layers of road structures, The Ministry of the Ukrainian SSR for Erection and Operation of Automotive Roads (MUSSREOAR), Kiev.

Nahm, J. J., Javanmard, K., Cowan, K. M. and Hale, A. H., 1994a, Slag mix mud conversion cementing technology: Reduction of mud disposal volumes and management of rig-site drilling wastes. *Journal of Petroleum Science and Engineering*, 11(1), 3–12.

Nahm, J. J., Romero, R. N., Hale, A. A., Keedy, C. R., Wyant, R. E., Briggs, B. R., Smith, T. R. and Lombardi, M. A., 1994b, Universal fluids improving cementing, *World Oil*, November, 67–72.

Nakamura, N., Sakai, M., Koibuchi, K. and Iijima, Y., 1986, Properties of high-strength concrete incorporating very finely ground granulated blast furnace slag. *2nd International Conference on the Use of Fly Ash, Silica Fume, Slag & Natural Pozzolans in Concrete*, ACI SP-91, 1361–1380.

Narang, K. C. and Chopra, S. K., 1983, Studies on alkaline activation of BF, steel and alloy slags. *Silicates Industrials*, 48(9), 175–182.

Neville, A., 1996, *Properties of Concrete* (New York: John Wiley and Sons, Inc.).

NIIZHB, 1983, Recommended design norms for structures made from slag alkaline cement concretes, The Research Institute for Concrete and Reinforced Concrete (NIIZHB), Moscow, USSR.

Occidental Chemical Corporation, Caustic Soda Handbook, 1992a, Occidental Chemical Corporation, Dallas, TX. 68p.

Occidental Chemical Corporation, Sodium Silicates Handbook, 1992b, Occidental Chemical Corporation, Dallas, TX, 20p.

Odler, I., 1991, Strength of cement. *Materials and Structure*, 24, 143–157.

Odler, I., 1998, Hydration, Setting and Hardening of Portland Cement, in Lea's *Chemistry of Cement and Concrete* (edited Hewlett) (London: Arnold), 241–289.

Odler, R., Skalny, J. and Brunauer, S., 1976, Properties of the system "clinker-lignosulfonate-carbonate". *6th International Congress on the Chemistry of Cement held in Moscow in 1974* (Moscow: Stroyizdat Publisher), Vol. 2, Book 2, 30–32.

Osbaeck, B., 1989, Ground blast furnace slags grinding methods, particle size distribution, and properties. *3rd International Conference on the Use of Fly Ash, Silica Fume, Slag & Natural Pozzolans in Concrete*, ACI SP-114, 1239–1263.

Osborn, E. F., DeVries, R. C., Gee, K. H. and Kraner, H. M., 1954, Optimum composition of blast-furnace slag as deduced from liquids data for the quaternary system $CaO–MgO–Al_2O_3–SiO_2$. *Journal of Metal*, 6(1), 33–45.

Osborne, G. J. and Singh, B., 1995, The durability of concretes made with blends of high alumina cement and ground blast furnace slag. *5th International Conference on the Use of Fly Ash, Silica Fume, Slag & Natural Pozzolans in Concrete*, ACI SP-153, Milwaukee, Wisconsin, USA, 885–909.

OST 67-10-84, 1984, Industry Standard. Technical Specification for Concretes, Heavyweight, Slag Alkaline, The Ministry of USSR for the Erection of Heavy Industry Enterprises, Moscow, USSR.

OST 67-11-84, 1984, Industry Standard. Technical Specification for Binder, Slag Alkaline. The Ministry of USSR for Erection of Heavy Industry Enterprises, Moscow, USSR.

OST 67-12-84, 1984, Industry Standard for Structures and Articles, Concrete and Reinforced Concrete Prefabricated from Slag Alkaline Binder Concrete, The Ministry of USSR for Erection of Heavy Industry Enterprises, Moscow, USSR.

Page, C. L., Short N. R. and El-Tarras, A., 1981, Diffusion of chloride ions in hardened cement pastes. *Cement and Concrete Research*, 11(3), 395–406.

Palacios, M. and Palomo, A., 2004, Alkali-activated fly ash matrices for lead immobilisation: a comparison of different leaching tests. *Advances in Cement Research*, 16(4), 137–144.

Palomo, A., Alonso, S., Fernández-Jiménez, A., Sobrados, I. and Sanz, J., 2004, Alkali activated of fly ashes, A NMR study of the reaction products. *Journal of American Ceramic Society*, 87(6), 1141–1145.

Palomo, A. and De la Fuente, J. I. L., 2003, Alkali-activated cementitous materials: alternative matrices for the immobilisation of hazardous wastes Part I. Stabilisation of boron. *Cement and Concrete Research*, 33(2), 281–288.

Palomo, A. and Glasser, F. P., 1992, Chemically bonded cementitious material based on metakaolin. *Bristish Ceramic Transactions and Journal*, 91, 107–112.

Palomo, A., Grutzeck, M. W. and Blanco, M. T., 1999, Alkali-activated fly ashes – a cement for the future. *Cement and Concrete Research*, 29(8), 1323–1329.

Palomo, A., Macias, A., Blanco, M. T. and Puertas, F., 1992, Physical chemical and mechanical characterization of geopolymers. *9th International Congress on the Chemistry of Cement*, New Delhi, India, V, 505–511.

Palomo, A. and Palacios, M., 2003, Alkali-activated cementitious materials: Alternative matrices for the immobilisation of hazardous wastes Part II. Stabilisation of chromium and lead. *Cement and Concrete Research*, 33(2), 289–295.

Pan, Q. and Zhang, C., 1999, Investigations on factors affecting performance of alkali-fly ash-slag binder. *Cement Engineering (in Chinese)*, 2, 1–3.

Parameswaran, P. S. and Chatterjee, A. K., 1986, Alkali activation of Indian blast furnace slag. *8th International Congress on the Chemistry of Cement*, Rio de Janeiro, Brazil, 4, 86–911.

Paschenko, A. A., Myasnikova, E. A., Serbin, V. P., Gumen, V. S., Evsyutin, Yu., R., Saldugey, M. M., Sanitsky, M. A., Serbin, V. P., Tokarchuk, V. V., Udachkin, I. B. and Chistyakov, V. V., 1991, *A Theory of Cement* (Kiev: Budivelnik Publisher).

Pavlik, V., 1994, Corrosion of hardened cement paste by acetic and nitric acids. Part I: Calculation of corrosion depth. *Cement and Concrete Research*, 24(3), 551–562.

Peng, J., 1982, New solid water glass-slag cements. *Cement (in Chinese)*, 6, 6–10.

Pera, J., 2001, Metakaolin and calcined clays. *Cement and Concrete Composites*, 23, p. iii.

Petropavlovsky, O. N., 1987, Slag Alkaline Binding Systems and Concretes Based on Steelmaking Slags, Ph.D. Thesis, Kiev Civil Engineering Institute, Kiev, USSR.

Petropavlovsky, O. N., 1990, The Structure Formation and Synthesis of Strength of the Slag Alkaline Binders Based on Steel-making Slags, Tsement, Leningrad, USSR, No. 11, pp. 5–7.

Pfeifer, D., McDonald, D. and Krauss, P., 1994, The rapid chloride test and its correlation to the 90-day chloride ponding test, *PCI Journal*, 38–47.

Popel, G. N., 1999, Synthesis of a mineral-like stone on alkaline aluminosilicate binders to produce the materials with the increased corrosion resistance. *2nd International Conference on Alkaline Cements and Concretes* (edited by Krivenko), Kiev, Ukraine, 208–219.

Popovics, S., 1998, *Strength and Related Properties of Concrete – A Quantitative Approach* (New York: John Wiley and Sons, Inc.).

Powers, T. C., 1958, Structure and Physical Properties of Hardened Portland Cement Pastes, *Journal of American Ceramic Society*, Vol. 41, No. 1, pp. 1–6.

Powers, T. C. and Brownyard, T. L., 1947, Studies of the physical properties of hardened Portland cement paste, Part 8 – The freezing of water in hardened Portland cement paste. *Journal of American Concrete Institute*, 43, 933–969.

Powers, T. C., Copeland, L. E., Hayes, J. C. and Mann, H. M., 1954. Permeability of portland cement paste. *ACI Journal Proceedings*, 51(3), 285–298.

Pu, X., Gan, C., Wu, L. and Chen, J., 1988, A study on new structural material advanced alkali-slag (JK) concrete. *Silicate Construction Products (in Chinese)*, 1, 6–11.

Pu, X., Gan, C., Wu, L. and Chen, J., 1989, Properties of alkali-slag (JK) concrete. *Bulletin of Chinese Ceramic Society (in Chinese)*, 1, 5–11.

Pu, X. and Chen, M., 1991, The preventive effect of silica fume on alkali-silica reactive expansion in alkali-slag concrete (JK) concrete. *International Symposium on Concrete Engineering*, Nanjing, P.R. China, 1197–1202.

Pu, X., Gan, C., He, O., Bai, G., Wu, L. and Chen, M., 1991, A study on durability of alkali-slag concrete (JK concrete). *International Symposium on Concrete Engineering*, Nanjing, P.R. China, 1144–1149.

Pu, X. and Yang, C., 1994, Study on alkali-silica reaction of alkali-slag concrete. *1st International Conference on Alkaline Cements and Concretes* (edited by Krivenko), Kiev, Ukraine, 2, 897–906.

Pu, X., Yang, C. and Gan, C., 1994, Research on Set-retarding of High and Super-high Strength Alkali-slag Cement and Concrete. *1st International Conference on Alkaline Cements and Concretes* (edited by Krivenko), Kiev, Ukraine, 2, 585–592.

Pu, X., Yang, C. and Liu, F., 1999, Studies on resistance of alkali activated slag concrete to acid attack. *2nd International Conference on Alkaline Cements and Concretes* (edited by Krivenko), Kiev, Ukraine, 717–721.

Purdon, A. O., 1940, The action of alkalis on blast-furnace slag. *Journal of the Society of Chemical Industry*, 59, 191–202.

Puertas, F., and Fernandez-Jimenez, A., 2003, Mineralogical and microstructural characterization of alkali-activated fly ash/slag pastes. *Cement and Concrete Composites*, 25, 287–293.

Puertas, F., Palomo, A., Fernandez-Jimenez, A., Jzquierdo, J. Z. and Granizo, M. L., 2003, Effect of superplasticizer on behaviour and properties of alkaline cements. *Advances in Cement Research*, 15(1), 23–28.

Puertas, F., Fernandez-Jimenez, A. and Blanco-Varela, M. T., 2004, Pore Solution in Alkali-activated slag cement pastes, relation to the composition and structure of calcium silicate hydrate. *Cement and Concrete Research*, 34(1), 139–148.

Putnis, A., 1992, *Introduction to Mineral Sciences*, Cambridge University Press, Cambridge, UK, p. 86.

Qian, G., Li, Y., Yi., F. and Shi, R., 2002, Improvement of metakaolin on radioactive Sr and Cs immobilization of alkali-activated slag matrix. *Journal of Hazardous Materials*, B92, 289.

Qian, G., Sun, D. D. and Tay, J. H., 2003a, Characterization of mercury- and zinc-doped alkali-activated slag matrix, Part I. Mercury. *Cement and Concrete Research*, 33(8), 1251–1256.

Qian, G., Sun, D. D. and Tay, J. H., 2003b, Characterization of mercury- and zinc-doped alkali-activated slag matrix, Part II. Zinc. *Cement and Concrete Research*, 33(8), 1257–1262.

Qian, J., Shi, C. and Wang, Z., 2001. Activation of blended cement containing fly ash. *Cement and Concrete Research*, 31(8), 1121–1127.

Querol, X., Alastuey, A., Lopez-Soler, A., Plana, F., Andres, J. M., Juan, R., Ferrer, P. and Ruiz, C. R., 1997, A fast method for recycling fly ash: microwave-assisted zeolite synthesis. *Environmental Science and Technology*, 31, 2527–2533.

Rajaokarivony-Andriambololona, Z., Thomassin, J. H., Baillif, P. and Touray, J. C., 1990, Experimental hydration of two synthetic glassy blast furnace slags in water and alkaline solutions (NaOH and KOH 0.1 N) at 40 °C: structure, composition and origin of the hydrated layer. *Journal of Materials Science*, 25, 2399–3410.

Raksha, V. A., 1975. An Investigation of Influence of Chemical Composition of Slags on Properties of Slag Alkaline Binders and Concretes. Ph.D. Thesis, Kiev Civil Engineering Institute, Kiev, USSR.

Ramachandran, V. S., Feldman, R. F. and Beaudoin, J. J., 1981, *Concrete Science – Treatise on Current Research* (London: Heyden & Son Ltd.).

RDIBS, 1988, Guide for manufacturing and use in construction of wall panels from lightweight slag alkaline binder concrete. *Research & Design Institute for Building Structures* (RDIBS), Kiev, USSR.

Regourd, M., 1980, Structure and behaviour of slag portland cement hydrates. *7th International Congress on the Chemistry of Cement*, Paris, France, 1, III-2/11-26.

Richardson, I. G. and Groves, G. W., 1992, Microstructure and microanalysis of hardened cement pastes involving ground blast-furnace slag. *Journal of Materials Science*, 27(22), 6204–6212.

Richardson, I. G., Brough, A. R., Brydson, R., Groves, G. W. and Dobson, C. M., 1993, Location of Aluminum in Substituted Calcium Silicate Hydrate (C–S–H) Gels as Determined by ^{29}Si and ^{27}Al NMR and EELS. *Journal of American Ceramic Society*, 76, pp. 2285–2288.

Richardson, I. G., Brough, A. R., Groves, G. W. and Dobson, C. M., 1994, The characterization of hardened alkali-activated blast furnace slag pastes and the nature of the calcium silicate hydrate (C–S–H) phase. *Cement and Concrete Research*, 24(5), 813–829.

RILEM 5-FMC Committee, 1985, Determination of the Fracture Energy of Mortar and Concrete by Beams of Three-Point Bend Tests on Notched Beam. *Materials and Structure*, 18(106), 285–290.

Rossler, M. and Odler, I., 1985, Investigations on the relationship between porosity, structure and strength of hydrated portland cement pastes, Part I. Effect of porosity. *Cement and Concrete Research*, 15(2), 320–330.

Roger, S. and Shi, C., 2004, Stabilization and Solidification of Hazardous, Radioactive and Mixed Wastes, CRC Press Inc., Boca, Florida.

Roper, H. F. and Auld, G. J., 1983, Characterization of a copper slag used in mine fill operation. *1st International Conference on the Use of Fly Ash, Silica Fume, Slag and Natural Pozzolans in Concrete*, ACI SP-79, American Concrete Institute, Detroit, Michigan, USA, 2, 1091–1110.

Rossi, G. and Forchielli, L., 1976, Porous structure and reactivity with lime of some natural Italian pozzolans. *Il Cemento*, 4, 215–221.

Rostami, H. and Brendley, W., 2003, Alkali ash material: a novel fly ash-based cement. *Environmental Science and Technology*, 37(15), 3454–3457.

Roy, A., Schilling, P. J., Eaton, H. C., Malone, P. G., Brabston, W. N. and Wakeley, L. D., 1992, Activation of ground-blast furnace slag by alkali-metal and alkaline-earth hydroxides. *Journal of American Ceramic Society*, 75(12), 3233–3240.

Roy, A., Schilling, P. J. and Eaton, H., 1996, Alkali activated class C fly ash cement, US Patent 5,565,028, October 15.

Roy, D. M., 1986, Mechanism of Cement Paste Degradation Due to Chemical and Physical Process. *Proceedings of 8th International Congress on the Chemistry of Cement*, Brazil, Vol. I, pp. 359–380.

Roy, D. M., 1987, New strong cement materials: chemically bonded ceramics. *Science*, 235, 651–658.

Roy, D. M., 1989, Hydration, microstructure, and chloride diffusion of slag-cement pastes and mortars. *3rd International Conference on the Use of Fly Ash, Silica Fume, Slag & Natural Pozzolans in Concrete*, ACI SP-114, Trondheim, Norway, 2, 1265–1281.

Roy, D. M., 1999, Alkali-activated cements: Opportunities and challenges. *Cement and Concrete Research*, 29(2), 249–254.

Roy, D. M. and Gouda, G. R., 1973, High strength generation in cementitious materials with very high strength. *Cement and Concrete Research*, 3, 807–820.

Roy, D. M., Gouda, G. R. and Brobowsky, A., 1972, Very high strength cement pastes prepared by hot-pressing and other high pressure techniques. *Cement and Concrete Research*, 2(3), 349–366.

Roy, D. M. and Idorn, G. M., 1982, Hydration, structure, and properties of blast furnace slag cements, mortars and concretes. *ACI Journal*, 79(6), 444–456.

Roy, D. M. and Idorn, G. M., 1985, Relationships between strength, pore structure and associated properties of slag-containing cementitious materials. *Materials Research Society Symposium*, 42, 133–142.

Roy, D. M. and Jiang, W., 1994, Microcharacteristics and properties of hardened alkali-activated cementitious materials. *Cement Technology*, American Ceramic Society, Columbus, USA, 40, 257–264.

Roy, D. M. and Langton, C. A., 1989, Characterization of cement-based ancient building materials for a repository in tuff, LA-11527-MS, Los Alamos National Lab., Los Alamos, NM, 1–100.

Roy, D. M. and Silsbee, M. R., 1992, Alkali-activated cementitious materials: An overview. *Materials Research Society Symposium*, 245, 153–164.

Roy, D. M. and Silsbee, M. R., 1994, Novel cements and cement products for applications in the 21st century. *Malhotra Symposium on Concrete Technology, Past, Present and Future*, SP-144, 349–382.

Roy, D. M. and Silsbee, M. R., 1995, Overview of slag microstructure and alkali-activated slag in concrete (Supplementary Paper). *2nd CANMET/ACI International Symposium on Advances in Concrete Technology*, Las Vegas, Nevada, USA, 700–715.

Roy, D. M., Jiang, W. and Silsbee, M. R., 2000, Chloride Diffusion in Ordinary, Blended, and Alkali-activated cement pastes and its relation to other properties. *Cement and Concrete Research*, 30(12), 1879–1884.

Roy, D. M. and Malek, R. I. A., 1993, Hydration of Slag Cement, Mineral Admixture. In Ghosh, Sarkar and Harsh (eds) *Cement and Concrete* (New Delhi, India: ABI Books Pty. Ltd.), pp. 84–117.

Royak, S. M., P'yachev, V. A. and Shkolnik, Ya. Sh., A Structure of blastfurnace slags and activity. *Tsement (in Russian)*, Leningrad, USSR, 8, 1978.

RSN 25-84, 1984, Republican Guidelines for Manufacturing and Use of Concrete and Reinforced Concrete Articles and Structures Based on Slag Alkaline Binders, The National Committee of the Uzbekistan Republic for Construction Tashkent, USSR.

RSN 336-84, 1984, Republican Building Norms for Production and Use of Slag Alkaline Binders, Concretes and Structures, The National Committee of the Ukrainian Republic of the USSR for Construction, Kiev, USSR.

RSN 344-87, 1987, Republican Building Norms for Recommended Consumptions of Ground Slag and Alkaline Activator for Making Slag Alkaline Cement Concretes, Precast and Cast-in-Situ Concrete, Reinforced Concrete Articles and Structures, The National Committee of the Ukrainian Republic of the USSR for Construction, Kiev, USSR.

RSN 354-90, 1991, The Use of Slag Alkaline Concrete in Monolithic Construction, Building Norms of Ukrainian Republic, State Committee of Ukrainian Republic for Construction, Kiev.

RST 5024-83, 1983, Republican Standard Technical Specification for Binder, Slag Alkaline, The National Committee of the Ukrainian Republic of the USSR for Construction, Kiev, USSR.

RST 5025-84, 1984, Republican Standard Technical Specification for Concretes, Heavyweight, The National Committee of the Ukrainian Republic of the USSR for Construction, Kiev, USSR.

RST 5026-84, 1984, Structures and Articles, Concrete, and Reinforced Concrete Prefabricated, Made from Slag Alkaline Binder Concrete. General Technical Requirements, Republican Standard, State Committee of Ukraine for Construction, Kiev.

RTU 153-65, 1966, Republican Technical Specifications for Manufacturing and of Precast Pilot-Scale Articles Made from Soil Silicate Concrete for Irrigation Construction, 1979, The State Committee of the Ukrainian Republic for Construction, Kiev, USSR.

Rumyna, G. V., 1974, Investigation of the Influence of Clay Minerals on Slag Alkaline Binder Concrete Properties. Ph.D. Thesis, Kiev Civil Engineering Institute, Kiev, USSR.

Russian Patent RU 213 4247, 1999, A method of producing hydrated sodium and potassium silicates.

Sabir, B. B., Wild, W. and Bai, J., 2001, Metakaolin and calcined clays as pozzolans for concrete: a review. *Cement and Concrete Composites*, 23, 441–454.

Sarvaranta, L. and Mikkola, E., 1994, Fiber mortar composites in fire conditions. *J. Fire Materials*, 18, 45–50.

Sato, K., Konishi, E. and Fukaya, K., 1986, Hydration of Blast Furnace Slag Particle. *8th International Congress on the Chemistry of Cement*, Rio de Janeiro, Brazil, 4, 96–103.

Satarin, V. I., 1976, Slag Portland cement (Principal Paper). *6th International Congress on the Chemistry of Cement held in Moscow in 1974* (Moscow: Stroiizadt Publisher), 3, 45–56.

Schilling, P. J., Roy, A., Eaton, H. C., Malone, P. G., and Brabston, W. N., 1994a, Microstructure, strength, and reaction products of ground granulated blast-furnace slag activated by highly concentrated NaOH solution. *Journal of Materials Research*, 9(1), 188–197.

Schilling, P. J., Butler, L. G., Roy, A. and Eaton, H. C., 1994b, 29Si and 27Al MAS-NMR of NaOH-activated blast furnace slag. *Journal of American Ceramic Society*, 77, 2363–2368.

Schramm, C. J., 1999, Microwave-dried amorphous alkali metal silicate powders and their use as builders in detergent composition, US Patent 5,961,663, Oct 5.

Schwiete, H. E. and Dolbor, F., 1963, The effect of the cooling conditions and the chemical composition on the hydraulic properties of haematitic slags. *Forschungsbericht des Landes*.

Scian, A. N., Lopez, J. M., Porto and Pereira, E., 1991, Mechanochemical activation of high aluimina cements – hydration characteristics. *Cement and Concrete Research*, 21(1), 51–60.

Sersale, R. and Frigione, G., 2004, Materials Activated with Alkaline Waste Water. *Proceedings 8th CANMET/ACI International Conference on Fly Ash, Silica Fume, Slag, and Natural Pozzolans*, Las Vegas, May 23–29, ed. V. M. Malhotra, American Concrete Institute, Michigan, U.S., 721–734.

Shan, W., Li, M. and Wei, R., 1983, Studies of granulated phosphorus slag as mineral admixtures. *Cement (in Chinese)*, P.R. China, 3, 23–27.

Sheikin, A. E., 1974, *Structure, Strength and Crack Resistance of a Cement Stone* (Moscow: Stroizdat Publisher).

Shen, X., Yan, S., Wu, X., Tang, M. and Yang, L., 1994, Immobilization of simulated high level wastes into AASC waste form. *Cement and Concrete Research*, 24(1), 133–138.

Shen, X., Yan, S., Wu, X., Tang, M. and Lu, C., 1996, Study on the Reaction Degree of Slag in High Level Radioactive Waste by Alkali Activated Slag Cement. *Bulletin of Chinese Ceramic Society*, 2(18), 4–7.

Shi, C., 1987, Activation of Granulated Phosphorus Slag, M.Sc. Thesis, Nanjing Institute of Technology, 1987, P.R. China.

Shi, C., 1988, Alkali-aggregate Reaction of alkali-slag cements. *Concrete and Cement Products (in Chinese)*, 4, 28–32.

Shi, C., 1992, Activation of Reactivity of Natural Pozzolan, Fly Ashes and Slag, Ph.D. Thesis, University of Calgary, Canada.

Shi, C., 1996a, Early microstructure development of activated lime-fly ash pastes. *Cement and Concrete Research*, 26(9), 1351–1359.

Shi, C., 1996b, Strength, pore structure and permeability of high performance alkali-activated slag mortars. *Cement and Concrete Research*, 26(12), 1789–1800.

Shi, C., 1997, Early hydration and microstructure development of alkali-activated slag pastes. *10th International Congress on the Chemistry of Cement*, Gothenburg, Sweden, 3ii099, 8pp.

Shi, C., 1998, Pozzolanic reaction and microstructure development of activated lime-fly ash pastes. *ACI Materials*, 95(5), 537–545.

Shi, C., 1999, Corrosion resistant cement made with steel mill by-products. *International Symposium on the Utilization of Metallurgical Slag*, Beijing, P.R. China, November 16–19, 171–178.

Shi, C., 2001, An overview on the activation of reactivity of natural pozzolan. *Canadian Journal of Civil Engineering*, 28(5), 778–786.

Shi, C., 2002, Characteristics and cementitious properties of ladle slag fines from steel production. *Cement and Concrete Research*, 3(32), 459–462.

Shi, C., 2003a, Corrosion resistance of alkali-activated slag cement. *Advances in Cement Research*, 15(2), 77–81.

Shi, C., 2003b, On the role and state of alkali ions during the hydration of alkali-activated slag cement. *Proceedings of the 11th International Congress on the Chemistry of Cement*, Durban, South Africa, pp. 2097–2105.

Shi, C., 2004a, Effect of mixing proportions of concrete on its electrical conductivity and the rapid chloride permeability test (ASTM C 1202 or ASSHTO T 277) results. *Cement and Concrete Research*, 34(3), 537–545.

Shi, C., 2004b, Steel slag – its production, processing, characteristics and cementitious properties. *Journal of Materials in Civil Engineering*, 16(3), 230–236.

Shi, C. and Day, R. L., 1993a, Acceleration of strength gain of lime-natural pozzolan cements by thermal activation. *Cement and Concrete Research*, 23(4), 824–832.

Shi, C. and Day, R. L., 1993b, Chemical activation of blended cement made with lime and natural pozzolans. *Cement and Concrete Research*, 23(6), 1389–1396.

Shi, C., Day, R. L. and Huizer, A., 1994, Chemical activation of natural pozzolan for low-cost masonry units. *Masonry International, Journal of the British Masonry Society*, 6, 1–3.

Shi, C. and Day, R. L., 1995a, Acceleration of the reactivity of fly ash by chemical activation. *Cement and Concrete Research*, 25(1), 15–21.

Shi, C. and Day, R. L., 1995b, Microstructure and reactivity of natural pozzolans, fly ash and blast furnace slag. *17th International Cement Microscopy Conference*, Calgary, Canada, 150–161.

Shi, C. and Day, R. L., 1995c, Chemical activation of lime-slag cement. *5th International Conference on Fly Ash, Silica Fume, Slag and Natural Pozzolans in Concrete*, ACI SP153-61. Milwaukee, Wisconsin, USA, 2, 1165–1177.

Shi, C. and Day, R. L., 1995d, A calorimetric study of early hydration of alkali-slag cements. *Cement and Concrete Research*, 25(6), 1333–1346.

Shi, C. and Day, R. L., 1996a, Alkali-slag Cements for The Solidification of Radioactive Wastes. In Gilliam and Wiles (eds) *Stabilization and Solidification of Hazardous, Radioactive, and Mixed Wastes*, ASTM STP 1240, American Society for Testing and Materials, Philadelphia, USA, 163–173.

Shi, C. and Day, R. L., 1996b, Factors affecting early hydration characteristics of alkali-slag cements. *Cement and Concrete Research*, 26(3), 439–448.

Shi, C. and Day, R. L., 1996c, Selectivity of alkaline activators for the activation of slags. *Cement, Concrete and Aggregate*, 18(1), 8–14.

Shi, C. and Day, R. L. 1999, Early strength development and hydration of alkali-activated blast furnace slag/fly ash blends. *Advances in Cement Research*, 11(4), 189–196.

Shi, C. and Day, R. L., 2000a, Pozzolanic reactions in the presence of chemical activators – Part I: Reaction kinetics. *Cement and Concrete Research*, 30(1), 51–58.

Shi, C. and Day, R. L., 2000b, Pozzolanic reactions in the presence of chemical activators – Part II: Reaction mechanisms. *Cement and Concrete Research*, 30(4), 607–613.

Shi, C. and Day, R. L., 2001, Comparison of different methods for enhancing reactivity of pozzolans. *Cement and Concrete Research*, 31(5), 813–818.

Shi, C., Day, R. L., Wu, X. and Tang, M., 1992a, Comparison of the microstructure and performance of alkali-slag and Portland cement pastes. *9th International Congress on the Chemistry of Cement*, New Delhi, India, III, 298–304.

Shi, C., Day, R. L., Wu, X. and Tang, M., 1992b, Uptake of metal ions by autoclaved cement pastes, in *Proceedings of Materials Research Society*. Materials Research Society, Boston, Vol. 245, 141–149.

Shi, C., Grattan-Bellew, P. E. and Stegemann, J. A., 1999, Conversion of A Waste Mud into a Pozzolanic Material. *Building and Construction Materials*, 13(5), 279–284.

Shi, C. and Li, Y., 1989a, Effect of the modulus of water glass on the activation of phosphorus slag. *Il Cemento*, 86(3), 161–168.

Shi, C. and Li, Y., 1989b, Investigation on some factors affecting the characteristics of alkali-phosphorus slag cement. *Cement and Concrete Research*, 19(4), 527–533.

Shi, C., Li, Y. and Tang, X., 1989a, A preliminary investigation on the activation mechanism of granulated phosphorus slag. *Journal of Southeast University (in Chinese)*, Nanjing, P.R. China, 19(1), 141–145.

Shi, C., Li, Y. and Tang, X., 1989b, Studies on the activation of phosphorus slag Supplementary Paper. *3rd International Conference on the Use of Fly Ash, Silica Fume, Slag and Natural Pozzolans in Concrete*, ACI SP-114, Trondheim, Norway, 657–666.

Shi, C., Shen, X., Wu, X. and Tang, M., 1994, Immobilization of radioactive wastes with portland and alkali-slag cement pastes, *Il Cemento*, 91, 97.

Shi, C. and Stegemann, J. A., 2000, Acid corrosion resistance of different cementing materials. *Cement and Concrete Research*, 30(6), 803–808.

Shi, C., Stegemann, J. A. and Caldwell, R., 1997a, An examination of interference in waste solidification through measurement of heat signature. *Waste Management*, 17, 249.

Shi, C., Stegemann, J. A. and Caldwell, R., 1997b Use of heat signature in solid-ification treatability studies. *Proceedings of 10th International Congress on the Chemistry of Cement*, Amarkai AB and Congrex Goteborg, Goteborg.

Shi, C., Stegemann, J. A. and Caldwell, R., 1998, Effect of supplementary cementing materials on the Rapid Chloride Permeability Test (AASHTO T 277 and ASTM C 1202) Results, ACI Materials, Vol. 95, 4, 389–394.

Shi, C., Tang, X. and Li, Y., 1991a, Thermal activation of phosphorus slag. *Il Cemento*, 88(4), 219–225.

Shi, C., Wu, X. and Tang, M., 1991b, Hydration of alkali-slag cements at 150°C. *Cement and Concrete Research*, 21(1), 91–100.

Shi, C., Wu, X. and Tang, M., 1993, Research on alkali-activated cementitious systems in China. *Advances in Cement Research*, 5(17), 1–7.

Shi, C. and Xie, P., 1998, Interface between Cement Paste and Quartz Sand in Alkali-Activated Slag Mortars, *Cement and Concrete Research*, 28(6), 887–896.

Shigemoto, N., Hayashi, H. and Miyaura, K., 1993, Selective formation of Na-X zeolite from coal fly ash by fusion with sodium hydroxide prior to hydrothermal reaction, *Journal of Materials Science*, 28, 4781–4786.

Shkolnick, Y. Sh., 1986, Physicochemical principles of the hydraulic activity of blast furnace slags. *8th International Congress on the Chemistry of Cement*, Rio de Janeiro, Brazil, 4, 133–136.

Shrivastava, O. P. and Glasser, F. P., 1985, Ion-exchange Properties of $Ca_5Si_6O_{18}H_2.4H_2O$. *Journal of Materials Science Letter*, 4, 1122.

Sikorsky, O. N., 1967, Method of design of properties of the soil-silicates concretes. *J.Sils'ke Budivnitstvo (in Ukrainian)*, Kiev, USSR, 8, 17–18.

Silverstrim, T. and Rostami, H., 1997. Fly Ash Cementitious Material and Method of Making A Product, US Patent 5,601,643.

Silverstrim, T. and Rostami, H., Xi, Y. and Martin, J., 1997, High perfor-mance characteristics of chemically activated fly ash (CAF), PCI/FHWA Inter-national Symposium on High Performance Concrete, New Orleans, Louisiana, 135–147.

Siemer, D. D., Olanrewaju, J., Scheetz, B. and Grutzeck, M. W., 2001, Development of Hydroceramic Waste Forms for *INEEL* Calcined Waste. *Ceramic Transactions*, 119, pp. 391–398.

Siemer, D. D., 2004, Hydroceramic Concrete, Stabilization and Solidification of Hazardous, Radioactive and Mixed Wastes (edited by Spence, R. and Shi, C.), CRC Press, Boca, Florida.

Singer, A., and Berkgatit, V., 1995, Cation exchange properties of hydrothermally treated coal fly ash. *Environmental Science and Technology*, 29, 748–753.

Singh, N., Rai, S. and Singh, N. B., 2001, Effect of sodium sulphate on the hydra-tion of granulated blast furnace slag blended Portland cement. *Indian Journal of Engineering and Materials Science*, 8(2), 110–113.

Skurchinskaya, Zh. V. and Belitsky, I. V., 1989, The regulation of setting processes in the slag-alkaline binders. *III National Scientific and Practical Conference on Slag-Alkaline Cements, Concretes and Structures* (edited by Glukhovsky), Kiev, USSR, I, 143–145.

Skvara, F., 1985, Alkali-activated slag cements. *J Stavivo (in Czech)*, Prague, Czechoslovakia, 63(1), 16–20.

Slota, R. J., 1987, Utilization of water glass as an activator in the manufacture of cementitious materials from waste by-products. *Cement and Concrete Research*, 17(5), 703–708.

Smithwick, R. W., 1982, A general analysis for mercury porosimetery. *Powder Technology*, 33, 201–209.

Smith, M. A. and Osborne, G. J., 1977, BFS/Fly ash cements. *World Cement Technology*, 8(4), 223–233.

Smolczyk, H. G., 1978, The effect of the chemistry of the slag on the strengths of blastfurnace cements. *Zement Kalk Gips*, 6, 294–296.

Smolczyk, H. G., 1980, Slag Structure and Identification of Slags. *7th International Congress on the Chemistry of Cement*, Paris, France, 1, III–I/4–16.

Song, S. and Jennings, H. M., 1999, Pore solution Chemistry of alkali-activated ground blast-furnace slag. *Cement and Concrete Research*, 29(2), 159–179.

Song, S., Sohn, D., Jennings, H. M. and Mason, T. O., 2000, Hydration of alkali-activated ground granulated blast furnace slag. *Journal of Materials Science*, 35, 249–257.

Spence, R. J. S., 1974, Lime and Surkhi Manufacture in India. In Spence (ed.) *Lime & Alternative Cements, One-day Seminar on Small-scale Manufacturing of Cementitious Materials*, Intermediate Technology Development Group, London, 18–21.

Spence, R. J. S. and Sakela, J. H., 1982, Lime-pozzolan as an Alternative Cementing Materials, Agrid Report, Strategies for Small Scale Mining and Mineral Industries, Regional Workshop, 3(8), 87–91.

Stade, H., 1989, On the reaction of C–S–H (di, poly) with alkali hydroxide. *Cement and Concrete Research*, 19(5), 802–810.

Stade, H. and Muller, D., 1987, On the coordination of Al in Ill-crystallized C–S–H phase by hydration of silicate and precipitation reactions at ambient temperature. *Cement and Concrete Research*, 17(4), 553–561.

Stark, J., Charchenko, I. and Krivenko, P., 2001, The utilization of cement kiln dust in slag-based binding systems. *J. Tsement I Ego Primenenie (Cement and Its Application)(in Russian)*, St. Petersburg, Russia, 2, 38–42.

Stark, J. and Wicht, B., 2001, *Dauerhaftigkeit von Beton* (Basel; Boston; Berlin: Birkhaeuser Verlag).

Stegemann, J. A., Caldwell, R. and Shi, C., 1995, Laboratory and Field Leaching of Solidified Wastes. *1st International Conference on Stabilization and Solidification*, Nancy, France, pp. 338–343.

Stegemann, J., Caldwell, R. and Shi, C., 1997, Variability of Field Solidified Wastes. *Journal of Hazardous Materials*, 52, 335–348.

Stegemann, J. A., Shi, C. and Caldwell, Robert, 1997, Response of various solidification systems to acid addition. Presented at the WASCON '97 – Putting Theory into Practice, June 4–6, The Netherlands.

Sugama, T. and Brothers, L. E., 2004, Sodium-silicate-activated slag for acid-resistant geothermal well cements. *Advances in Cement Research*, 16(2), 77–87.

Sun, S., 1983, Investigations on Steel Slag Cements. In Sun (ed.) *Collections of Achievements on the Treatment and Applications of Metallurgical Industrial Wastes* (Beijing: Chinese Metallurgical Industry Press), 1, pp. 1–71.

Sun, S., Wang, J., Zu, G. and Li, Y., 1993, Steel Slag Cement. In Lu (ed.) *Pretreatment and Reclamation of Dusts, Sludges and Scales in Steel Plants*, McMaster University, Canada, 253–260.

Sun, S. and Yuan, 1983, Study of steel slag cement. *Silicates Industrielles*, 2, 31–34.

Sun, W., Lui, M. and Wei, R., 1984, A study on the use of granulated phosphorus slag as a cement replacement. *Cement (in Chinese)*, 3, 23–28.

Sun, W., Mandel, J. A. and Said, S., 1986, Study of the interface strength in steel fiber-reinforced cement-based composites. *ACI Journal*, 83, 597–605.

Sun, W., Zhang, Y., Lin, W. and Liu, Z., In situ monitoring of the hydration process of K-PS geopolymer cement with ESEM. *Cement and Concrete Research*, 34, 935–940.

Swamy, R. N., 1993, Fly ash and slag, standards and specifications – help or hindrance. *Materials and Structure*, 26, 600–613.

Swanepoel, J. C. and Strydom, C. A., 2002, Utilisation of fly ash in a geopolymeric material. *Applied Geochemistry*, 17, 1143–1148

Synder, R., Esfandi, E. and Surapaneni, S., 1996, Control of concrete sewer corrosion via the Crown Spray Process. *Water Environment Research*, 68(3), 338–347.

Takemoto, K. and Uchikawa, H., 1980, Hydration of pozzolanic cement. *7th International Congress on the Chemistry of Cement*, Paris, France, 1, IV–2/–29.

Talling, B., 1989, Effect of curing conditions on alkali-activated slags. *3rd International Conference on the Use of Fly Ash, Silica Fume, Slag & Natural Pozzolans in Concrete*, ACI SP-114, Trondheim, Norway, 2, 1485–1500.

Talling, B. and Brandstetr, J., 1989, Present and future of alkali-activated slag concrete. *3rd International Conference on the Use of Fly Ash, Silica Fume, Slag & Natural Pozzolans in Concrete*, ACI SP-114, Trondheim, Norway, 1519–1546.

Talling, B. and Brandstetr, J., 1993, Clinker-Free Concrete Based on Alkali-Activated Slag. In Sarkar and Harsh (eds) *Mineral Admixtures in Cement and Concrete* (New Delhi, India: ABI Books Pty. Ltd.).

Tamas, F. D., Sarkar, A. K. and Roy, D. M., 1976, Effect of variables upon the silylation products of hydrated cements. In *Hydraulic Cement Pastes: Their Structure and Properties, Conference held at Sheffield on 8–9 April 1976*, Cement and Concrete Association, Wexham Springs, Slough SL3 6PL, U.K., 55–72.

Tang, L. and Nilsson, L.-O., 1993, Chloride binding capacity and binding isotherms of OPC pastes and mortars. *Cement and Concrete Research*, 23, 247–253.

Tang, M., 1973, Investigation of Mineral Compositions of Steel Slags for Cement Production, Research Report, Nanjing Institute of Chemical Technology, Nanjing, P.R. China.

Tang, M., 1994, Optimum mix design for alkali-activated slag-high calcium fly ash concrete. *Journal of Shenyang Architectural and Civil Engineering Institute*, 10(4), 315–321.

Tang, M. and Han, S., 1981, Effect of $Ca(OH)_2$ on alkali-silica reaction. *Journal of Chinese Silicate Society*, 9(2), 160–166.

Tang, M., Han, S. and Zhen, S., 1989, A rapid method for identification of alkali reactivity of aggregates. *Cement and Concrete Research*, 13(2), 417–422.

Tang, X., 1986, Portland Phosphorus Slag Cements, Research Report, Nanjing Institute of Technology, Nanjing, China.

Tang, X., and Shi, C., 1988, Alkali-phosphorus slag binders. *Silicate Construction Products (in Chinese)*, 1, 28–32.

Tango, C. E. S. and Vaidergorin, E. Y. L., 1992, Some studies on the activation of blastfurnace slag in cements without clinker. *9th International Congress on the Chemistry of Cement*, New Delhi, India, III, 101–107.

Taplin, J. H., 1962, The temperature coefficient of the rate of hydration of β-dicalcium silicate. *4th International Symposium on the Chemistry of Cement*, Washington, USA, 1, 263–266.

Taylor, H. F. W., 1987, A method for predicting alkali ion concentrations in cement pore solutions. *Advances in Cement Research*, 1(1), 5–16.

Taylor, H. F. W., 1990, *Cement Chemistry* (London: Academic Press).

Taylor, H. F. W., 1993, Nanostructure of C-S-H: Current Status. *Advanced Cement Based Materials*, 1, 38–46.

Teoreanu, I., 1991, The interaction mechanism of blast-furnace slags with water: The role of activating agents. *Il Cemento*, 8(2), 91–97.

Teoreanu, I., Georgescu, M. and Puri, A., 1980, Hydrated phases in slag-water-activator systems. *7th International Congress on the Chemistry of Cement*, Paris, France, 2, III-99-111.

The Dow Chemical Company, The Caustic Soda Solution Handbook, 1994, The Dow Chemical Company, Midland, Michigan, USA.

The Dow Chemical Company, METSO Sodium Metasilicates for Industrial and Commercial Cleaning, 1993, 12, The PQ Corporation.

The PQ Corporation, PQ Sodium Silicates – Liquids and Solids, 1994, 16, The PQ Corporation.

Thomas, M. D. A. and Innis, F. A., 1998, Effect of slag on expansion due to alkali-aggregate reaction in concrete. *ACI Materials Journal*, 95(6), 716–724.

Timkovich, V Yu., 1986, Genesis of structure and strength of the slag alkaline cements and concretes, Ph.D. Thesis, Kiev Civil Engineering Institute, Kiev, USSR.

Tritthart, J., 1989, Chloride binding in cement – II. The influence of the hydroxide concentration in the pore solution of hardened cement paste on chloride binding. *J. Cement and Concrete Research*, 19(5), 683–691.

TSNIIOMTP, 1985a, Concreting technique, design of structures made from slag alkaline cement concretes, Central Scientific Research and Design Institute for Management, Mechanization and Technical Assistance to Construction (TSNI-IOMTP), Manual of practice, Moscow, USSR.

TSNIIOMTP, 1985b, Recommendations for concreting technique, design of structures made from slag alkaline cement concretes, Central Scientific Research and Design Institute for Management, Mechanization and Technical Assistance for Construction (TSNIIOMTP), Moscow, USSR.

TU 67-Ukr181-74, 1974, Technical Specification, for Binder, Slag Alkaline, The Ministry of the Ukrainian Republic for Erection of Heavy Industry Enterprises, Kiev, USSR.

TU-33 UzSSR-03-82, 1982, Technical Specification for Concrete and Reinforced Slabs for Linings of Channels and Chutes of Irrigation Systems Made from Slag Alkaline Binder, Tashkent, USSR.

TU 204 UzbSSR 1-83, 1983, Technical Specification for Slag Alkaline Binder Based on Granulated Non-ferrous Slags, Ministry of the Uzbekistan Republic for Housing and Public Utilities, Tashkent, USSR.

TU 33 UzbSSR 01-84, 1984, Technical Specification for Articles Made from Lightweight Concrete on Slag Alkaline Binders, Glavsredazirsovhozstroi, Ministry for Irrigation and Water Supply of the USSR, Tashkent, USSR.

TU 67-629-84, 1984, Technical Specifications for Heat-Resistant Slag Alkaline Binder Concretes. Kiev, USSR.

TU 65-484-84, 1984, Technical Specifications for Wood-based Concrete Made from Slag Alkaline Binder, Ministry of USSR for Industrial Construction, Kiev, USSR.

TU 67-648-84, 1984, Technical Specifications for Granulated Non-ferrous Slags for Slag Alkaline Binders, Ministry of USSR for Construction of Heavy Industry Enterprises, Moscow, USSR.

TU 6-18-35-85, 1985, Technical Specifications for a Mixture of Salts (Industrial Alkali-containing Wastes from Production of Sodium Sulfide), Ministry of USSR for Chemical Industry, Kiev, USSR.

TU 14-14-145-85, 1985, Technical Specifications for a Mixture of Alkalis, Solid (Wastes from Cleaning Metallic Castings in Alkaline Solutions), Ministry for Ferrous Metallurgy of the Ukrainian Republic, Kiev, USSR.

TU 65 UkrSSR 196-86, 1987, Technical Specifications for Wall Panels, One-layered, Made from Lightweight Concrete on Slag Alkaline Binder for Industrial Buildings, Kiev, USSR.

TU 6-16-29-45-86, 1986, Technical Specification for An Alkaline Liquid Extract (Waste from Air Purification from Carbon Dioxide), Ministry for Chemical Industry of the USSR, Kiev, USSR.

TU 14-11-228-87, 1987, Technical Specifications for Slag Alkaline Binder from Silicomanganese Granulated Slag, Ministry of the USSR for Ferrous Metallurgy, Moscow, USSR.

TU 203.1-1-88, 1988, Technical Specifications for High-volume Blocks of Series BKR-2EK on Slag Alkaline Binder, Ministry of USSR for Housing-Civil Engineering, Moscow, USSR.

TU 37.002.0442-88, 1988, Technical Specification for a Mixture of Alkalis, Liquid (Waste from Cleaning Metallic Castings in Alkaline Solutions, Ministry of the USSR for Automotive Industry, Kiev, USSR.

TU 67-1010-88, 1989, Technical Specification for Boards, Mineral-Wood-Aggregate, Slag Alkaline Binder. Ministry of USSR for Construction in the Regions of Ural and Western Siberia, Moscow, USSR.

TU 67-1019-89, 1989, Technical Specification for Concretes, Heavyweight, Slag Alkaline, Moscow, USSR.

TU 67-1020-89, 1989, Technical Specification for Binder, Slag Alkaline, Ministry of the USSR for Construction in the Regions of Ural and Western Siberia, Moscow, USSR.

TU 67-1021-89, 1989, Technical specification for Structures and Articles, Concrete and Reinforced Concrete and Prefabricated from Slag Alkaline Binder Concrete. Ministry of the USSR for Construction in the Regions of Ural and Western Siberia, Moscow, USSR.

TU 234 UkrSSR 212-89, 1989, Technical Specification for Concrete Blocks for Foundation Walls Made from Slag Alkaline Binder, Ministry of the Ukrainian Republic for Construction, Kiev, USSR.

TU 12.0175921.003-89, 1989, Technical Specifications for Stones, Concrete, for Walls, Made from Metallurgical Granulated Slags and Slag Alkaline Binders, Kiev, USSR.

TU 7- BelSSR 5-90, 1990, Technical Specification for Binder, Slag Alkaline, Made from Cupola/Iron/Granulated Slag, The National Committee of the Belarussian Republic of the USSR for Construction, Minsk, USSR.

TU 559-10.20-001-90, 1990, Technical Specifications for Binder, Slag Alkaline, Made from Ferronickel Granulated Slag, State Enterprise Zhitomiroblagrostroi, Zhitomir, USSR.

TU 14-11-228-90, 1990, Technical Specifications for Binder, Slag Alkaline, Made from Silicomanganese Granulated Slag, Ministry of Metallurgy of the USSR, Kiev, USSR.

TU 6-46-91, 1991, Technical Specification for Soda-Sulfate Liquid Extract (Waste from Autoclave Treatment of Barite Concentrate with Soda), Ministry for Chemical Industry of the USSR, Kiev, USSR.

TU 10.20 UkrSSR 169-91, 1991, Technical Specification for Slag Alkaline Binder Made with Sulfate-Containing Compounds of Alkali Metals, The Ukrainian Cooperative State Corporation on Agricultural Construction, Kiev, USSR.

TU 14-113 UzbSR 11-91, 1991, Technical Specification for Slags, Electrosteelmaking for making a Slag Alkaline Cement, Bekabad, USSR.

TU 10.15 UzSSR 04-91, 1991, Technical Specification for Slag Alkaline Binder Based on Electrothermophosphorus and Electrosteelmelting Slags., The State Cooperative Enterprise "Uzagrostroi" and Tashkent Architectural and Construction Institute, Tashkent, USSR.

TU 10.20. 186-92, 1992, Technical Specification for Heavyweight Slag Alkaline Binder Concrete made with Sulfate-Containing Compounds of Alkali Metals, Kiev, USSR.

TU U V.2.7-16403272.006-98, 1998, Technical Specification for Binders, Alkaline, for Special Purposes: Geocements, Kiev, USSR.

Uchikawa, H., 1986, Blended and Special Cements: Effect of Blending Component on Hydration and Structure Formation. *Proceedings of the 8th International Congress on the Chemistry of Cement and Concrete, Principal Report* (Brazil), 1, 249–280.

US Bureau of Reclamation, 1975, Concrete Manual (8th edn), Denver, Colorado, 1975.

USEPA, 1993, *Technology Resource Document — Solidification/Stabilization and Its Application to Waste Materials*, EPA/530/R-93/012, June 1993.

USEPA, 1996, *Innovative Treatment Technologies: Annual Status Report, 8th Edition*, EPA/542/R-96/010, November 1996.

Usherov-Marshak, A. V., Krivenko, P. V. and Pershina, L. A., 1998, The role of solid phase basicity on heat evolution during hardening of cements. *Cement and Concrete Research*, 28(9), 1289–1296.

USSR Standard 10181.1-81, 1981, Measurement of workability/Consistence of Concrete Mixtures.

GOST 10181.1-81, 1981, National Standard for Concrete Mixes. Determination of workability/consistence, Moscow, USSR.

Vail, J. G., 1928, Soluble Silicates in Industry, American Chemical Society Monograph Series, The American Catalog Company, Inc., New York, USA.

Vail, J. G. and Wills, J. H., 1952, Soluble Silicates – Their Properties and Uses, Vol. 1. Chemistry, American Chemical Society Monograph Series (New York: Reinhold Publishing Corporation).

Van Jaarsveld, J. G. S., van Deventer J. S. J. and Lorenzen, L., 1996, The potential use of geopolymeric materials to immobilise toxic metals, Part I. theory and applications. *Minerals Engineering*, 10, 659–669.

Van Jaarsveld, J. G. S. and Van Deventer, J. S. J., 1989, Effect of alkali-metal activator on the properties of fly ash-based geopolymers. *Ind. Eng. Chem Res.*, 38, 3932–3941.

Voinovitch, I. A. and Dron, R., 1976, Action of different activators on the hydration of slag. *J.Silicates Industriels*, 41(9), 209–212.

Volyansky, A. A. 1958, *Strength and Deformability of Lightweight Concretes from Alevrolites of the Donbass* (Kiev: GosStroyizdat UkrSSR Publisher).

VSN 2-97-77, 1978, Code of practice for manufacturing technology for prefabricated reinforced slag alkaline cement concrete, Building Norms (Provisional), State Scientific-Research Institute for Erection of Pipelines (VNIIST), Moscow, USSR.

VSN 65.12-83, 1983, Building Norms for Concrete and Reinforced Concrete Structures Made from Slag Alkaline Binder Concrete. Manual for design and manufacturing (Provisional), Moscow, USSR.

VSN 67-247-83, 1984, Manual of Practice for the Use of Concretes Made from Slag Alkaline Binder in Aggressive Mineral and Organic Environments, Tchelyabinsk, USSR.

VTU 183-74, 1974, Technical Specification (Provisional) for Slag Alkaline Binder-Made Materials, Kiev, USSR.

Wachsmuth, F., Geiseler, J., Fix, W., Koch, K. and Schwerdtfeger, K., 1980, Contribution to the structure of BOF-steel slags and its influence in their volume stability. *C. I. M. M. International Symposium on Metallurgical Slags*, Paper 5.8, Halifax, Canada, 1–18.

Wang, S. D., 2000a, The role of sodium during the hydration of alkali-activated slag. *Advances in Cement Research*, 12(2), 65–69.

Wang S. D., 2000b, Alkali-activated slag: Hydration process and development of microstructure. *Advances in Cement Research*, 12(4), 163–172.

Wang, S. D. and Zhao, Z., 1990, Hydration features of new phosphorus slag cement. *Journal of Chinese Silicate Society*, 18(4), 379–384.

Wang, S. D., 1992, A new method to estimate the hydraulic activity of slag - the reactivity method. *Journal of the Chinese Ceramic Society*, 20(2), 196–200.

Wang, S. D. and Scrivener, K. L., 1993, Microchemistry of alkaline activation of slag. *3rd Beijing International Symposium on Cement and Concrete*, International Academic Publishers, Beijing, P.R. China, 2, 1047–1053.

Wang, S. and Scrivener, K. L., 1995, Hydration products of alkali activated slag cement. *Cement and Concrete Research*, 25(3), 561–571.

Wang, S., Scrivener, K. L. and Pratt, P. L., 1994, Factors affecting the strength of alkali-activated slag. *Cement and Concrete Research*, 24(6), 1033–1043.

Wang, S., Pu, X., Scrivener, K. L. and Pratt, P. L., 1995, Activated slag cement and concrete. A review of properties and problems. *Advances in Cement Research*, 7(27), 93–102.

Wang S. and Scrivener, K. L., 2003, 29Si and 27Al study of alkali-activated slag. *Cement and Concrete Research*, 33(5), 769–774.

Wang, Y. and Lin, D., 1983, The steel slag blended cement. *Silicates Industrielles*, 6, 121–126.

Weldes, H. H. and Lange, K. R., 1969, Properties of soluble silicates. *Industry and Engineering Chemistry*, 61(4), 29–44.

Wills, J. H., 1982. A Short History of the Manufacture of Soluble Silicates in the United States, in Falcone (ed.) *Soluble Silicates* (Washington, D.C.: American Chemical Society).

Wu, C., Zhang, Y. and Hu, Z., 1993, Properties and application of alkali-slag cement. *Journal of the Chinese Ceramic Society*, 21(2), 176–181.

Wu., D., Pei, Y. and Huang, B., 1996, Slag/Mud Mixtures Improve Cementing Operations in China, Oil & Gas Journal (Dec. 23), pp. 95–100.

Wu, Q., 1999, Effect of the ratio of water glass on properties of alkali-activated slag cement. *Cement Engineering (in Chinese)*, 5, 10–11.

Wu, X., 1984, A study on mineral phases in phosphorus slag. *Journal of Yinnan Building Materials (in Chinese)*, 4, 5–14.

Wu, X., Langton, C. A. and Roy, D. M., 1983, Hydration of slag cement at early stage. *J. Cement and Concrete Research*, 13(2), 277–286.

Wu. X., Yan, S., Sheng, X. and Tang, M., 1991, Alkali-activated cement based radioactive waste forms. *Cement and Concrete Research*, 21, 16.

Wu, X., Zhu, H., Hou, X. and Li, H., 1999, Study on steel slag and fly ash composite Portland cement. *Cement and Concrete Research*, 29, 1103–1106.

Wu, Z., and Naik, T., 2003, Chemically activated Blended Cements, *ACI Materials*, 100(5), 434–440.

Xi, Y. and Jennings, H. M., 1992, Relationships between Microstructure and Creep and Shrinkage of Cement Pastes. In Skalny (ed.) *Materials Science of Concrete III*, American Ceramic Society, 37–70.

Xie, Z. and Xi, Y., 2001. Hardening mechanisms of an alkali-activated Class F fly ash. *Cement and Concrete Research*, 31(9), 1245–1249.

Xu, B. and Pu, X., 1999, Study on solid alkaline AAS cement. *2nd International Conference on Alkaline Cements and Concretes* (edited by Krivenko), Kiev, Ukraine, 58–63.

Xu, H. and Van Deventer, J. S. J., 2000, The Geopolymerisation of Alumino-silicate Minerals. *International Journal of Mineral Processing*, 59(3), 247–266.

Xu, Z., 1988, Relationship between the Composition, Structure and Mechanical Properties of Low-Porosity Cementitious Materials and Its Modification Approaches, Ph.D. Thesis, Nanjing Institute of Chemical Technology, P.R. China.

Xu, Z., Deng, Y., Wu, X. and Tang, M., 1993, Influence of various hydraulic binders on performance of very low porosity cementitious system. *Cement and Concrete Research*, 23(2), 462–470.

Xu, Z., Tang, M. and Beaudoin, J. J., 1992, Relationship between composition, structure and mechanical properties of very low porosity system. *Cement and Concrete Research*, 22(2), 187–195.

YB/T 022-92, 1992, Steel Slag for Use in Cement, Ministry of Metallurgical Industry of China.

Yang, C., 1997, Alkali-Aggregate Reaction of Alkaline Cement Systems, Ph.D. Thesis, Chongqing Jiangzhu University, Chongqing, P.R. China.

Yang, C., Pu, X. and Wu, F., 1999, Studies on alkali-silica reaction (ASR) expansions of alkali activated slag cement mortars. *2nd International Conference on Alkaline Cements and Concretes* (edited by Krivenko), Kiev, Ukraine, 101–108.

Ye, G. and Burstrom, E., 1995, Utilization and stabilization of steelmaking slags. AFR-Report 57, Swedish Waste Research Council.

Ye, G. and Liao, J., 1999, The Slag Air-Quenching Technology. *Proceedings of International Symposium on the Utilization of Metallurgical Slag*, Beijing, 434–437.

Young, J. F., 1991, Macrodefect-free cement: A review. *Materials Research Society Symposium*, 179, 101–122.

Yin, Y., 1985, *University Chemistry Handbook* (Shandong: Shandong Science and Technology Press), P.R. China.

Yu, S. and Wang, W., 1990, Hardening mechanism of clinker-free sodium silicate slag cement. *Journal of Chinese Silicate Society (in Chinese)*, 18(2), 104–109.

Yuan, R., Gao, Q. and Ouyang, S., 1987, Study on structure and latent hydraulic activity of slag and its activation mechanism. *Journal of Wuhan University of Technology (in Chinese)*, P.R. China, 3, 297–303.

Zachariasen, W. H., 1932, The atomic arrangement in glass. *Journal of the American Chemical Society*, 54, 3841–3851.

Zahrada, J., 1986, Effect of the storage time, silicate modulus, and density of a sodium silicate solution on the hydraulic activity of slag-alkali-binder. *J Stavivo (in Czech)*, Prague, Czechoslovakia, 64(1), 23–26.

Zhang, C., Zhang, L. and Wei, Q., 1997, Limit of SO_3 in alkali-activated fly ash products. *Comprehensive Uses of Fly Ashes*, 4, 19–21.

Zhang, X. and Groves, G. W., 1990, The Alkali-silica reaction in OPC/silica glass mortar with particular reference to pessimum effects. *Advances in Cement Research*, 3(9), 9–13.

Zhong, B. and Yang, N., 1993, Hydration characteristics of water glass-activated slag cement. *Bulletin of Chinese Ceramic Society*, 23(6), 4–8.

Zhou, H., Wu, X., Xu, Z. and Tang, M., 1993, Kinetic study on hydration of alkali-activated slag. *Cement and Concrete Research*, 23(6), 1253–1258.

Zhu, G., Sun, S., Wu, G. and Hu, J., 1989, Quality of steelmaking slag and assessing method for its cementing property. *2nd International Symposium on Cement and Concrete*, Beijing, P.R. China, 2, 295–302.

Zhu, G. and Sun, S., 1999, Current situation and trend of iron and steel slag utilization in P.R. China. *International Symposium on the Utilization of Metallurgical Slag*, Beijing, P.R. China, 10–15.

Zhu, X., Wang, S. and Qi, W., 2001, Research on pump delivered alkali-slag concrete and its application, *Concrete (in Chinese)*, 4, 16–18.

Zivica, V., 1993, Alkali-silicate admixture for cement composites incorporating pozzolan or blast furnace slag. *Cement and Concrete Research*, 23(5), 1215–1222.

Index

9 780367 863630